# Matrix Theory and Applications with MATLAB®

## Darald J. Hartfiel

**CRC Press**
Boca Raton   London   New York   Washington, D.C.

## Library of Congress Cataloging-in-Publication Data

Hartfiel, D.J.
   Matrix theory and applications with MATLAB / Darald J. Hartfiel.
      p.   cm.
   Includes bibliographical references and index.
   ISBN 1-58488-108-9 (alk. paper)
      1. Matrices—Data processing. 2. MATLAB. I. Title.

QA188 .H37 2000
512.9′434—dc21                                                                 00-046817

**Visit the CRC Press Web site at www.crcpress.com**

No claim to original U.S. Government works
International Standard Book Number 1-58488-108-9
Library of Congress Card Number 00-046817
Printed in the United States of America        2  3  4  5  6  7  8  9  0
Printed on acid-free paper

# Preface

This text is intended for a basic course in matrix theory and applications. "Basic" here means that the material chosen is what is most often seen in these courses, and the presentation stresses insight and understanding.

There is no common definition of "understanding"; however, there is some agreement of its consequences. A person that understands material should be able to do the following:

- fill in any missing pieces of the material that were not given,

- adjust the material to cover like problems,

- extend the material beyond what was seen, and perhaps

- create by adding new work to the material.

In this text, there are some places where results are given for special cases, e.g., $2 \times 2$ matrices rather than $n \times n$ matrices. This was done so that the notation required was simple and the idea of the proof was easier to glean. Of course, understanding this case should mean that the general proof can be seen as well. Still, writing out the proof can take some time in dealing with subscripts and such. There are also a few places where the first two consequences come into play. Understanding is important, and it seems that when students don't understand, they resort to memorization of material that means nothing to them.

There are some special features of the text, which are described as follows:

1. *Optional subsections.* At the end of each section, a subsection entitled Optional covers applications of the material in the section. Its intent is to show how the material in the subsection is used.

2. *MATLAB\* subsections.* Also at the end of most sections, there is a subsection entitled MATLAB. These subsections discuss the various commands we use in MATLAB to do the computations described in the sections. Of course, learning requires that some problems be done by hand. However, for larger problems, some kind of software

---

*MATLAB is a registered trademark of The MathWorks, Inc. For product information, please contact:

The MathWorks, Inc.
3 Apple Hill Drive
Natick, MA 01760-2098
Tel: 508 647-7000
Fax: 508-647-7001
E-mail: info@mathworks.com
Web: www.mathworks.com

is required (as it saves considerable time). We used MATLAB since it is the software of choice in this area.

Code for the pictures and other graphics used in the text are also given in these subsections. Code for some algorithms used are given as well. This is important for several reasons. First, the code can be used to get color pictures rather than the black and whites shown in the text. Second, with a little effort, code can be adjusted or extended to handle other problems. Third, this code is important since not everyone uses MATLAB on a daily basis. It is nice to be able to review code a bit to bring the work back to mind.

Actually, the students can work through the Optional and MATLAB subsections themselves.

3. *Visuals.* Many professors believe that pictures are important to learning, and studies on the hemispheres of the brain support that view. This text supports much of the verbal material with pictures. In fact, there are 129 pictures or drawings in this text. Exercises involving drawing and pictures are also given.

4. *Examples.* Much of the theory given in the text is supported by examples, 115 in fact. This serves several purposes. Some students learn better by looking at examples, although there is always the problem of mimicking here rather than working from basic ideas. And, some professors may choose to cover some of the material by discussing and showing examples rather than by discussing and proving.

5. *Exercises.* There are 450 exercises in the text designed to help students learn the material in the sections: practice calculations, applying results, completing proofs, and such.

6. *Order.* The first 7 chapters of this text represent basic matrix theory. Beyond that, the chapters can be taken in any order. These latter chapters are short and perhaps a bit more advanced.

In conclusion, I would like to thank those students who, over the years, provided feedback on how they felt they learned material. It was helpful.

In addition, I would like to thank my wife, Faye, for typing and working with me on this manuscript. Both of us thank John MacKendrick from MacKichan Software for his help in typesetting problems.

And, I would like to thank my editor, Bob Stern, for his advice and help on producing this text.

Darald J. Hartfiel

# Author's Page

Darald J. Hartfiel received his Ph.D. degree in mathematics from the University of Houston in 1969. Since then, he has been on the faculty of Texas A&M University.

Professor Hartfiel has written 97 research papers, mostly in matrix theory and related areas. He is the author of two books, one of which is a monograph.

# Contents

# 1
# Review of Matrix Algebra

In this book we will assume some basic background in matrices, linear equations, and determinants as this material is usually studied in previous courses.

What is assumed is reviewed in this chapter. In reviewing, a few remarks and examples are usually enough to bring the work back to mind. Observing the technical notations used in the book and working through a few problems will also help.

Almost all of the work in this book can be done using either the set $\mathcal{R}$ of real numbers or the set $\mathcal{C}$ of complex numbers. Any exceptions (when we work only in $\mathcal{R}$ or in $\mathcal{C}$) will be stated.

In linear algebra and matrix theory, it is traditional to refer to numbers as *scalars* (real or complex). We will use Greek letters $\alpha, \beta, \ldots$ to denote scalars.

## 1.1 Matrices, Systems of Linear Equations, Determinants

Since we use complex numbers in this book, it may be helpful to give a brief review of them. A complex number is written in the form $a + bi$ where $a$ (the real part) and $b$ (the imaginary part) are real numbers and $i = \sqrt{-1}$. If the complex number has imaginary part 0, we simply write $a$ for $a + 0i$. Since complex numbers commute, $bi = ib$, so we can also write complex numbers as $a + ib$.

Complex numbers can be plotted in the complex plane by finding $a$ on the real axis ($x$-axis) and $b$ on the imaginary axis ($y$-axis) and plotting $a + bi$ at $(a, b)$. (See Figure 1.1.)

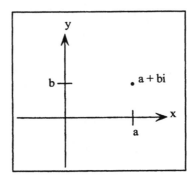

FIGURE 1.1.

The computing rules are as those for real numbers, using $i^2 = -1$ to simplify. For example,

$$(2 + 3i)(4 + 5i) =$$
$$2 \cdot 4 + 2 \cdot 5i + 3i \cdot 4 + 3i \cdot 5i =$$
$$8 + 10i + 12i - 15 = -7 + 22i.$$

If

$$z = a + bi$$

its conjugate is

$$\bar{z} = a - bi.$$

Calculation shows that if $w = c + di$, then

$$\overline{z + w} = \bar{z} + \bar{w} \text{ and } \overline{zw} = \bar{z}\,\bar{w}.$$

Since $z\bar{z} = a^2 + b^2$ is a real number, we can simplify a fraction by multiplying its numerator and its denominator by the conjugate of the denominator. For example,

$$\frac{3 + 2i}{4 + 5i} = \frac{3 + 2i}{4 + 5i} \cdot \frac{4 - 5i}{4 - 5i} = \frac{22 - 7i}{41} = \frac{22}{41} - \frac{7}{41}i.$$

We also use that the absolute value (also called the modulus) of $z$ is

$$|z| = \sqrt{a^2 + b^2}.$$

Viewed in the complex plane, it is seen (Figure 1.2) that $|z|$ is the distance between $z$ and 0.

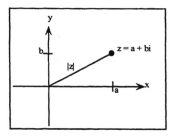

FIGURE 1.2.

By direct calculation, it can be seen that

$$|z + w| \leq |z| + |w|, \quad |zw| = |z|\,|w|, \quad \text{and} \quad |z| = (z\overline{z})^{\frac{1}{2}}.$$

Finally, recall that

$$e^{ib} = \cos b + i \sin b,$$

and

$$e^{a+ib} = e^a e^{ib}.$$

For example, $e^{2+i\frac{\pi}{3}} = e^2 \left(\cos\frac{\pi}{3} + i \sin\frac{\pi}{3}\right) = e^2 \left(\frac{1}{2} + \frac{\sqrt{3}}{2}i\right) = 3.69 + 6.40i$ (rounded to the hundredths decimal place).

### 1.1.1 Matrix Algebra

A matrix is an $m \times n$ array of numbers placed in $m$ rows and $n$ columns. In general, we exhibit a matrix as

$$\begin{bmatrix} a_{11} & a_{12} & \cdots & a_{1n} \\ a_{21} & a_{22} & \cdots & a_{2n} \\ & & \cdots & \\ a_{m1} & a_{m2} & \cdots & a_{mn} \end{bmatrix}$$

or more compactly as $[a_{ij}]$, depicting the entry $a_{ij}$ in row $i$ and column $j$.

$$\begin{array}{c} \phantom{i-}\quad\quad\quad\quad j \\ i- \left[ \begin{array}{ccccc} & & - & & \\ & & - & & \\ - & - & - & a_{ij} & - \\ & & - & & \\ & & - & & \end{array} \right] \end{array}$$

The *size* of this matrix is $m \times n$. If $m = n$, the matrix is often called *square*.

We use capital letters to denote matrices and the corresponding lower case letters for entries. Thus, we write

$$A = [a_{ij}], \ B = [b_{ij}], \ \text{etc.}$$

If a matrix is $1 \times n$ or $m \times 1$, we call it a vector and simply write

$$x = [x_i].$$

The $m \times 1$ vector, $e_i$, called a unit vector, defined by

$$e_i = \left[ \begin{array}{c} 0 \\ \dots \\ 0 \\ 1 \\ 0 \\ \dots \\ 0 \end{array} \right] \leftarrow i\text{-th row,}$$

appears throughout the text.

Recall that the arithmetic of matrices is like the arithmetic of numbers, except

i. Matrices do not necessarily commute under multiplication.

ii. Matrices need not have multiplicative inverses.

We will again see this as we give a brief review of the algebra (arithmetic) of matrices.

Developing the algebra of matrices, for $m \times n$ matrices $A$ and $B$, define addition as

$$A + B = [a_{ij} + b_{ij}].$$

And if $\alpha$ is a scalar, define scalar multiplication as

$$\alpha A = [\alpha a_{ij}].$$

Thus,

$$\begin{bmatrix} 1 & 0 \\ 2 & -1 \end{bmatrix} + \begin{bmatrix} 2 & -1 \\ -2 & 0 \end{bmatrix} = \begin{bmatrix} 3 & -1 \\ 0 & -1 \end{bmatrix}$$

and

$$2\begin{bmatrix} 0 & 1 \\ -2 & 3 \end{bmatrix} = \begin{bmatrix} 0 & 2 \\ -4 & 6 \end{bmatrix}.$$

Recall that equality between matrices of the same size means that all corresponding entries are equal. Using this, the following properties for $m \times n$ matrices and scalars are easily seen.

(a) $A + B = B + A$

(b) $(A + B) + C = A + (B + C)$

(c) The matrix 0, all of whose entries are 0, satisfies

$$A + 0 = 0 + A = A.$$

(d) For each matrix $A$, the matrix $-A = [-a_{ij}]$ satisfies

$$A + (-A) = (-A) + A = 0.$$

(e) $\alpha(A + B) = \alpha A + \alpha B$

(f) $(\alpha + \beta)A = \alpha A + \beta A$

(g) $(\alpha\beta)A = \alpha(\beta A)$

(h) $1A = A$

It may be helpful to demonstrate one of these results.

**Proof (e).** By direct computation,

$$\begin{aligned}
\alpha(A + B) &= \alpha([a_{ij}] + [b_{ij}]) \\
&= \alpha[a_{ij} + b_{ij}] \\
&= [\alpha(a_{ij} + b_{ij})] \\
&= [\alpha a_{ij} + \alpha b_{ij}] \\
&= [\alpha a_{ij}] + [\alpha b_{ij}] \\
&= \alpha[a_{ij}] + \alpha[b_{ij}] \\
&= \alpha A + \alpha B.
\end{aligned}$$

This verifies (e). ∎

Let $A$ be an $m \times r$ matrix and $B$ an $r \times n$ matrix. The product $AB$ is defined entrywise as the $m \times n$ matrix $C$ where

$$c_{ij} = a_{i1}b_{1j} + a_{i2}b_{2j} + \cdots + a_{ir}b_{rj}$$

$$= \sum_{k=1}^{r} a_{ik}b_{kj}$$

for $1 \le i \le m$ and $1 \le j \le n$.

Computing this product in terms of the rows of $A$, called *forward multiplication*, yields the rows of $AB$; e.g., to compute the $i$-th row of $AB$, we multiply the $i$-th row of $A$ and the matrix $B$ as in

$$a_i \begin{bmatrix} b_1 \\ \cdots \\ b_r \end{bmatrix} = a_{i1}b_1 + \cdots + a_{ir}b_r$$

where $a_i$ is the $i$-th row of $A$ and $b_1, \ldots, b_r$ the rows of $B$. (Here we can think of taking row $a_i$, tilting it forward or vertically, so its entries are against the corresponding rows of $B$, and multiplying through.)

Computing columns of the product, called *backward multiplication*, yields for the $j$-th column of $AB$,

$$[a_1 \cdots a_r]\, b_j = b_{1j}a_1 + \cdots + b_{rj}a_r$$

where $b_j$ is the $j$-th column of $B$ and $a_1, \ldots, a_r$ are the columns of $A$. (Here we can think of taking a column $b_j$ of $B$, tilting it backward, or horizontally, so its entries are against the corresponding columns of $B$, and multiplying through.) Viewing a product as a backward multiplication will allow us to see (as obvious) many matrix results. It is useful.

An example may help.

**Example 1.1** *Let* $I = \begin{bmatrix} 1 & 0 \\ 0 & 1 \end{bmatrix}$. *We show that* $IA = A$ *for any* $2 \times 2$ *matrix* $A$. *Here, if* $a_1$ *and* $a_2$ *are the rows of* $A$, *then by forward multiplication*

$$IA = \begin{bmatrix} 1 & 0 \\ 0 & 1 \end{bmatrix} \begin{bmatrix} a_1 \\ a_2 \end{bmatrix} = \begin{bmatrix} a_1 \\ a_2 \end{bmatrix} = A.$$

*To see this, note that multiplying* $A$ *by the first row of* $I$, *we tilt vertically and multiply through*

$$\begin{matrix} 1 \\ 0 \end{matrix} \begin{bmatrix} a_1 \\ a_2 \end{bmatrix}$$

$$\text{(do mentally)}$$

$$\begin{bmatrix} 1 & 0 \end{bmatrix} \begin{bmatrix} a_1 \\ a_2 \end{bmatrix} \qquad = \qquad [a_1]$$

*obtaining the first row of A. Doing the same for the second row of I, we have*

$$
\begin{array}{c}
\begin{array}{c} 0 \\ 1 \end{array}
\left[ \begin{array}{c} a_1 \\ a_2 \end{array} \right] \\
\textit{(do mentally)}
\end{array}
$$

$$
\nearrow \qquad\qquad\qquad\qquad\qquad \searrow
$$

$$
\begin{bmatrix} 0 & 1 \end{bmatrix} \left[ \begin{array}{c} a_1 \\ a_2 \end{array} \right] \qquad = \qquad [a_2]
$$

*obtaining the second row of A.*

  *We now show $AI = A$ by using backward multiplication. Here, if $a_1$ and $a_2$ denote the columns of A, we have*

$$
\begin{array}{c}
\begin{array}{cc} 1 & 0 \end{array} \\
\left[ \begin{array}{cc} a_1 & a_2 \end{array} \right] \\
\textit{(do mentally)}
\end{array}
$$

$$
\nearrow \qquad\qquad\qquad\qquad\qquad \searrow
$$

$$
\begin{bmatrix} a_1 & a_2 \end{bmatrix} \left[ \begin{array}{c} 1 \\ 0 \end{array} \right] \qquad = \qquad [a_1]
$$

*and*

$$
\begin{array}{c}
\begin{array}{cc} 0 & 1 \end{array} \\
\left[ \begin{array}{cc} a_1 & a_2 \end{array} \right] \\
\textit{(do mentally)}
\end{array}
$$

$$
\nearrow \qquad\qquad\qquad\qquad\qquad \searrow
$$

$$
\begin{bmatrix} a_1 & a_2 \end{bmatrix} \left[ \begin{array}{c} 0 \\ 1 \end{array} \right] \qquad = \qquad [a_2] \ .
$$

*Thus,*

$$
\begin{bmatrix} a_1 & a_2 \end{bmatrix} \begin{bmatrix} 1 & 0 \\ 0 & 1 \end{bmatrix} = \begin{bmatrix} a_1 & a_2 \end{bmatrix} \ .
$$

A square matrix $T$ is *upper triangular* if $t_{ij} = 0$ whenever $i > j$. It is *lower triangular* if $t_{ij} = 0$ when $i < j$. When we say that $T$ is *triangular*, $T$ can be either upper triangular of lower triangular. If $T_1$ and $T_2$ are $n \times n$ upper triangular matrices, then, using forward multiplication, we see that $T_1 T_2$ is upper triangular. (Observe that if $r$ is the $i$-th row of $T_1$ and $c$ the $j$-th column of $T_2$ and $i > j$, then the nonzero entries of $r$ correspond to 0 entries in $c$ and so $rc = 0$. The companion result for lower triangular matrices is also true.

  A square matrix $D$ is a *diagonal* matrix if $D$ is both upper triangular and lower triangular. A diagonal matrix is often written $D = diag(d_{11}, \ldots, d_{nn})$, simply identifying the main diagonal entries $d_{11}, \ldots, d_{nn}$ of $D$. An *identity matrix* $I$ is a diagonal matrix with 1's on the main diagonal. Note that, if the products are defined, $IA = A$ by forward multiplication and $AI = A$ by backward multiplication.

  If $A$ is an $n \times n$ matrix, then

$$
A^\circ = I
$$

and if $k$ is a positive integer,

$$A^k = A \cdots A$$

where $A$ appears here as a factor $k$ times.

Additional properties of the product, for all matrices in which the expressed sums and products are defined, follow.

(i) $(AB)C = A(BC)$

(j) $A(B+C) = AB + AC$

(k) $(B+C)A = BA + CA$

(l) $A(\alpha B) = \alpha(AB)$ (Often in computing, a scalar $\alpha$ is caught between matrices. This property assures $\alpha$ can be pulled out and placed in front of the product.)

An additional computation may be helpful.

**Proof (l).** Suppose $A$ is an $m \times r$ matrix and $B$ is an $r \times n$ matrix. Then

$$A(\alpha B) = [a_{ij}](\alpha[b_{ij}]) = [a_{ij}][\alpha b_{ij}] = \left[\sum_{k=1}^{r} a_{ik}(\alpha b_{kj})\right]$$

$$= \left[\sum_{k=1}^{r} \alpha(a_{ik}b_{kj})\right] = \alpha\left[\sum_{k=1}^{r} a_{ik}b_{kj}\right] = \alpha(AB).$$

This verifies the result. ∎

The definition of matrix multiplication allows the writing of systems of linear equations

$$a_{11}x_1 + a_{12}x_2 + \cdots + a_{1n}x_n = b_1$$
$$a_{21}x_1 + a_{22}x_2 + \cdots + a_{2n}x_n = b_2$$
$$\cdots$$
$$a_{m1}x_1 + a_{m2}x_2 + \cdots + a_{mn}x_n = b_m$$

in compact form as the matrix equation

$$Ax = b$$

where $A = [a_{ij}]$, $x = \begin{bmatrix} x_1 \\ \cdots \\ x_n \end{bmatrix}$, and $b = \begin{bmatrix} b_1 \\ \cdots \\ b_m \end{bmatrix}$.

If $A$ is a square matrix and there is a square matrix $B$ which satisfies the *inverse equation*

$$AX = XA = I$$

then $B$ is an *inverse* of $A$ and $A$ is *nonsingular*. (Some books use the word *invertible*.) If $A$ has no inverse, it is *singular*. If $B$ and $C$ are inverses for $A$, then

$$B = BI = B(AC) = (BA)C = IC = C$$

so there can be at most one inverse for $A$. (There may be none.) We denote this inverse, when it exists, as $A^{-1}$.

Properties of the inverse are

(m) $T^{-1}$ is upper triangular when $T$ is upper triangular and nonsingular. The companion result for lower triangular matrices also holds.

(n) $\left(A^{-1}\right)^{-1} = A$, when $A$ is nonsingular.

(o) $(AB)^{-1} = B^{-1}A^{-1}$, when both $A$ and $B$ are nonsingular. This can be extended, by induction, to $(A \cdots A_k)^{-1} = A_k^{-1} \cdots A_1^{-1}$. Thus, $(A^m)^{-1} = \left(A^{-1}\right)^m$, or simply $A^{-m}$, when $A$ is nonsingular and $m$ a positive integer.

In problems involving inverses, the inverse equation is often used. We show this in the following computation.

**Proof (o).** Note that by replacing parentheses

$$(AB)\left(B^{-1}A^{-1}\right) = A\left(BB^{-1}\right)A^{-1} = A^{-1}IA = I$$

and similarly $\left(B^{-1}A^{-1}\right)(AB) = I$. Since $B^{-1}A^{-1}$ satisfies the inverse equation for $AB$, $(AB)^{-1} = B^{-1}A^{-1}$.  ∎

Finally, for an $m \times n$ matrix $A$, we define

$$\bar{A} = \begin{bmatrix} \overline{a_{11}} & \overline{a_{12}} & \cdots & \overline{a_{1n}} \\ \overline{a_{21}} & \overline{a_{22}} & \cdots & \overline{a_{2n}} \\ & & \cdots & \\ \overline{a_{m1}} & \overline{a_{m2}} & \cdots & \overline{a_{mn}} \end{bmatrix} \quad \text{and} \quad A^t = \begin{bmatrix} a_{11} & a_{21} & \cdots & a_{m1} \\ a_{12} & a_{22} & \cdots & a_{m2} \\ & & \cdots & \\ a_{1n} & a_{2n} & \cdots & a_{mn} \end{bmatrix}$$

called the *conjugate* and the *transpose* of $A$, respectively. Using these, we define

$$A^H = \left(\bar{A}\right)^t,$$

called the *conjugate transpose* of $A$.

**Example 1.2** *If* $A = \begin{bmatrix} 1+i & 2-3i \\ 2 & 4+2i \end{bmatrix}$, *then*

$$A^H = \bar{A}^t = \begin{bmatrix} \overline{1+i} & \overline{2-3i} \\ \overline{2} & \overline{4+2i} \end{bmatrix}^t$$

$$= \begin{bmatrix} 1-i & 2+3i \\ 2 & 4-2i \end{bmatrix}^t = \begin{bmatrix} 1-i & 2 \\ 2+3i & 4-2i \end{bmatrix}.$$

Matrices which satisfy

$$A = A^H$$

are called *Hermitian*, or *symmetric* if $A$ has real entries. (In the latter case, $A^H = A^t$.)

Properties of the conjugate transpose are

(p) $\left(A^H\right)^H = A$ for any $m \times n$ matrix $A$.

(q) $(A^H)^{-1} = \left(A^{-1}\right)^H$ if $A$ is nonsingular.

(r) $(A+B)^H = A^H + B^H$ for any $m \times n$ matrices $A$ and $B$.

(s) $(AB)^H = B^H A^H$ for any $m \times r$ matrix $A$ and $r \times n$ matrix $B$. This can be extended to $(A_1 \cdots A_k)^H = A_k^H \cdots A_1^H$, so, in terms of reversing the order of the products, it is like the inverse of a product.

Another demonstration of a computation shows how to use this notation.

**Proof (r).** By direct computation,

$$(A+B)^H = ([a_{ij}] + [b_{ij}])^H = [a_{ij} + b_{ij}]^H = \left[\overline{a_{ij} + b_{ij}}\right]^t$$
$$= \left[\bar{a}_{ij} + \bar{b}_{ij}\right]^t = \left[\bar{a}_{ji} + \bar{b}_{ji}\right] \quad (ij\text{-th entry is } \bar{a}_{ji} + \bar{b}_{ji})$$
$$= [\bar{a}_{ji}] + \left[\bar{b}_{ji}\right] = [\bar{a}_{ij}]^t + \left[\bar{b}_{ij}\right]^t = A^H + B^H.$$

Thus the result is established. ∎

It is sometimes useful to do matrix arithmetic on submatrices which make up the matrix. By partitioning the rows and columns, a matrix $A$ can be partitioned into *submatrices* $A_{ij}$ (sometimes called *blocks*) say

$$A = \begin{bmatrix} A_{11} & A_{12} & \cdots & A_{1r} \\ A_{21} & A_{22} & \cdots & A_{2r} \\ & & \cdots & \\ A_{p1} & A_{p2} & \cdots & A_{pr} \end{bmatrix}.$$

For example, if we partition

$$A = \begin{array}{cc} & \begin{array}{cccc} 1 & 2 & \quad 3 & 4 \end{array} \\ \begin{array}{c} 1 \\ 2 \\ \\ 3 \end{array} & \left[\begin{array}{cc|cc} 2 & 3 & -5 & 13 \\ 7 & 4 & 2 & 0 \\ \hline 5 & 16 & -2 & 8 \end{array}\right] \end{array},$$

then we can write $A = \begin{bmatrix} A_{11} & A_{12} \\ A_{21} & A_{22} \end{bmatrix}$ where $A_{11} = \begin{bmatrix} 2 & 3 \\ 7 & 4 \end{bmatrix}$, $A_{12} = \begin{bmatrix} -5 & 13 \\ 2 & 0 \end{bmatrix}$, $A_{21} = \begin{bmatrix} 5 & 16 \end{bmatrix}$, and $A_{22} = \begin{bmatrix} -2 & 8 \end{bmatrix}$.

If $B$ is a matrix, partitioned as is $A$, then addition can be done using the submatrices, that is,

$$A + B = [A_{ij} + B_{ij}].$$

And, if the expressed matrix sums and products of the blocks are defined,

$$AB = \left[\sum_{k=1}^{r} A_{ik} B_{kj}\right].$$

Note that this partitioned arithmetic is exactly like that for the entry arithmetic previously described.

**Example 1.3** *Some examples of partitioned arithmetic follow.*

(a) *Let* $A = \begin{bmatrix} A_{11} & A_{12} \\ A_{21} & A_{22} \end{bmatrix}$, $B = \begin{bmatrix} B_{11} & B_{12} \\ B_{21} & B_{22} \end{bmatrix}$ *where A and B are* $4 \times 4$
*matrices and all submatrices are* $2 \times 2$ *matrices. Then*

$$A + B = \begin{bmatrix} A_{11} + B_{11} & A_{12} + B_{12} \\ A_{21} + B_{21} & A_{22} + B_{22} \end{bmatrix}$$

*and*

$$AB = \begin{bmatrix} A_{11}B_{11} + A_{12}B_{21} & A_{11}B_{12} + A_{12}B_{22} \\ A_{21}B_{11} + A_{22}B_{21} & A_{21}B_{12} + A_{22}B_{22} \end{bmatrix}.$$

(b) *Let A be* $m \times r$ *matrix and B an* $r \times n$ *matrix. If* $B = [b_1 b_2 \dots b_n]$,
*where* $b_k$ *is the k-th column of B, then*

$$AB = [Ab_1 Ab_2 \dots Ab_n].$$

*If* $A = \begin{bmatrix} a_1 \\ a_2 \\ \dots \\ a_m \end{bmatrix}$, *where* $a_k$ *is the k-th row of A, then* $AB = \begin{bmatrix} a_1 B \\ a_2 B \\ \dots \\ a_m B \end{bmatrix}.$

## 1.1.2  Systems of Linear Equations

To solve a system of linear equations

$$a_{11}x_1 + a_{12}x_2 + \cdots + a_{1n}x_n = b_1$$
$$a_{21}x_1 + a_{22}x_2 + \cdots + a_{2n}x_n = b_2$$
$$\cdots$$
$$a_{m1}x_1 + a_{m2}x_2 + \cdots + a_{mn}x_n = b_m$$

,

we simplify it to a system in *row echelon form* (staggered rows), such as

$$\circledast x_1 + *x_2 + \cdots + *x_n = *$$
$$\circledast x_2 + \cdots + *x_n = *$$
$$\cdots$$
$$\circledast\, x_n = *.$$

Here the $\circledast$'s are nonzero scalars and the $*$'s are arbitrary scalars. Using this form, the scalars $\circledast$ are called *pivots* and the variables corresponding to them are called *pivot variables*. All other variables are called *free variables*.

To solve the simplified system, we set each free variable equal to an arbitrary scalar. We then solve for the pivot variables, starting with the last equation and working up, in terms of the free variables. The solutions are then expressed using only the free variables. This method is called *back substitution*.

To simplify a system we can use the following elementary operations.

a. Interchange: $R_i \leftrightarrow R_j$, interchange equations $i$ and $j$.

b. Scale: $\alpha R_i$, multiply equation $i$ by a nonzero scalar $\alpha$.

c. Add: $\alpha R_j + R_i$, add $\alpha$ times equation $j$ to equation $i$.

It can be shown that applying an elementary operation to a system of linear equations will not change its solution set.

Operation (a) can be applied to a system to obtain a nonzero coefficient of $x_1$ in equation 1. Then operation (c) can be applied to eliminate $x_1$ from equations 2 through $m$. And, this method can then be applied to the system with the 1-st equation deleted. Continuing, we obtain a row echelon form.

An example will help recall the method, called *Gaussian elimination*.

**Example 1.4** *Solve*

$$1x_1 + 1x_2 + 1x_3 + 1x_4 = \phantom{-}10$$
$$2x_1 + 2x_2 + 0x_3 + 4x_4 = \phantom{-}44$$
$$3x_1 + 3x_2 + 7x_3 - 1x_4 = -18.$$

*Since the arithmetic will only take place on the constants, we do the operations on the corresponding augmented matrix*

$$
\begin{array}{cccc}
x_1 & x_2 & x_3 & x_4 \\
\end{array}
$$

$$
\left[
\begin{array}{cccc|c}
1 & 1 & 1 & 1 & 10 \\
2 & 2 & 0 & 4 & 44 \\
3 & 3 & 7 & -1 & -18 \\
\end{array}
\right].
$$

*Applying* $-2R_1 + R_2$ *and* $-3R_1 + R_3$ *yields*

$$
\begin{array}{cccc}
x_1 & x_2 & x_3 & x_4 \\
\end{array}
$$

$$
\left[
\begin{array}{cccc|c}
1 & 1 & 1 & 1 & 10 \\
0 & 0 & -2 & 2 & 24 \\
0 & 0 & 4 & -4 & -48 \\
\end{array}
\right].
$$

*Now we use* $2R_2 + R_3$ *to get*

$$
\begin{array}{cccc}
x_1 & x_2 & x_3 & x_4 \\
\end{array}
$$

$$
\left[
\begin{array}{cccc|c}
1 & 1 & 1 & 1 & 10 \\
0 & 0 & -2 & 2 & 24 \\
0 & 0 & 0 & 0 & 0 \\
\end{array}
\right].
$$

*This says that* $x_1$ *and* $x_3$ *are pivot variables while* $x_2$ *and* $x_4$ *are free. Set*

$$x_2 = \alpha$$
$$x_4 = \beta, \text{ arbitrary constants.}$$

*Solving for the pivot variables, in terms of the free variables, we have, from the second equation*

$$-2x_3 + 2\beta = 24$$

*so*

$$x_3 = -12 + \beta.$$

*And, from the first equation,*

$$x_1 + \alpha + x_3 + \beta = 10$$

*so*

$$x_1 = 22 - \alpha - 2\beta.$$

*Thus*

$$x = \begin{bmatrix} x_1 \\ x_2 \\ x_3 \\ x_4 \end{bmatrix} = \begin{bmatrix} 22 - \alpha - 2\beta \\ \alpha \\ -12 + \beta \\ \beta \end{bmatrix}$$

$$= \begin{bmatrix} 22 \\ 0 \\ -12 \\ 0 \end{bmatrix} + \alpha \begin{bmatrix} -1 \\ 1 \\ 0 \\ 0 \end{bmatrix} + \beta \begin{bmatrix} -2 \\ 0 \\ 1 \\ 1 \end{bmatrix}.$$

*Since $\alpha$ and $\beta$ can be any scalars, there are infinitely many solutions.*

If, while applying Gaussian elimination, we scale each pivot so it is 1 and then apply add operations to obtain 0's below and above the pivot, the method is called *Gauss-Jordan*. This method requires about $\frac{1}{6}$ more arithmetic operations than does Gaussian elimination, so it isn't used for large problems.

The row echelon form obtained by using Gauss-Jordan is called the reduced row echelon form (rref), and rref does appear as a command on most calculators and in computer software.

If $A$ is a nonsingular matrix, its inverse can be found by solving

$$AX = I. \tag{1.1}$$

Note the solution is $X = A^{-1}$. (To solve, multiply the equation through $A^{-1}$.)

If $X = [x_1 \ldots x_n]$, where $x_j$ is the $j$-th column of $X$, then by partitioned arithmetic,

$$Ax_1 = e_1, \ldots, Ax_n = e_n. \tag{1.2}$$

These equations can be solved simultaneously, using the augmented matrix

$$[A \mid I].$$

If in the process of applying Gaussian elimination, or Gauss-Jordan, to this matrix, a row of 0's is encountered in the first block, then, as given in the exercises, $A$ is singular.

An example will help bring the method back to mind.

**Example 1.5** *Let $A = \begin{bmatrix} 1 & 1 \\ 1 & 3 \end{bmatrix}$. The augmented matrix for $AX = I$ is $[A \mid I]$. Applying Gauss-Jordan, we start with*

$$\begin{bmatrix} 1 & 1 & 1 & 0 \\ 1 & 3 & 0 & 1 \end{bmatrix}.$$

*Using* $-R_1 + R_2$, *we have*

$$\left[\begin{array}{cc|cc} 1 & 1 & 1 & 0 \\ 0 & 2 & -1 & 1 \end{array}\right].$$

*To obtain 1's on the pivots, we apply* $\frac{1}{2}R_2$, *yielding*

$$\left[\begin{array}{cc|cc} 1 & 1 & 1 & 0 \\ 0 & 1 & -\frac{1}{2} & \frac{1}{2} \end{array}\right].$$

*Getting 0's above the second pivot, we use* $-R_2 + R_1$. *So we have*

$$\left[\begin{array}{cc|cc} 1 & 0 & \frac{3}{2} & -\frac{1}{2} \\ 0 & 1 & -\frac{1}{2} & \frac{1}{2} \end{array}\right].$$

*Thus, solving (1.1) or equivalently (1.2), we get*

$$A^{-1} = \left[\begin{array}{cc} \frac{3}{2} & -\frac{1}{2} \\ -\frac{1}{2} & \frac{1}{2} \end{array}\right].$$

*As can be seen, this matrix is in the second block of the augmented matrix above.*

### 1.1.3   Determinant

Recall that if $A$ is a $1 \times 1$ matrix, say $A = [a_{11}]$, then

$$\det A = a_{11}.$$

If $A$ is an $n \times n$ matrix, with $n > 1$, we use the following inductive definition.

   i. Let $A_{ij}$ denote the matrix obtained from A by deleting row $i$ and column $j$.

   ii. Set $c_{ij} = (-1)^{i+j} \det A_{ij}$. This number is called the *ij-th cofactor* of $A$. (The $\det A_{ij}$ is called the *ij*-th *minor* of $A$.)

   Using this notation, we define ·

$$\det A = a_{11}c_{11} + a_{12}c_{12} + \cdots + a_{1n}c_{1n}$$
$$= a_{11}\left[(-1)^{1+1}\det A_{11}\right]$$
$$+ a_{12}\left[(-1)^{1+2}\det A_{12}\right] + \cdots + a_{1n}\left[(-1)^{1+n}\det A_{1n}\right].$$

**Example 1.6** *Applying the definition,*

$$(a)\ \det\left[\begin{array}{cc} a & b \\ c & d \end{array}\right] = a\left[(-1)^{1+1}\det A_{11}\right] + b\left[(-1)^{1+2}\det A_{12}\right]$$
$$= ad - bc.$$

*(b)* $\det \begin{bmatrix} 2 & 3 & -1 \\ 0 & 4 & -2 \\ -3 & 1 & 0 \end{bmatrix} = 2 \left[ (-1)^{1+1} \det \begin{bmatrix} 4 & -2 \\ 1 & 0 \end{bmatrix} \right] +$

$3 \left[ (-1)^{1+2} \det \begin{bmatrix} 0 & -2 \\ -3 & 0 \end{bmatrix} \right] + (-1) \left[ (-1)^{1+3} \det \begin{bmatrix} 0 & 4 \\ -3 & 1 \end{bmatrix} \right] =$

$2 \cdot 1 \cdot 2 + 3 \cdot (-1) \cdot (-6) + (-1) \cdot 1 \cdot 12 = 10.$

Actually, the determinant can be expanded along any row or column as given below. The proof, a bit intricate, is outlined in the exercises.

(a) $\det A = \sum\limits_{k=1}^{n} a_{ik} c_{ik}$ (the i-th row expansion)

(b) $\det A = \sum\limits_{k=1}^{n} a_{kj} c_{kj}$ (the j-th column expansion)

**Example 1.7** *Let* $A = \begin{bmatrix} 1 & 4 & 3 \\ -2 & 0 & 0 \\ 3 & -1 & 0 \end{bmatrix}$. *Expanding the determinant along the third column, to make use of the numerous 0's there, we have*

$$\det A = 3 \left[ (-1)^{1+3} \det \begin{bmatrix} -2 & 0 \\ 3 & -1 \end{bmatrix} \right] + 0 \cdot c_{23} + 0 \cdot c_{33}$$

$$= 6.$$

Some easy consequences of (a) and (b) follow.

(c) If $T$ is a triangular matrix, then $\det T = t_{11} t_{22} \cdots t_{nn}$. As an example, expanding along the first columns,

$$\det \begin{bmatrix} t_{11} & t_{12} & t_{13} \\ 0 & t_{22} & t_{23} \\ 0 & 0 & t_{33} \end{bmatrix} = t_{11} \det \begin{bmatrix} t_{22} & t_{23} \\ 0 & t_{33} \end{bmatrix} = t_{11} t_{22} t_{33}.$$

(d) If a row of $A$ is a scalar multiple of another row of $A$, then $\det A = 0$. For example,

$$\det \begin{bmatrix} a & b \\ \alpha a & \alpha b \end{bmatrix} = a \alpha b - b \alpha a = 0.$$

(e) If $b_1, b_2, \ldots, b_r$ are $n \times 1$ vectors, then

$$\det \left[ \left( \sum_{k=1}^{r} b_k \right) a_2 \ldots a_n \right] = \sum_{k=1}^{r} \det [b_k \, a_2 \ldots a_n]$$

where $a_k$ is the $k$-th column of $A$. (The result also holds when $\sum_{k=1}^{r} b_k$ is the $j$-th column instead of the first one.) As an example, expanding along the first column,

$$\det \left[\begin{bmatrix} a \\ c \end{bmatrix} + \begin{bmatrix} b \\ d \end{bmatrix} \; \begin{matrix} e \\ f \end{matrix} \right]$$

$$= \det \begin{bmatrix} a+b & e \\ c+d & f \end{bmatrix}$$

$$= (a+b)\,f - (c+d)\,e$$

$$= af - ce + bf - de$$

$$= \det \begin{bmatrix} a & e \\ c & f \end{bmatrix} + \det \begin{bmatrix} b & e \\ d & f \end{bmatrix}.$$

The next property shows that all determinant results about rows hold equally for columns.

(f) $\det A = \det A^t$, $\det A^H = \overline{\det A}$  Observe that

$$\det \begin{bmatrix} a & b \\ c & d \end{bmatrix}^t = \det \begin{bmatrix} a & c \\ b & d \end{bmatrix} = ad - bc = \det \begin{bmatrix} a & b \\ c & d \end{bmatrix}.$$

A group of results, showing how the determinant behaves when elementary operations are performed on the matrix, follows.

(g) If two rows of $A$ are interchanged, obtaining $B$, then $\det B = -\det A$. For example,

$$\det \begin{bmatrix} a & b \\ c & d \end{bmatrix} = ad - bc \text{ while}$$

$$\det \begin{bmatrix} c & d \\ a & b \end{bmatrix} = bc - ad.$$

(h) If any row of $A$ is multiplied by a scalar $\alpha$, obtaining $B$, then $\det B = \alpha \det A$. (So we can pull out a scalar if it appears as a factor in all of the entries of a row.) Observe that in the $2 \times 2$ case,

$$\det \begin{bmatrix} a & b \\ \alpha c & \alpha d \end{bmatrix} = a\alpha d - b\alpha c$$

$$= \alpha\,(ad - bc)$$

$$= \alpha \det \begin{bmatrix} a & b \\ c & d \end{bmatrix}.$$

(i) If a scalar multiple of a row of $A$ is added to another row, obtaining $B$, then $\det B = \det A$. For example, expanding along the last row, using the transpose of (e),

$$\det \begin{bmatrix} a & b \\ c+a & d+b \end{bmatrix} = \det \begin{bmatrix} a & b \\ c & d \end{bmatrix} + \det \begin{bmatrix} a & b \\ a & b \end{bmatrix}$$

$$= \det \begin{bmatrix} a & b \\ c & d \end{bmatrix}.$$

(j) If interchange and add operations are applied to $A$ to obtain a row echelon form $E$, then $\det A = (-1)^t \det E$ where $t$ is the number of times the interchange operation was used. An example will demonstrate the result.

$$\text{Let } A = \begin{bmatrix} 1 & 0 & 1 \\ 2 & 0 & -1 \\ -1 & 1 & 2 \end{bmatrix}. \text{ Applying } -2R_1 + R_2 \text{ and } R_1 + R_3,$$

$$\det A = \det \begin{bmatrix} 1 & 0 & 1 \\ 0 & 0 & -3 \\ 0 & 1 & 3 \end{bmatrix}.$$

Applying $R_2 \leftrightarrow R_3$ yields

$$\det A = -\det \begin{bmatrix} 1 & 0 & 1 \\ 0 & 1 & 3 \\ 0 & 0 & -3 \end{bmatrix}$$

$$= -(-3)$$
$$= 3.$$

One of the strongest results about determinants concerns the product of matrices. It can be used to derive several useful results about matrices.

(k) If $B$ is also an $n \times n$ matrix, then $\det(AB) = (\det A)(\det B)$.

The key ideas for the proofs follow.

**Proof.** For (c), if $T$ is upper triangular, expand the determinant along column 1, and continue this on the subsequent cofactors. The lower triangular case is handled similarly.

Property (d) is proved by induction on $n$. For $n = 2$, the property can be checked directly. Assuming the property for $n = r$, to show the property for $n = r + 1$, expand $\det B$ about a row other than the two which are scalar multiples and use the induction hypothesis on the cofactors.

Property (e) is proved by expanding the determinant along the first column and rearranging.

Property (g) is proved as was (d) .

For property (h), if row $i$ was multiplied by $\alpha$, then expanding $\det B$ along row $i$ yields

$$\begin{aligned}
\det B &= \alpha a_{i1}c_{i1} + \cdots + \alpha a_{in}c_{in}\\
&= \alpha \left(a_{i1}c_{i1} + \cdots + a_{in}c_{in}\right)\\
&= \alpha \det A.
\end{aligned}$$

To prove property (i), use (e) and (d). For property (j), use (g) and (i). And, the proof of (k) is outline in the exercises.  ∎

The *adjoint of* $A$ is defined, by using cofactors, as

$$\operatorname{adj} A = [c_{ij}]^t .$$

**Example 1.8**  *Let* $A = \begin{bmatrix} a & b & c \\ d & e & f \\ g & h & i \end{bmatrix}$ . *Then*

$$\begin{aligned}
\operatorname{adj} A &= \begin{bmatrix} c_{11} & c_{12} & c_{13} \\ c_{21} & c_{22} & c_{23} \\ c_{31} & c_{32} & c_{33} \end{bmatrix}^t\\[2mm]
&= \begin{bmatrix} ei-fh & -(di-fg) & dh-eg \\ -(bi-ch) & ai-cg & -(ah-bg) \\ bf-ce & -(af-cd) & ae-bd \end{bmatrix}^t\\[2mm]
&= \begin{bmatrix} ei-fh & -bi+ch & bf-ce \\ -di+fg & ai-cg & -af+cd \\ dh-eg & -ah+bg & ae-bd \end{bmatrix} .
\end{aligned}$$

Three properties of the adjoint are listed below:

(l)  $A\,(\operatorname{adj} A) = (\operatorname{adj} A)\,A = (\det A)\,I.$

Thus if $\det A \neq 0$, from the inverse equation,

(m)  $A^{-1} = \frac{1}{\det A} \operatorname{adj} A.$  For example, let $A = \begin{bmatrix} a & b \\ c & d \end{bmatrix}$. Then $\operatorname{adj} A = \begin{bmatrix} d & -b \\ -c & a \end{bmatrix}$ and $A^{-1} = \frac{1}{\det A} \operatorname{adj} A = \begin{bmatrix} \frac{d}{ad-bc} & \frac{-b}{ad-bc} \\ \frac{-c}{ad-bc} & \frac{a}{ad-bc} \end{bmatrix}$. Checking, we see that $AA^{-1} = I.$

If $A$ is nonsingular and $b$ is an $n \times 1$ vector, we have Cramer's Rule.

(n)  The solution to $Ax = b$ has as its i-th entry, $x_i = \frac{\det A_i}{\det A}$, where $A_i$ is the matrix obtained from $A$ by replacing column $i$ by $b$.  As an example, if $\begin{bmatrix} 1 & 2 \\ 3 & 4 \end{bmatrix} x = \begin{bmatrix} 1 \\ 1 \end{bmatrix}$, we have

$$x_1 = \frac{\det \begin{bmatrix} 1 & 2 \\ 1 & 4 \end{bmatrix}}{\det \begin{bmatrix} 1 & 2 \\ 3 & 4 \end{bmatrix}} = \frac{2}{-2} = -1,$$

$$x_2 = \frac{\det \begin{bmatrix} 1 & 1 \\ 3 & 1 \end{bmatrix}}{\det \begin{bmatrix} 1 & 2 \\ 3 & 4 \end{bmatrix}} = \frac{-2}{-2} = 1.$$

Finally, the determinant determines nonsingularity.

(o) $A$ is nonsingular if and only if $\det A \neq 0$.

**Proof.** We argue a few of these results leaving the others as exercises.
For (l), we do the $2 \times 2$ case which can be extended to the $n \times n$ case. Here

$$A \operatorname{adj} A = \begin{bmatrix} a_{11} & a_{12} \\ a_{21} & a_{22} \end{bmatrix} \begin{bmatrix} c_{11} & c_{21} \\ c_{12} & c_{22} \end{bmatrix}$$

$$= \begin{bmatrix} a_{11}c_{11} + a_{12}c_{12} & a_{11}c_{21} + a_{12}c_{22} \\ a_{21}c_{11} + a_{22}c_{12} & a_{21}c_{21} + a_{22}c_{22} \end{bmatrix}$$

$$= \begin{bmatrix} \det A & 0 \\ 0 & \det A \end{bmatrix}$$

noting that the off diagonal entries are determinants of matrices with duplicate rows. For example, the $1, 2$-entry is an expansion along the 2nd row of the matrix obtained from $A$ by replacing the 2nd row with the 1st row,

$$a_{11}c_{21} + a_{12}c_{22} = \det \begin{bmatrix} a_{11} & a_{12} \\ a_{11} & a_{12} \end{bmatrix} = 0.$$

Similarly, $(\operatorname{adj} A)A = (\det A)I$.

For (m), if $A$ is nonsingular, $AA^{-1} = I$. Thus, $\det A \det A^{-1} = 1$, and so $\det A \neq 0$. Now, by (l), $A(\frac{1}{\det A} \operatorname{adj} A) = (\frac{1}{\det A} \operatorname{adj} A) A = I$. So $A^{-1} = \frac{1}{\det A} \operatorname{adj} A$.

In (o), if $\det A \neq 0$, from (m), $A^{-1} = \frac{1}{\det A} \operatorname{adj} A$. So $A$ is nonsingular. On the other hand, if $A$ is nonsingular, $\det A \neq 0$ as argued in part (m).

This concludes the proof. ∎

### 1.1.4 ˙ Optional (Ranking)

Suppose, in a tournament, four tennis players, named 1, 2, 3, and 4, play each other exactly once. We draw a directed graph with vertices 1, 2, 3, 4 and arcs from $i$ to $j$ if $i$ beats $j$. Define $A = [a_{ij}]$ where

$$a_{ij} = \begin{cases} 1 \text{ if there is an arc from i to j} \\ 0 \text{ otherwise.} \end{cases}$$

If $A^2 = \left[a_{ij}^{(2)}\right]$, then we can show that

$$a_{ij}^{(2)} = \text{number of secondary wins from i to j}$$
$$\text{i.e., the number of players k where i}$$
$$\text{beats k and k beats j.}$$

To rank the players, set $B = A + A^2$. The sum of the $i$-th row entries in $B$ gives that player's total wins and secondary wins. This total is used to rank the players.

For example, if the outcome of the tournament, in digraph form, is given in Figure 1.3,

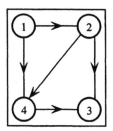

FIGURE 1.3.

then $A = \begin{bmatrix} 0 & 1 & 1 & 1 \\ 0 & 0 & 1 & 1 \\ 0 & 0 & 0 & 0 \\ 0 & 0 & 1 & 0 \end{bmatrix}$ and $A^2 = \begin{bmatrix} 0 & 0 & 2 & 1 \\ 0 & 0 & 1 & 0 \\ 0 & 0 & 0 & 0 \\ 0 & 0 & 0 & 0 \end{bmatrix}$.

Thus $B = A + A^2 = \begin{bmatrix} 0 & 1 & 3 & 2 \\ 0 & 0 & 2 & 1 \\ 0 & 0 & 0 & 0 \\ 0 & 0 & 1 & 0 \end{bmatrix}$. Summing the rows of $B$ yields

$\begin{bmatrix} 6 \\ 3 \\ 0 \\ 1 \end{bmatrix}$. Thus, the ranking is 1: Player 1; 2: Player 2; 3: Player 4; and 4: Player 3.

Expressions such as $A + A^2$, with some refinements, have been used to determine the power of pro football teams. What has been shown above should be considered as a starting point rather than a finished product.

## 1.1.5  MATLAB (Solving $Ax = b$)

Some of the basics of the MATLAB package are given in Appendix A. These basics include how to calculate answers to the computations in this section. We will add to this some additional remarks about solving $Ax = b$.

1. Solving $Ax = b$: If $A$ is $n \times n$ and nonsingular, $A \backslash b$ will provide a solution to $Ax = b$ or indicate it is having a problem. When this occurs, sometimes the mathematics problem can be redescribed to eliminate the difficulty.

   If $A$ is $m \times n$, we can solve $Ax = b$ using the augmented matrix $B = [A \,|\, b]$ and finding the reduced row echelon form. An example follows.

   $B = \begin{bmatrix} 1 & 1 & 1; & 1 & 1 & 1; & 1 & 1 & 1 \end{bmatrix}$;
   rref($B$)

   $$\text{ans} = \begin{bmatrix} 1 & 1 & | & 1 \\ 0 & 0 & | & 0 \\ 0 & 0 & | & 0 \end{bmatrix}$$

   Now, we can write out the solution, $x = \begin{bmatrix} 1 - \alpha \\ \alpha \end{bmatrix} = \begin{bmatrix} 1 \\ 0 \end{bmatrix} + \alpha \begin{bmatrix} -1 \\ 1 \end{bmatrix}$.

2. Least squares solving $Ax = b$: If $A$ is $m \times n$ with $m \neq n$, $A \backslash b$ provides a least-squares solution to $Ax = b$. A least-squares solution is not always a solution. (We study least-squares solutions in Chapter 7 and Chapter 8.) An example follows.

   $A = \begin{bmatrix} 1 & 1; & 1 & 1; & 1 & 1 \end{bmatrix}$;
   $b = \begin{bmatrix} 1; & 2; & 3 \end{bmatrix}$;
   $A \backslash b$

   $$\text{ans} = \begin{bmatrix} 2 \\ 0 \end{bmatrix}$$

If MATLAB is having a solving problem, a warning is given. Warnings here usually indicate that in computing, some 'small' number was assumed to be 0. The 0 didn't occur due to rounding and consequently was set to 0. And had it not been set to 0, the answer may be very different.

For more, type in *help mldivide*.

## Exercises

1. Express in the form $a + bi$.
   - (a) $(3 - 2i)(-4 + 5i)$
   - (b) $(2 - 3i)^2$
   - (c) $\frac{2-5i}{3+2i}$
   - (d) $|4 - 3i|$
   - (e) $e^{(2+3i)t}$, where $t$ is a real parameter

2. Prove that if $z = a + bi$ and $w = c + di$, then

   (a) $\overline{z + w} = \overline{z} + \overline{w}$.     (b) $\overline{zw} = \overline{z}\,\overline{w}$.
   (c) $|z + w| \le |z| + |w|$.    (d) $|zw| = |z|\,|w|$.
   (e) $|z| = (z\overline{z})^{\frac{1}{2}}$.

3. Let $A$, $B$ be $m \times n$ matrices and $\alpha, \beta$ scalars. Prove the following.

   (a) $A + B = B + A$    (b) $(A + B) + C = A + (B + C)$
   (c) $A + 0 = A$    (d) $A + (-A) = 0$
   (e) $(\alpha\beta) A = \alpha (\beta A)$    (f) $(\alpha + \beta) A = \alpha A + \beta A$

4. Compute expressions for the following.

   (a) $\begin{bmatrix} 0 & 1 \\ 2 & 0 \end{bmatrix} \begin{bmatrix} b_1 \\ b_2 \end{bmatrix}$ by forward multiplication. (Here $b_1$ and $b_2$ are $1 \times 2$ vectors.)

   (b) $[a_1 a_2] \begin{bmatrix} 0 & 1 \\ -1 & 1 \end{bmatrix}$ by backward multiplication. (Here $a_1$ and $a_2$ are $2 \times 1$ vectors.)

5. Compute by backward multiplication.

$$[p_1 \; p_2 \; p_3] \begin{bmatrix} \lambda_1 & 0 & 0 \\ 0 & \lambda_2 & 0 \\ 0 & 0 & \lambda_3 \end{bmatrix}$$

6. Two parts:

   (a) Let $T = \begin{bmatrix} t_{11} & t_{12} & t_{13} \\ 0 & t_{22} & t_{23} \\ 0 & 0 & t_{33} \end{bmatrix}$. If $T$ is nonsingular, use the inverse equation $(XT = I)$ and forward multiplication to show that $T^{-1}$ is upper triangular and that its main diagonal is $t_{11}^{-1}, t_{22}^{-1}, t_{33}^{-1}$.

   (b) Extend this result to nonsingular upper triangular matrices in general. (Hint: Start with the last row of $X$.)

7. Let $A$, $B$, and $C$ be $n \times n$ matrices.

   (a) If $AB = AC$, then $B$ need not be $C$. Give an example of this where none of $A$, $B$, or $C$ is 0. Also, explain what arithmetic property, for real numbers, is missing from the arithmetic of matrices that causes this to occur.

   (b) Do the same for: $AB = 0$ doesn't imply $A = 0$ or $B = 0$.

8. Let $A$, $B$, and $C$ be matrices. Assuming all multiplications are defined, prove the following.

(a) $(AB)C = A(BC)$

(b) $A(B+C) = AB + AC$

(c) $(B+C)A = BA + CA$

9. Compute $A^{-1}$, if it exists, by solving $AX = I$, using the augmented matrix.

(a) $A = \begin{bmatrix} 2 & 1 \\ 1 & 1 \end{bmatrix}$   (b) $A = \begin{bmatrix} 1 & -1 \\ -1 & 1 \end{bmatrix}$

10. Prove that if $A$ and $B$ are $n \times n$ matrices and $AB = I$, then $B = A^{-1}$.

11. Let $A$ be an $n \times n$ nonsingular matrix. Prove the following.

(a) $\left(A^{-1}\right)^{-1} = A$

(b) $\left(A^m\right)^{-1} = \left(A^{-1}\right)^m$ , $m$ a positive integer

(c) $\left(A^H\right)^{-1} = \left(A^{-1}\right)^H$

12. Let $A$ be an $m \times n$ matrix. Prove that $\left(A^H\right)^H = A$.

13. Let $A$ be an $m \times r$ matrix and $B$ an $r \times n$ matrix. Prove that $(AB)^H = B^H A^H$.

14. Solve the following.

$$x_1 - x_2 + x_3 + x_4 = 6$$
$$2x_1 + x_2 + 3x_3 - x_4 = 5$$

Indicate, as in Example 1.4, all operations done in solving. Also identify all free variables and pivot variables. Use

(a) Gaussian elimination.

(b) Gauss-Jordan.

15. Consider the system of linear equations

$$ax_1 + bx_2 = e \qquad\qquad (1.3)$$
$$cx_1 + dx_2 = f$$

where a, b, c, d, e, and f are constants.

(a) Apply $\delta R_1 + R_2$ to (1.3) to obtain

$$ax_1 + bx_2 = e \qquad\qquad (1.4)$$
$$(c + \delta a)x_1 + (d + \delta b)x_2 = f + \delta e.$$

Show that if $(\alpha, \beta)$ is a solution to (1.3), it is a solution to (1.4) and vice versa.

(b) Repeat (a) for $\delta R_1$, where $\delta \neq 0$.

16. Consider the matrix equation

$$\begin{bmatrix} a & b \\ c & d \end{bmatrix}\begin{bmatrix} x_1 & y_1 \\ x_2 & y_2 \end{bmatrix} = \begin{bmatrix} e & g \\ f & h \end{bmatrix}$$

where a, b, c, d, e, f, g, and h are constants. To solve, we equate the corresponding columns to get

$$\begin{array}{lll} ax_1 + bx_2 = e & & ay_1 + by_2 = g \\ cx_1 + dx_2 = f & \text{and} & cy_1 + dy_2 = h. \end{array} \tag{1.5}$$

Explain how to solve these equations simultaneously.

17. Write out the expression for

$$\det \begin{bmatrix} a & b & c \\ d & e & f \\ g & h & i \end{bmatrix}.$$

18. Compute $\det A$ for

(a) $A = \begin{bmatrix} 1 & 2 & 2 \\ 1 & -1 & -1 \\ 3 & 1 & 4 \end{bmatrix}$ by expanding along the 2nd column.

(b) $A = \begin{bmatrix} 1 & 2 & 0 & 3 \\ -2 & 1 & 2 & 5 \\ 1 & 3 & 0 & 0 \\ 2 & 1 & 0 & 0 \end{bmatrix}$ by any row or column expansion.

19. Let $A$ be an $n \times n$ matrix. Prove by induction.

(a) $\det A = \det A^t$

(b) If A has two rows that are scalar multiples of each other, then $\det A = 0$.

20. Let $A = \begin{bmatrix} 2 & 0 & 1 \\ 0 & -1 & 3 \\ 2 & 3 & 0 \end{bmatrix}$. Compute adj $A$.

21. If $A$ has entries which are rational numbers, are the entries in adj $A$ rational numbers? Explain.

22. Prove that if $A$ is a $3 \times 3$ nonsingular matrix, then $Ax = b$ has as its solution $x$ where $x_i = \frac{\det A_i}{\det A}$. (Hint: Use that $x = A^{-1}b = \frac{1}{\det A}(\text{adj } A)b$ and write out the expressions for $x_1$, $x_2$, and $x_3$.)

23. Let $A$ be an $n \times n$ matrix. Prove by induction that expanding the determinant along any row yields the same result as expanding about the first row. (Hint: For the general step, expand the determinant along the $i$-th row and then those cofactors along the 1st row. Show this is the same as expanding about the first row and then those cofactors along the $i$-th row of $A$.)

24. Let $A$ be an $n \times n$ matrix and $E$ a row echelon form for $A$. Prove that if $A$ is singular, $E$ has a row of 0's and vice versa.

25. Let $A$ and $B$ be a $3 \times 3$ matrices. Prove that $\det(AB) = \det A \det B$ using the following outline.

    (a) Prove that if an elementary operation is done on $A$ and on $AB$, obtaining $\hat{A}$ and $\widehat{AB}$, then $\hat{A}B = \widehat{AB}$. (Hint: Write $AB = \begin{bmatrix} a_1 B \\ a_2 B \\ a_3 B \end{bmatrix}$.)

    (b) Apply interchange and add operations to $A$ and $AB$ to obtain $E$ and $\widehat{AB}$, respectively, where $E$ is a row echelon form. Then, by (a), $EB = \widehat{AB}$. If $t$ interchange operations were used, then

    $$\det(AB) = (-1)^t \det(EB)$$

    $$= (-1)^t \det \begin{bmatrix} e_{11}b_1+ & e_{12}b_2+ & e_{13}b_3 \\ & e_{22}b_2+ & e_{23}b_3 \\ & & e_{33}b_3 \end{bmatrix}.$$

    Using properties (d) and (e), continue to get

    $$\det(AB) = (-1)^t \det \begin{bmatrix} e_{11} & b_1 \\ e_{22} & b_2 \\ e_{33} & b_3 \end{bmatrix}$$

    $$= (-1)^t e_{11}e_{22}e_{33} \det B$$

    $$= (-1)^t \det E \det B.$$

    Now use that $\det A = (-1)^t \det E$ to finish the work.

26. Let $A = \begin{bmatrix} B & c \\ d & \alpha \end{bmatrix}$ where $B$ is a square matrix and $\alpha$ a scalar. Using partitioned arithmetic, compute $A^2$. (Be sure all multiplications used are defined.)

27. Let $A = \begin{bmatrix} I & B \\ 0 & I \end{bmatrix}$ be a partitioned matrix where both $I$'s are $n \times n$. Using partition arithmetic, and the inverse equation, find the corresponding partitioned form of $A^{-1}$.

28. The equation

$$.26x_1 + .35x_2 = .09$$
$$.54x_1 + .70x_2 = .16$$

has solution $x = (-1, 1)^t$. Show that $\hat{x} = (-1.51, \ 1.38)^t$ nearly solves the equation by showing that $b - Ax$ is small.

29. (Optional) Rank the players for the tournament graph given in Figure 1.4.

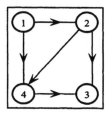

FIGURE 1.4.

30. (MATLAB) Let $A = \begin{bmatrix} 1 & 2 & 1 \\ 2 & -1 & 0 \\ 0 & 1 & 3 \end{bmatrix}$ and $B = \begin{bmatrix} 0 & 1 & -1 \\ 1 & 3 & 1 \\ 2 & 1 & 1 \end{bmatrix}$. Compute the following.

   (a) $A + B$    (b) $A - B$
   (c) $AB$        (d) $3A$
   (e) $A^{-1}B$   (f) $BA^{-1}$
   (g) $A^{-1}$    (h) $A^6$
   (i) $\det A$    (j) rref $A$

31. (MATLAB) Let $A$ and $B$ be as in Exercise 30 and $b = (1, 0, 1)^t$.

   (a) Solve $Ax = b$.

   (b) Solve $Ax = b$ using *format long*. (To extend the display of the answer on the screen type in *format long*. This will provide answers to about 15 digits. To return to standard format, type in *format short*.)

32. (MATLAB) Solve

$$1x_1 + 1x_2 + 0x_3 + 1x_4 = 7$$
$$1x_1 + 0x_2 + 1x_3 + 0x_4 = 4$$
$$0x_1 + 1x_2 + 1x_3 + 1x_4 = 10$$

(a) By using the reduced row echelon form.

(b) By using $A \backslash b$.

33. (MATLAB)  Solve $\begin{bmatrix} 1 & 0 & 1 \\ 1 & 1 & 0 \\ 0 & 1 & 0 \end{bmatrix} X = \begin{bmatrix} 1 & -1 \\ 2 & 0 \\ 1 & 3 \end{bmatrix}$ for $X$.

# 2
# Introduction to Vector Spaces

It has been observed that there are many algebraic systems which, in terms of arithmetic properties, are just like $R^2$ and $R^3$. These systems often arise in mathematical work. In this chapter, we give a general study of these systems.

As we go through this chapter, we will see very little direct application (for example model building) of it. The reason for this is that this chapter introduces concepts and techniques (tools, so to speak) which are then used throughout matrix theory and applications. These tools are important to learn.

## 2.1 Vector Spaces

In this section, we study algebraic systems having arithmetic properties like those of $R^2$ and $R^3$. These algebraic systems are called vector spaces. The general (abstract) definition of a vector space follows.

**Definition 2.1** *A vector space is a nonempty set $V$ with elements called vectors , together with a set of numbers, called scalars. The set of numbers can be $\mathcal{R}$ or $\mathcal{G}$. (When we need emphasis, we can use the words real vector space or complex vector space to distinguish the two cases.)*

  *(a) On $V$ there is an operation, called vector addition, that combines any pair of vectors $x$ and $y$ into a vector, denoted by $x + y$, called their sum. This addition must satisfy the following properties.*

i. $x + y = y + x$ for all vectors $x$ and $y$.

ii. $(x + y) + z = x + (y + z)$ for all vectors $x$, $y$ and $z$.

iii. There is a unique vector, denoted by $0$, such that $0 + x = x$ for all vectors $x$.

iv. For any vector $x$, there corresponds a unique vector, denoted by $-x$, such that $x + (-x) = 0$.

(b) On $V$ there is an operation, called scalar multiplication, that combines a scalar $\alpha$ and a vector $x$ into a vector, denoted by $\alpha x$, called their product. This scalar multiplication must satisfy the properties below.

v. $\alpha(x + y) = \alpha x + \alpha y$ for all vectors $x$ and $y$ and scalars $\alpha$.

vi. $(\alpha + \beta)x = \alpha x + \beta x$ for all vectors $x$ and scalars $\alpha$ and $\beta$.

vii. $\alpha(\beta x) = (\alpha\beta)x$ for all vectors $x$ and scalars $\alpha$ and $\beta$.

viii. $1x = x$ or all vectors $x$.

We should remark for clarity that when we talk about scalars, we mean the scalars (from either $\mathcal{R}$ or $\mathcal{C}$) of the vector space. (The properties given above can be recalled since they are the basic arithmetic properties of $R^2$ and $R^3$, four involving $+$, and four involving a mix of $+$ and scalar multiplication.)

We intend to develop a theory (a collection of results) about vector spaces. Results will be stated about vector spaces, in general, and thus proofs can only use the properties listed in the definition of a vector space. To show how this is done, we provide a proof of a lemma extending the properties of a vector space.

**Lemma 2.1**  *The following are also properties of a vector space.*

(a) $0x = 0$

(b) $\alpha 0 = 0$

(c) $-1x = -x$

(d) $\alpha x = 0$ implies $\alpha = 0$ or $x = 0$

**Proof (a).**  Using that the scalar $0$ satisfies $0 + 0 = 0$, we have

$$0x = (0 + 0)x = 0x + 0x. \qquad (2.1)$$

Now, $0x$ is a vector and thus has an additive inverse, $-(0x)$. Adding this vector to the left and right sides of (2.1) and simplifying using the properties

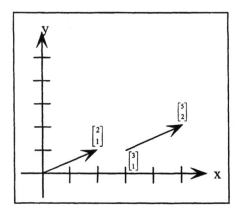

FIGURE 2.1.

of a vector space, yields

$$-(0x) + 0x = -(0x) + (0x + 0x)$$
$$-(0x) + 0x = (-(0x) + 0x) + 0x$$
$$0 = 0 + 0x$$
$$0 = 0x$$

the desired result. ∎

As mentioned previously, there are many vector spaces. A few of these follow.

**Example 2.1** *Trivial Space: Let $V = \{0\}$ with addition and scalar multiplication defined by the tables*

| + | 0 |
|---|---|
| 0 | 0 |

*and*

| · | 0 |
|---|---|
| $\alpha$ | 0 |

*for all scalars $\alpha$. This is a vector space.*

**Example 2.2** *Vector spaces $R^2$ and $R^3$ (This example is helpful in developing a geometric view of vector calculations.): We will develop this material in $R^2$. The generalization to $R^3$ should be clear.*

*Recall from calculus, a geometric vector from a point $x = (x_1, x_2)^t$ to a point $y = (y_1, y_2)^t$, written $\overrightarrow{xy}$, is a directed line segment from $x$ to $y$. The inclination of such a vector is $(y_1 - x_1, y_2 - x_2)^t$. Two geometric vectors are equal (equivalent) if they have the same inclination. For example, in the diagram (Figure 2.1), the geometric vectors are equal.*

*Two geometric vectors can be added by finding any two equivalent vectors with the same initial point and adding those vectors by the parallelogram law. Alternately we can take equivalent vectors that are appended at end and initial points and complete the triangle for the sum.*

*Any point x in $R^2$ can be associated with the geometric vector from the origin to x. With this association, arithmetic in $R^2$ can be envisioned in terms of geometric vectors. For example,*

1. *Scalar multiplication:* $2\begin{bmatrix} 2 \\ 1 \end{bmatrix} = \begin{bmatrix} 4 \\ 2 \end{bmatrix}$ *can be seen by drawing the geometric vector from* $\begin{bmatrix} 0 \\ 0 \end{bmatrix}$ *to* $\begin{bmatrix} 2 \\ 1 \end{bmatrix}$ *and by scaling this corresponding geometric vector by 2. (See Figure 2.2.)*

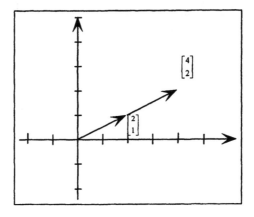

FIGURE 2.2.

2. *Addition:* $\begin{bmatrix} 1 \\ 4 \end{bmatrix} + \begin{bmatrix} 4 \\ 1 \end{bmatrix} = \begin{bmatrix} 5 \\ 5 \end{bmatrix}$ *can be seen by adding the geometric vectors corresponding to* $\begin{bmatrix} 1 \\ 4 \end{bmatrix}$ *and* $\begin{bmatrix} 4 \\ 1 \end{bmatrix}$ *by the parallelogram rule, or appending (an end to an initial) and completing the triangle. (See Figure 2.3.)*

*Appending is also useful in seeing subtraction. Note in Figure 2.4, that a geometric vector equivalent to x − p can be seen by beginning at the point p, going to 0 (to obtain a geometric vector equivalent to −p), then proceeding to the point x as diagrammed (adding x to −p). Completing the triangle gives a geometric vector for x − p. Note that to find the corresponding point*

$$\begin{bmatrix} x_1 - p_1 \\ x_2 - p_2 \end{bmatrix}$$

*in $R^2$, we need to start this vector at the origin.*

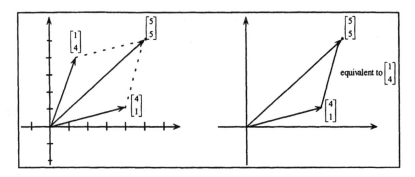

FIGURE 2.3.

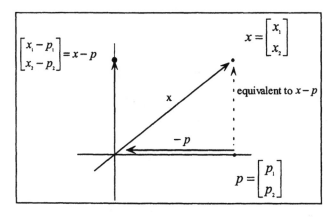

FIGURE 2.4.

*We also use the following parametric descriptions from calculus.*

1. *Line:  The line through points a and b, a ≠ b, is given by*

$$x = ta + (1 - t)\, b, \text{ where } -\infty < t < \infty. \tag{2.2}$$

2. *Segment:  The segment between points a and b, a ≠ b, is given by*

$$x = ta + (1 - t)\, b, \text{ where } 0 \le t \le 1. \tag{2.3}$$

We use (2.2) and (2.3) as the equations of lines through, and segments between, vectors $a$ and $b$, $a \ne b$, in any vector space $V$.

**Example 2.3** *Matrix Space and Euclidean Space:  Let*

$$R^{m \times n} = \{A : A \text{ is an } m \times n \text{ matrix with entries in } R\}.$$

*Using real scalars, matrix addition, and scalar multiplication, $R^{m \times n}$ is a vector space. (Since the vectors are matrices, we can call $R^{m \times n}$ a matrix* space.) $C^{m \times n}$ is defined similarly, using complex numbers in vectors as well as for the scalars.

For simplicity of notation, set

$$R^m = R^{m \times 1} \text{ and } C^m = C^{m \times 1}.$$

These are the classical real or complex Euclidean m-spaces, respectively. We use the words Euclidean m-space, or simply the symbol $E^m$, to denote either $R^m$ or $C^m$.

**Example 2.4** *Polynomial space: Let*

$$P_n = \left\{ \begin{array}{l} p : p(t) = a_{n-1}t^{n-1} + a_{n-2}t^{n-2} + \cdots + a_0 \\ \text{where } a_{n-1}, a_{n-2}, \ldots, a_0 \text{ are real scalars} \\ \text{and } t \text{ a real variable} \end{array} \right\}.$$

Using real scalars, the usual addition (adding coefficients of like terms) and scalar multiplication (multiplying the coefficients by the scalar), $P_n$ is a vector space.

We should also recall here that two polynomials are equal if and only if the coefficients of their corresponding terms are equal, e.g., if

$$a_2t^2 + a_1t + a_0 = b_2t^2 + b_1t + b_0,$$

then $a_2 = b_2$, $a_1 = b_1$, and $a_0 = b_0$.

**Example 2.5** *Function space: Let $[a, b]$ be an interval of real numbers and*

$$C[a, b] = \{f : f \text{ is a real continuous function on } [a, b]\}.$$

Using real scalars, the usual definition of addition and scalar multiplication, namely

$$(f + g)(t) = f(t) + g(t) \text{ and}$$
$$(\alpha f)(t) = \alpha(f(t)),$$

$C[a, b]$ is a vector space.

It is also helpful to recall that two functions, $f$ and $g$, are equal (written $f = g$) if and only if

$$f(t) = g(t)$$

for all $t$.

For the open interval $(a, b)$, $C(a, b)$ is defined similarly.

Most vector spaces arise inside the larger vector spaces given in the examples above. The definition below describes such sets.

**Definition 2.2**  *Let $V$ be a vector space and $W$ a nonempty subset of $V$. Then $W$ is a subspace of $V$ provided that $W$ is*

   *i. Closed under addition: if $x, y \in W$ then $x + y \in W$.*

   *ii. Closed under scalar multiplication: if $x \in W$ and $\alpha$ any scalar, then $\alpha x \in W$.*

*(The addition and scalar multiplication is as that in $V$.)*

We leave it as an exercise to show that, using this definition, geometrically a subspace in $R^3$ must be one of the following:

(a) The set $\{0\}$.

(b) A line through the origin.

(c) A plane through the origin.

(d) $R^3$ itself.

(See illustrations of each in Figure 2.5.)

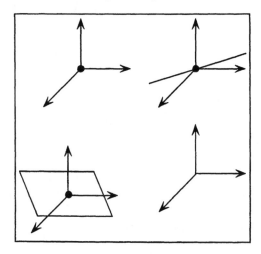

FIGURE 2.5.

We now show that sets which are subspaces are actually vector spaces.

**Theorem 2.1**  *Every subspace is a vector space.*

**Proof.** Suppose $W \subseteq V$ and satisfies the definition of a subspace. We show that $W$ satisfies the definition of a vector space.

By the definition of a subspace, $W \neq \emptyset$ and properties (a) and (b) of the definition of a vector space hold. Thus, we need only verify properties (i) through (viii) of the definition of a vector space. We do a sample of these.

**Property i.** Let $x, y \in W$. Then $x + y = y + x$ since this is true in $V$. (The addition table for $W$ is a subtable of the addition table for $V$.)

**Property iv.** Let $x \in W$. By Lemma 2.1, $-1x = -x$. Thus by (b) of the definition of a subspace, we know that $-x \in W$. Hence, this verifies property (iv). ∎

An example showing how to apply this theorem follows.

**Example 2.6**  *We show that the set $W \subseteq R^{2\times2}$ of symmetric matrices is a vector space.*

*Here we simply check the properties of the definition of a subspace. Clearly, $W \neq \emptyset$.*

*Closure of addition:* Let $\begin{bmatrix} a & b \\ b & c \end{bmatrix}, \begin{bmatrix} d & e \\ e & f \end{bmatrix} \in W$. Then $\begin{bmatrix} a & b \\ b & c \end{bmatrix} +$ $\begin{bmatrix} d & e \\ e & f \end{bmatrix} = \begin{bmatrix} a+d & b+e \\ b+e & c+f \end{bmatrix}$. *Since this sum is symmetric, the sum is in $W$.*

*Closure of scalar multiplication:* Let $\alpha$ be a scalar and

$$\begin{bmatrix} a & b \\ b & c \end{bmatrix} \in W.$$

*Then*

$$\alpha \begin{bmatrix} a & b \\ b & c \end{bmatrix} = \begin{bmatrix} \alpha a & \alpha b \\ \alpha b & \alpha c \end{bmatrix}.$$

*Thus, the product is symmetric and hence in $W$.*

*Having verified the properties of a subspace, it follows that $W$ is a subspace and thus a vector space.*

In the remaining work, we will show how axes can be put in a vector space. (This goal will help unify the work.) To get the idea of how this is done, note that $R^3$ has axes determined by $e_1 = \begin{bmatrix} 1 \\ 0 \\ 0 \end{bmatrix}$, $e_2 = \begin{bmatrix} 0 \\ 1 \\ 0 \end{bmatrix}$,

$e_3 = \begin{bmatrix} 0 \\ 0 \\ 1 \end{bmatrix}$. In some sense (which we describe later), these vectors point out different dimensions. And, any $x \in R^3$ can be reached using them. For example, if $x = (2, 3, 4)^t$, then

$$x = 2e_1 + 3e_2 + 4e_3,$$

and the coordinates 2, 3, 4 tell how $x$ is reached; i.e., go 2 units on the axis determined by $e_1$, then 3 units in the direction of the axis determined by $e_2$, etc.

To obtain axes in an arbitrary vector space, we look for vectors with two special properties. (i) We must be able to reach any vector using them, and (ii) these vectors must point out different dimensions. The first property is mathematically described below.

**Definition 2.3**  *Let $S = \{x_1, \ldots, x_m\}$ be a nonempty subset of a vector space $V$. If*

$$x = \alpha_1 x_1 + \cdots + \alpha_m x_m$$

*for some scalars $\alpha_1, \ldots, \alpha_m$, then $x$ is a linear combination of $x_1, \ldots, x_m$. (So $x$ can be reached by going $\alpha_1$ units on the axis determined by $x_1$, etc.) The set of all linear combinations of $x_1, \ldots, x_m$, the set that $S$ spans, is called the* span *of $S$. That is,*

$$\text{span } S = \left\{ \begin{array}{l} x : x = \alpha_1 x_1 + \cdots + \alpha_m x_m \\ \textit{for some scalars } \alpha_1, \ldots, \alpha_m \end{array} \right\}$$

*(So $S$ is the set of all reachable vectors.)*

**Example 2.7**  *If we view $R^3$, the span of a nonzero vector, say $x_1$, is a line. (See Figure 2.6.) The span of the two noncollinear vectors, say $x_1$*

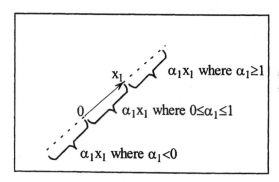

FIGURE 2.6.

*and $x_2$ illustrated in Figure 2.7, is a plane.*
    *And if we have three noncoplanar vectors, they span $R^3$.*

As we might expect, spans provide subspaces.

**Theorem 2.2**  *Let $V$ be a vector space and $S = \{x_1, \ldots, x_m\}$ a nonempty subset of $V$. Then* span $S$ *is a subspace of $V$.*

**Proof.**  To show span $S$ is a subspace, we need to verify each property of the definition of a subspace.

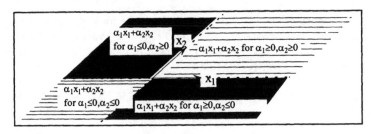

FIGURE 2.7.

Closure of addition: Let $x, y \in$ span $S$. Then

$$x = \alpha_1 x_1 + \cdots + \alpha_m x_m \text{ and}$$
$$y = \beta_1 x_1 + \cdots + \beta_m x_m$$

for some scalars $\alpha_1, \ldots, \alpha_m$ and $\beta_1, \ldots, \beta_m$. Adding we get

$$x + y = (\alpha_1 + \beta_1)x_1 + \cdots + (\alpha_m + \beta_m)x_m$$

a linear combination of $x_1, \ldots, x_m$. Thus, $x + y \in$ span $S$.

Closure of scalar multiplication: left as an exercise. ∎

We conclude this section by demonstrating that just a few vectors can span $R^{2\times 2}$.

**Example 2.8** *Let* $E_{11} = \begin{bmatrix} 1 & 0 \\ 0 & 0 \end{bmatrix}$, $E_{12} = \begin{bmatrix} 0 & 1 \\ 0 & 0 \end{bmatrix}$, $E_{21} = \begin{bmatrix} 0 & 0 \\ 1 & 0 \end{bmatrix}$, *and* $E_{22} = \begin{bmatrix} 0 & 0 \\ 0 & 1 \end{bmatrix}$. *We show these matrices span* $R^{2\times 2}$.

*To do this, we take an arbitrary matrix* $A$ *in* $R^{2\times 2}$, *say* $A = \begin{bmatrix} a & b \\ c & d \end{bmatrix}$. *We need to show that* $A$ *is a linear combination of* $E_{11}, E_{12}, E_{21},$ *and* $E_{22}$. *Thus, set*

$$\alpha_1 E_{11} + \alpha_2 E_{12} + \alpha_3 E_{21} + \alpha_4 E_{22} = A.$$

*Equating corresponding entries, we have a solution, namely*

$$\alpha_1 = a, \ \alpha_2 = b, \ \alpha_3 = c, \text{ and } \alpha_4 = d.$$

*Hence,* $A \in$ span $\{E_{11}, E_{12}, E_{21}, E_{22}\}$ *and since* $A$ *was arbitrarily chosen,*

$$\text{span}\,\{E_{11}, E_{12}, E_{21}, E_{22}\} = R^{2\times 2}.$$

## 2.1.1   Optional (Geometrical Description of the Solutions to $Ax = b$)

In this optional, we give a geometrical view of the solution set of $Ax = b$. To see this, we consider a small system,

$$
\begin{array}{rrrcr}
2x & +4y & -6z & = & 2 \\
-3x & -6y & +9z & = & -3
\end{array} .
$$

As a matrix equation we can write this system as

$$
\begin{bmatrix} 2 & 4 & -6 \\ -3 & -6 & 9 \end{bmatrix} \begin{bmatrix} x \\ y \\ z \end{bmatrix} = \begin{bmatrix} 2 \\ -3 \end{bmatrix}.
$$

To solve, we apply Gaussian elimination to get

$$
\begin{bmatrix} x \\ y \\ z \end{bmatrix} = \begin{bmatrix} 1 - 2\beta + 3\alpha \\ \beta \\ \alpha \end{bmatrix}
$$

where $\alpha$ and $\beta$ are arbitrary. Thus, there are infinitely many solutions, one for each pair of chosen $\alpha$ and $\beta$.

Actually, the solution set is more than an infinite set, it has shape. Note that we can pull out $\alpha$ and $\beta$ from the vector and write it as

$$
\begin{bmatrix} 1 \\ 0 \\ 0 \end{bmatrix} + \alpha \begin{bmatrix} 3 \\ 0 \\ 1 \end{bmatrix} + \beta \begin{bmatrix} -2 \\ 1 \\ 0 \end{bmatrix}.
$$

If we graph all vectors of the form $\alpha \begin{bmatrix} 3 \\ 0 \\ 1 \end{bmatrix} + \beta \begin{bmatrix} -2 \\ 1 \\ 0 \end{bmatrix}$, we get

$$
\operatorname{span} \left\{ \begin{bmatrix} 3 \\ 0 \\ 1 \end{bmatrix}, \begin{bmatrix} -2 \\ 1 \\ 0 \end{bmatrix} \right\},
$$

a plane through the origin. The graph of the solution set then is a translation, by $\begin{bmatrix} 1 \\ 0 \\ 0 \end{bmatrix}$, of that plane, as depicted in Figure 2.8.

Although $Ax = b$ is a more general equation than that just studied, its solutions (providing there are any) can be described in the form

$$
x = x_0 + \alpha_1 x_1 + \cdots + \alpha_m x_m
$$

where $x_0, \ldots, x_m$ are vectors and $\alpha_1, \ldots, \alpha_m$ the free variables. Thus, if $W = \operatorname{span} \{x_1, \ldots, x_m\}$, then the solution set is

$$
x_0 + W = \{x : x = x_0 + w \text{ where } w \in W\},
$$

a translation by $x_0$ of the subspace $W$. Such sets are called *affine spaces* (or linear manifolds). Thus, the set of solutions to a system of linear equations has that kind of shape.

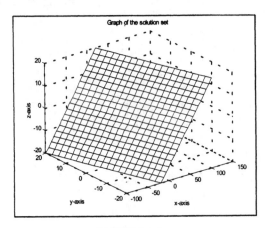

FIGURE 2.8.

## 2.1.2    MATLAB (Graphics)

The basics of graphing a function can be found in Appendix A.

**Code for Graphing** $x = 1 - 2y + 3z$

```
y = linspace (−20,  20,  20);
z = linspace (−20,  20,  20);
[y, z] = meshgrid (y, z);              % More points give a finer grid.
mesh (1 − 2 * y + 3 * z, y, z)
xlabel('x-axis'), ylabel('y-axis'),    % Labels axes.
       zlabel('z-axis')
title ('Graph of  the solution set')   % Gives graph a title.
```

For more information on graphing, type in *help mesh*.

# Exercises

1. Prove (b), (c), and (d) of Lemma 2.1.

2. Two parts.

   (a) Draw geometric vectors corresponding to $(1, 2)^t$, $(2, 1)^t$, $-2 (1, 2)^t$, and $2 (1, 2)^t - (2, 1)^t$.

   (b) Using Figure 2.9, show (i) a vector equivalent to $a - b$ using $a$ and $b$ and (ii) $a - b$ originating at the origin.

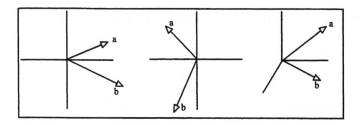

FIGURE 2.9.

3. Four parts.

(a) Give the 0 vector for each of the following vector spaces.
  i. $R^4$    ii. $R^{2\times 3}$
  iii. $P_3$    iv. $C[0,1]$

(b) Give the form of an arbitrary vector in the following.
  i. $R^3$    ii. $R^{2\times 3}$
  iii. $P_3$

(c) Give $-x$ for the following $x$'s.

  i. $x = \begin{bmatrix} 1 \\ -1 \\ 0 \end{bmatrix}$

  ii. $x = \begin{bmatrix} 1 & -1 \\ -1 & 0 \end{bmatrix}$
  iii. $x = t^2 + 2t - 1$
  iv. $x = \sin t$

(d) Identify, by context, which symbols are vectors and which are scalars.
  i. $0x = 0$    ii. $x + (-x)$
  iii. $A + 0 = 0$    iv. $\alpha f(t) + \beta g(t) = 0$

4. Let $f, g, h \in C[a, b,]$ and $\alpha, \beta$ scalars. Prove the following.

(a) $(f + g) + h = f + (g + h)$
(b) $\alpha(f + g) = \alpha f + \alpha g$

5. Find the parametric equation of the line determined by the following.

(a) $x = \begin{bmatrix} 1 \\ 0 \\ 1 \end{bmatrix}, y = \begin{bmatrix} 1 \\ -1 \\ 2 \end{bmatrix}$

(b) $A = \begin{bmatrix} 1 & 1 \\ 0 & 1 \end{bmatrix}, B = \begin{bmatrix} -1 & 1 \\ 1 & 1 \end{bmatrix}$

6. Prove that each of the following is a subspace.

   (a) $W = \left\{ x : x = (x_1, x_2, x_3)^t \text{ and } x_1 + x_2 = x_3 \right\}$
   (b) $W = \{ p : p(t) = at^2 + bt + c \text{ and } a + b + c = 0 \}$
   (c) $W = \{ A : A \in R^{2 \times 2} \text{ and } A \text{ is upper triangular} \}$
   (d) $W = \{ f : f \in C[-1, 1] \text{ and } f(0) = 0 \}$
   (e) $W = \{ A : A \in R^{2 \times 2} \text{ and } a_{11} + a_{12} = 0, a_{21} + a_{22} = 0 \}$
   (f) $C_1(-\infty, \infty) = \{ f : f \in C(-\infty, \infty) \text{ and } f \text{ is differentiable} \}$
       (Use known calculus results.)
   (g) $W = \{ f : f \in C_1(-\infty, \infty) \text{ and } f' - f = 0 \}$
   (h) $W = \{ x : x \text{ is a function on the nonnegative integers and } x(k+1) + x(k) = 0 \text{ for all } k. \}$ (Use that the set of functions defined on the nonnegative integers is a vector space.)

7. Show that the following subsets of $R^{2 \times 2}$ are not subspaces.

   (a) $W = \{ A : A \text{ is the singular matrix} \}$
   (b) $W = \{ A : A \text{ is the nonsingular matrix} \}$

8. Prove that if $W$ is a subspace of a vector space $V$, then $0 \in W$.

9. Prove Theorem 2.1, property (ii).

10. Graph the solution set of $y = x^2$ in $R^2$. Is the solution set, namely $\{ (x, y) : y = x^2 \}$ a vector space?

11. Explain why, geometrically, a subspace in $R^3$ must be $\{0\}$, a line through the origin, a plane through the origin, or $R^3$. (The explanation may be a bit rough, so support it with drawings.)

12. Decide if $x$ is a linear combination of $y$ and $z$.

   (a) $x = (1, -1, 2)^t$, $y = (2, 0, -1)^t$, $z = (1, 0, 1)^t$
   (b) $x = \begin{bmatrix} 1 & 1 \\ 1 & 0 \end{bmatrix}$, $y = \begin{bmatrix} 0 & 1 \\ 1 & 1 \end{bmatrix}$, $z = \begin{bmatrix} 1 & 0 \\ 0 & 1 \end{bmatrix}$
   (c) $x = t + 1$, $y = 2t - 3$, $z = 4$

13. Three parts. Prove the following.

   (a) $S = \left\{ (1, 1, 0)^t, (1, -2, -1)^t, (-1, 2, 2)^t \right\}$ spans $R^3$.
   (b) $S = \left\{ \begin{bmatrix} 1 & 0 \\ 0 & 0 \end{bmatrix}, \begin{bmatrix} 0 & 0 \\ 0 & 1 \end{bmatrix}, \begin{bmatrix} 0 & 1 \\ 1 & 0 \end{bmatrix} \right\}$ spans the symmetric matrices in $R^{2 \times 2}$.

(c) $S = \{t - 1, t + 1\}$ spans $P_2$.

14. Decide if the given set spans the given vector space.

    (a) $\left\{(1,1,1)^t, (-1,1,0)^t, (0,1,1)^t\right\}, R^3$

    (b) $\{1 + t, t + t^2\}, F_2$

    (c) $\left\{\begin{bmatrix} 1 & 1 \\ 1 & 0 \end{bmatrix}, \begin{bmatrix} 0 & 1 \\ 1 & 1 \end{bmatrix}, \begin{bmatrix} 1 & 0 \\ 1 & 1 \end{bmatrix}\right\}, R^{2\times2}$

15. Draw span $\left\{(1,1,0)^t, (1,1,1)^t\right\}$ in $R^3$. Find two other vectors that span the same space.

16. Prove Theorem 2.2, part b.

17. Let $V$ and $W$ be vector spaces with the same set of scalars. Define on $V \times W$,

$$(v_1, w_1) + (v_2, w_2) = (v_1 + v_2, w_1 + w_2) \text{ and}$$
$$\alpha(v_1, w_1) = (\alpha v_1, \alpha w_1)$$

where $v_1, v_2 \in V$; $w_1, w_2 \in W$, and $\alpha$ a scalar. Prove that $V \times W$ with this $+$ and scalar multiplication is a vector space.

18. Let $V$ be a vector space and $U$, $W$ subspaces of $V$. Prove that each of the following is a subspace of $V$.

    (a) $U \cap W$

    (b) $U + W$, $U + W = \{x = u + w \text{ where } u \in U, w \in W\}$

19. (MATLAB) Solve

$$\begin{aligned} x_1 + x_2 + x_3 &= 3 \\ x_1 \phantom{{}+x_2} - x_3 &= 0 \end{aligned}$$

by using the rref command. Graph the solution set as done in the example in Optional.

## 2.2  Dimension

In this section we continue the study, started in the last section, of finding vectors that form axes in a vector space. We now mathematically describe the second special property (vectors pointing out different dimensions) needed for such sets.

We first describe the property algebraically, so we can calculate. Later in this section, we will show that our algebraic description is what we want geometrically.

**Definition 2.4** *Let $V$ be a vector space and $S = \{x_1, \ldots, x_m\}$ a nonempty subset of $V$. If the* pendent equation

$$\alpha_1 x_1 + \cdots + \alpha_m x_m = 0$$

*has only the trivial solution, $\alpha_1 = \cdots = \alpha_m = 0$, then $S$ is* linearly independent. *The set $S$ is* linearly dependent *if the pendent equation has a nontrivial solution.*

Alternately (as often used in other books), we say that vectors $x_1, \ldots, x_m$ are linearly independent or linearly dependent if the "set" of these vectors is linearly independent or linearly dependent .

The following example shows how decisions about linearly independent and linearly dependent sets are made.

**Example 2.9** *Decide if*

$$\begin{bmatrix} 1 & 1 \\ 1 & 0 \end{bmatrix}, \begin{bmatrix} 1 & 1 \\ 0 & 1 \end{bmatrix}, \begin{bmatrix} 1 & 0 \\ 1 & 1 \end{bmatrix}$$

*are linearly independent or linearly dependent.*

*To do this, we solve the corresponding pendent equation,*

$$\alpha_1 \begin{bmatrix} 1 & 1 \\ 1 & 0 \end{bmatrix} + \alpha_2 \begin{bmatrix} 1 & 1 \\ 0 & 1 \end{bmatrix} + \alpha_3 \begin{bmatrix} 1 & 0 \\ 1 & 1 \end{bmatrix} = \begin{bmatrix} 0 & 0 \\ 0 & 0 \end{bmatrix}.$$

*Equating corresponding entries, we have*

$$
\begin{aligned}
\alpha_1 &+ \alpha_2 &+ \alpha_3 &= 0 \\
\alpha_1 &+ \alpha_2 & &= 0 \\
\alpha_1 & &+ \alpha_3 &= 0 \\
&\alpha_2 &+ \alpha_3 &= 0
\end{aligned}.
$$

*Solving, by say Gaussian elimination, yields only*

$$\alpha_1 = 0, \ \alpha_2 = 0, \ \alpha_3 = 0.$$

*Thus,* $\begin{bmatrix} 1 & 1 \\ 1 & 0 \end{bmatrix}, \begin{bmatrix} 1 & 1 \\ 0 & 1 \end{bmatrix}, \begin{bmatrix} 1 & 0 \\ 1 & 1 \end{bmatrix}$ *are linearly independent.*

We now attach some geometry to our definition. First, let $x$ be a vector. Observe that if $x = 0$, then $\alpha 0 = 0$ has nontrivial solutions (Any $\alpha \neq 0$ will do.), so $\{0\}$ is linearly dependent. If $x \neq 0$, then $\alpha x = 0$ implies $\alpha = 0$ by Lemma 2.1 (d), so $\{x\}$ is linearly independent. (Actually, a single such vector generates an axis.)

Now consider a set of vectors $S = \{x_1, \ldots, x_m\}$ where $m \geq 2$. If

$$x_k \notin \operatorname{span} S \backslash \{x_k\},$$

then $x_k$ is not reachable using the vectors in $S \backslash \{x_k\}$. (Recall that $S \backslash \{x_k\}$ is the set $S$ with the vector $x_k$ removed.) Thus, we say $x_k$ is *independent* in $S$. If

$$x_k \in \text{span } S \backslash \{x_k\},$$

$x_k$ is reachable using the vectors in $S \backslash \{x_k\}$, and so we say $x_k$ is *dependent* in $S$.

We intend to show that $S$ is a linearly independent set if and only if each vector in $S$ is independent in $S$. Thus in $R^3$ such sets would appear as shown in Figure 2.10.

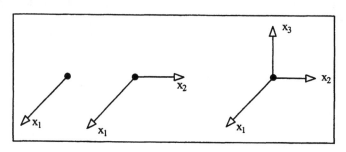

FIGURE 2.10.

We need the following lemma.

**Lemma 2.2** *Let $V$ be a vector space and $S = \{x_1, \ldots, x_m\}$ a subset of $V$ containing at least two vectors. Then $S$ is linearly dependent if and only if $S$ contains a dependent vector.*

**Proof.** We need to argue two parts for the biconditional.

Part a. Suppose $S$ contains a dependent vector. Without loss of generality (the vectors in $S$ can be reindexed), we suppose $x_1$ is dependent in $S$. Thus

$$x_1 = \beta_2 x_2 + \cdots + \beta_m x_m$$

for some scalars $\beta_2, \ldots, \beta_m$. Now by rearranging,

$$1x_1 - \beta_2 x_2 - \cdots - \beta_m x_m = 0$$

and so the pendent equation has a nontrivial solution, namely

$$(1, -\beta_2, \ldots, -\beta_m).$$

Thus, $S$ is linearly dependent.

Part b. The converse implication is left as an exercise. ∎

Another form (actually the contrapositive) of this lemma says that $S = \{x_1, \ldots, x_m\}$ is linearly independent if and only if each $x_k \in S$ is independent in $S$.

As we expect, dependent vectors can be removed from a set without affecting the span.

**Theorem 2.3** *Let $S = \{x_1, \ldots, x_m\}$ be a subset of a vector space $V$. If $x_k$ is dependent in $S$, then*

$$\operatorname{span} S = \operatorname{span} S \backslash \{x_k\}.$$

**Proof.** For simplicity of notation, we re-index the vectors in $S$ so that $x_k$ becomes $x_1$.

We show that $\operatorname{span} S = \operatorname{span} S \backslash \{x_1\}$, which is an equality of sets argument. Thus, let $x \in \operatorname{span} S$. Then

$$x = \beta_1 x_1 + \cdots + \beta_m x_m \tag{2.4}$$

for some scalars $\beta_1, \ldots, \beta_m$. Since $x_1$ is dependent in $S$, we can write

$$x_1 = \gamma_2 x_2 + \cdots + \gamma_m x_m. \tag{2.5}$$

for some scalars $\gamma_2, \ldots, \gamma_m$.

Substituting (2.5) into (2.4), we have

$$x = (\beta_1 \gamma_2 + \beta_2) x_2 + \cdots + (\beta_1 \gamma_m + \beta_m) x_m$$

which says that $x \in \operatorname{span} S \backslash \{x_1\}$. Thus $\operatorname{span} S \subseteq \operatorname{span} S \backslash \{x_1\}$.

Now let $x \in \operatorname{span} S \backslash \{x_1\}$. Then

$$x = \beta_2 x_2 + \cdots + \beta_m x_m$$

for some scalars $\beta_2, \ldots, \beta_m$ Writing

$$x = 0 x_1 + \beta_2 x_2 + \cdots + \beta_m x_m$$

shows that $x \in \operatorname{span} S$. Thus $\operatorname{span} S \backslash \{x_1\} \subseteq \operatorname{span} S$. ∎

We give an example showing how we can use this theorem.

**Example 2.10** *Identify the shape of* $\operatorname{span} S$ *where* $S = \{x_1, x_2, x_3, x_4\}$ *and*

$$x_1 = \begin{bmatrix} 1 \\ 0 \\ 1 \end{bmatrix}, \ x_2 = \begin{bmatrix} 1 \\ 1 \\ 0 \end{bmatrix}, \ x_3 = \begin{bmatrix} 2 \\ 1 \\ 1 \end{bmatrix}, \ x_4 = \begin{bmatrix} 0 \\ -1 \\ 1 \end{bmatrix}.$$

*To find dependent vectors in $S$, we consider the pendent equation*

$$\alpha_1 x_1 + \alpha_2 x_2 + \alpha_3 x_3 + \alpha_4 x_4 = 0. \tag{2.6}$$

*Putting this into an augmented matrix and finding a row echelon form yields*

$$\left[\begin{array}{cccc|c} 1 & 1 & 2 & 0 & 0 \\ 0 & 1 & 1 & -1 & 0 \\ 0 & 0 & 0 & 0 & 0 \end{array}\right].$$

*Thus $\alpha_3$ and $\alpha_4$ are free. Setting $\alpha_3 = 0$, $\alpha_4 = -1$ and solving, we get $(1, -1, 0, -1)$, and by plugging into (2.6),*

$$x_1 - x_2 = x_4.$$

*Similarly, setting $\alpha_3 = -1$, $\alpha_4 = 0$ yields the solution $(1, 1, -1, 0)$, so by (2.6)*

$$x_1 + x_2 = x_3.$$

*(In general, if all free variables $\alpha_k$ are set to 0, except, say $\alpha_i$, which is set to -1, then we see that $x_i$ is a linear combination of the vectors corresponding to pivot variables.)*
    *Thus*

$$\text{span } S = \text{span } S \backslash \{x_4\} = \text{span } S \backslash \{x_3, x_4\}.$$

*Now, if all free variables are set to 0, the resulting equation (2.6) contains only vectors corresponding to pivot variables, namely*

$$\alpha_1 x_1 + \alpha_2 x_2 = 0.$$

*A row echelon form for this equation can be obtained from the augmented matrix above by deleting the columns corresponding to the free variables,*

$$\left[\begin{array}{cc|c} 1 & 1 & 0 \\ 0 & 1 & 0 \\ 0 & 0 & 0 \end{array}\right].$$

*Thus, the only solution to this equation is $\alpha_1 = \alpha_2 = 0$ and so $x_1$, $x_2$ are linearly independent. So*

$$\text{span } S = \text{span } \left\{ \begin{bmatrix} 1 \\ 0 \\ 1 \end{bmatrix}, \begin{bmatrix} 1 \\ 1 \\ 0 \end{bmatrix} \right\},$$

*which is a plane.*

As demonstrated in the example, we have the following.

**Corollary 2.1** *Let $A$ be an $m \times n$ matrix and $E$ a row echelon form of $A$. Let $S = \{a_1, \ldots, a_n\}$ where $a_i$ is the i-th column of $A$.*

(a) *The columns of $A$ corresponding to the columns of $E$ that don't contain pivots are dependent in $S$. And, they can be removed from $S$ without affecting the span.*

(b) *The columns of $A$ corresponding to the columns of $E$ that contain pivots are linearly independent vectors.*

We now describe those sets which can be used to form axes.

**Definition 2.5** *Let $V$ be a vector space and $S$, $S = \{x_1, \ldots, x_n\}$, a nonempty subset of $V$. The set $S$ is a basis for $V$ if*

*i. $S$ is linearly independent and*

*ii. span $S = V$.*

*In addition, we consider the set $S$ as ordered, so $x_1$ is the first vector, $x_2$ the second vector, etc. in $S$. (Now each vector, say $x_i$, in $S$ determines an axis in $V$, namely span $\{x_i\}$.)*

**Example 2.11** *Some vector spaces with bases follow.*

(a) *$R^n$ has as a basis $\{e_1, \ldots, e_n\}$. (There are others.)*

(b) *$R^{m \times n}$ has as a basis $\{E_{ij} | 1 \le i \le m$ and $1 \le j \le n\}$ where $E_{ij}$ is the matrix having a 1 in the $ij$-th position and 0's elsewhere. (There are others.)*

(c) *$P_n$ has as a basis $\{1, t, \ldots, t^{n-1}\}$. (There are others.)*

Let $V$ be a vector space that has a basis, say $S = \{x_1, \ldots, x_n\}$. We now show how coordinates are attached to vectors in $V$. (This is somewhat like the calculus problem of attaching of polar coordinates to points in $R^2$.)

For any $x \in V$ we can write

$$x = \alpha_1 x_1 + \cdots + \alpha_n x_n \tag{2.7}$$

for some scalars $\alpha_1, \ldots, \alpha_n$. Note that these scalars must be unique since if

$$x = \beta_1 x_1 + \cdots + \beta_n x_n \tag{2.8}$$

for some scalars $\beta_1, \ldots, \beta_n$, then by subtracting (2.8) from (2.7), we have

$$0 = (\alpha_1 - \beta_1) x_1 + \cdots + (\alpha_n - \beta_n) x_n.$$

Since $S$ is linearly independent,

$$\alpha_1 - \beta_1 = 0, \ldots, \alpha_n - \beta_n = 0 \text{ or}$$
$$\alpha_1 = \beta_1, \ldots, \alpha_n = \beta_n.$$

Thus, we can define the $S$-coordinates for $x$ as

$$[x]_S = \begin{bmatrix} \alpha_1 \\ \cdots \\ \alpha_n \end{bmatrix}.$$

(Of course, the coordinates depend on $S$.) We see that a basis gives a coordinate system, which we call the $S$-coordinate system, with axes determined $x_1, \ldots, x_n$. The vector $x$ is located by proceeding $\alpha_1$ units along the axis determined by $x_1$ to get $\alpha_1 x_1$, then $\alpha_2$ units parallel to the axis determined by $x_2$ to get $\alpha_1 x_1 + \alpha_2 x_2$, etc. (Geometrically we would add by appending.) We label the axes as $y_1, \ldots, y_n$, respectively. (See Figure 2.11.)

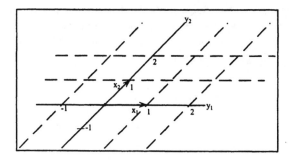

FIGURE 2.11.

A particular example follows.

**Example 2.12**  *Note that $R^2$ has $S = \left\{(1,1)^t, (-1,1)^t\right\}$ as a basis. To find the coordinates of $(1,3)^t$ with respect to this basis, we set*

$$\begin{bmatrix} 1 \\ 3 \end{bmatrix} = \alpha_1 \begin{bmatrix} 1 \\ 1 \end{bmatrix} + \alpha_2 \begin{bmatrix} -1 \\ 1 \end{bmatrix}$$

*and solve for $\alpha_1$, $\alpha_2$. This gives $\alpha_1 = 2$, $\alpha_2 = 1$. Thus,*

$$\left[\begin{bmatrix} 1 \\ 3 \end{bmatrix}\right]_S = \begin{bmatrix} 2 \\ 1 \end{bmatrix}.$$

*To locate $\begin{bmatrix} 1 \\ 3 \end{bmatrix}$ with respect to $S$, we move 2 units along the $y_1$-axis followed by 1 unit in the direction of the $y_2$-axis, as shown in Figure 2.12.*

The number of axes, or vectors in a basis, gives the dimension of a vector space. To show this, we need a technical result.

**Lemma 2.3**  *Let $V$ be a vector space having a basis of $n$ vectors. The number of vectors in any linearly independent set of $V$ cannot exceed $n$.*

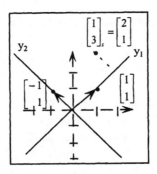

FIGURE 2.12.

**Proof.** We will prove this lemma for Euclidean $n$-space, which is more insightful, leaving the general argument as an exercise.

Let $S = \{y_1, \dots, y_m\}$ be any set of vectors in $E^n$ where $m > n$. We show that $S$ must be linearly dependent. For this, consider the pendent equation

$$\alpha_1 y_1 + \cdots + \alpha_m y_m = 0. \tag{2.9}$$

Using backward multiplication, write (2.9) as the matrix equation

$$[y_1 \dots y_m] \begin{bmatrix} \alpha_1 \\ \cdots \\ \alpha_m \end{bmatrix} = 0.$$

Note that the coefficient matrix is $n \times m$, and thus if we compute a row echelon form (say by Gaussian elimination) for the augmented matrix $[y_1 \dots y_m \,|\, 0]$, there is a free variable. From this it follows that there are $\infty$-many solutions to (2.9) and hence $S$ is linearly dependent. ∎

A consequence of the lemma follows.

**Theorem 2.4** *Let $V$ be a vector space having a basis. Then all bases of $V$ contain the same number of vectors.*

**Proof.** Let $B_1 = \{x_1, \dots, x_p\}$ and $B_2 = \{y_1, \dots, y_q\}$ be bases for $V$. Using Lemma 2.3, noting the $B_1$ is linearly independent, it follows that $p \leq q$. Using Lemma 2.3, with $B_2$ linearly independent, yields $q \leq p$. Thus, $p = q$, the desired result. ∎

By using this theorem, we can define the dimension of a vector space as we intended, counting vectors in a basis.

**Definition 2.6** *Let $V$ be a vector space.*

   i. If $V = \{0\}$, then $\dim V = 0$.

ii. If $V$ has a basis, say $\{x_1, \ldots, x_r\}$, then $\dim V = r$.

iii. In all other cases, $\dim V = \infty$.

**Example 2.13** *Applying the definition, we can see that*

(a) $\dim R^n = n$, $\dim C^n = n$.

(b) $\dim R^{m \times n} = mn$, $\dim C^{m \times n} = mn$.

(c) $\dim P_n = n$.

And, interestingly, as given in an exercise

(d) $\dim C\,(+\infty, \infty) = \infty$.

A more complicated example follows.

**Example 2.14** *Let $A$ be an $m \times n$ matrix. Define the null space of $A$ as*

$$N\,(A) = \{x : Ax = 0\}\,.$$

*It is left as an exercise to show that $N(A)$ is a subspace.*
*We compute the dimension of $N(A)$ for a small example. For this, let*
$A = \begin{bmatrix} 2 & -4 & 2 \\ -3 & 6 & -3 \end{bmatrix}$. *To find the null space we solve*

$$Ax = 0.$$

*This yields*

$$x = \begin{bmatrix} 2\beta - \alpha \\ \beta \\ \alpha \end{bmatrix}$$

*where $\alpha$, $\beta$ are free variables. Factoring these scalars out of $x$ yields*

$$x = \alpha \begin{bmatrix} -1 \\ 0 \\ 1 \end{bmatrix} + \beta \begin{bmatrix} 2 \\ 1 \\ 0 \end{bmatrix}$$

*and so $N\,(A) = \mathrm{span} \left\{ \begin{bmatrix} -1 \\ 0 \\ 1 \end{bmatrix}, \begin{bmatrix} 2 \\ 1 \\ 0 \end{bmatrix} \right\}.$*

*If we set $x = 0$ and observe the last two entries of the vectors in the equation*

$$0 = \alpha \begin{bmatrix} -1 \\ 0 \\ 1 \end{bmatrix} + \beta \begin{bmatrix} 2 \\ 1 \\ 0 \end{bmatrix},$$

*it is clear that $\alpha = \beta = 0$, so* $\begin{bmatrix} -1 \\ 0 \\ 1 \end{bmatrix}, \begin{bmatrix} 2 \\ 1 \\ 0 \end{bmatrix}$ *are linearly independent.*

*Hence,* $\left\{ \begin{bmatrix} -1 \\ 0 \\ 1 \end{bmatrix}, \begin{bmatrix} 2 \\ 1 \\ 0 \end{bmatrix} \right\}$ *is a basis for $N(A)$ and so*

$$\dim N(A) = 2.$$

In general,

$$\dim N(A) = \quad \begin{array}{l} \text{number of free variables} \\ \text{in the solution of } Ax = 0. \end{array} \tag{2.10}$$

And, a basis for this subspace can be found from the solution, as in the example, by using the vectors which are coefficients of the free variables.

Finally, we point out a result useful in establishing when linearly independent sets actually form bases.

**Corollary 2.2** *Let $V$ be a vector space with $\dim V = n$. Then, any set of $n$ linearly independent vectors in $V$ is a basis of $V$.*

**Proof.** Suppose $x_1, \dots, x_n$ are linearly independent vectors in $V$. To show $x_1, \dots, x_n$ forms a basis for $V$, we need only show that

$$\text{span}\{x_1, \dots, x_n\} = V.$$

Let $x \in V$. Then by Lemma 2.3, $x_1, \dots, x_n, x$ are linearly dependent. Thus there is a nontrivial solution $(\beta_1, \dots, \beta_n, \beta)$ to the pendent equation

$$\alpha_1 x_1 + \cdots + \alpha_n x_n + \alpha x = 0.$$

Note that $\beta = 0$ implies that

$$\alpha_1 x_1 + \cdots + \alpha_n x_n = 0$$

has a nontrivial solution, namely $(\beta_1, \dots, \beta_n)$, which contradicts that $x_1, \dots, x_n$ are linearly independent. Thus, $\beta \neq 0$.

Since $\beta \neq 0$, we can solve

$$\beta_1 x_1 + \cdots + \beta_n x_n + \beta x = 0$$

for $x$, yielding

$$x = \frac{-\beta_1}{\beta} x_1 + \cdots + \frac{-\beta_n}{\beta} x_n.$$

Thus $x \in \text{span}\{x_1, \dots, x_n\}$. And, since $x$ was chosen arbitrarily,

$$\text{span}\{x_1, \dots, x_n\} = V,$$

which is what we wanted to prove. ■

An example showing how this corollary can be used follows.

**Example 2.15** *Can we write any polynomial in $P_3$ in the form*

$$a + b(t-1) + c(t-1)^2?$$

*Note that $1, t-1, (t-1)^2$ are linearly independent in $P_3$. Since $\dim P_3 = 3$, these vectors also form a basis for $P_3$. Hence, the answer is yes.*

### 2.2.1   Optional (Dimension of Convex sets)

Using vector space notions, we can describe basic geometrical objects in $R^n$.

1. *Parallelepiped*: To describe a parallelogram in $R^2$, let $x$, $y$ be linearly independent vectors. (See Figure 2.13.) Then $\{\alpha x + \beta y$ where $0 \le \alpha \le 1,\ 0 \le \beta \le 1\}$ describes all points in the parallelogram with sides $x$ and $y$. For a parallelepiped in $R^3$, let $x, y, z$ be linearly independent. Then the description is $\{\alpha x + \beta y + \gamma z$ where $0 \le \alpha \le 1, 0 \le \beta \le 1,\ 0 \le \gamma \le 1\}$.

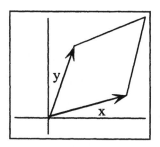

FIGURE 2.13.

In $R^n$ a parallelepiped determined from linearly independent vectors $x_1, \dots, x_n$ is $\{\alpha_1 x_1 + \cdots + \alpha_n x_n$ where $0 \le \alpha_1 \le 1, \dots, 0 \le \alpha_n \le 1\}$.

2. *Pyramid*: Using the same technique as in 1, we have that if $x$ and $y$ are linearly independent in $R^2$ then $\{\alpha x + \beta y$ where $0 \le \alpha,\ 0 \le \beta,$ and $\alpha + \beta \le 1\}$ is a triangle with vertices $0$, $x$, and $y$. (See Figure 2.15.)

In $R^n$, for $x_1, x_2, \dots, x_n$ linearly independent, $\{\alpha x_1 + \alpha_2 x_x + \cdots + \alpha_n x_n$ where $0 \le \alpha_1,\ 0 \le \alpha_2, \dots,\ 0 \le \alpha_n$ and $\alpha_1 + \alpha_2 + \cdots + \alpha_n \le 1\}$ describes a pyramid.

A nonempty set $S$ in a vector space $V$ is *convex* if for each $x, y \in S$, the segment between $x$ and $y$, namely

$$\alpha x + (1-\alpha) y$$

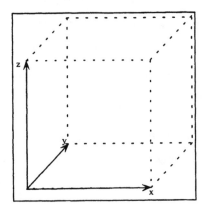

FIGURE 2.14.

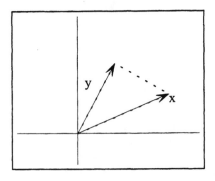

FIGURE 2.15.

where $0 \le \alpha \le 1$, is in $S$. The disc in Figure 2.16 is convex, and the L is not.

A subspace $W$ of $V$ is clearly convex and so are its translations. And, parallelepipeds and pyramids are convex.

For any convex set $S$, define the dimension of $S$ as follows:

i. $\dim S = 0$ if $S = \{x_0\}$, i.e., $S$ contains a single point.

ii. $\dim S = r$ if $r > 0$ is the largest integer such that $S$ contains $r + 1$ vectors $x_0, x_1, \dots ,, x_r$ for which $x_1 - x_0, x_2 - x_0, \dots , x_r - x_0$ are linearly independent. Note in Figure 2.17 that a line has dimension 1, a disc dimension 2, and a pyramid dimension 3.

iii. $\dim S = \infty$ otherwise.

Since a vector space $V$ itself is convex, it has a dimension as described above. If the vector space dimension of $V$ is $n$, then $V$ has $n$ linearly independent vectors, say $x_1, \dots , x_n$. Taking $x_0 = 0$, we have that the

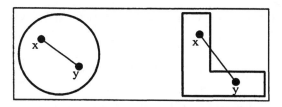

FIGURE 2.16.

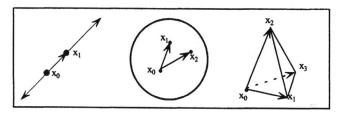

FIGURE 2.17.

convex dimension of $V$ is at least $n$. And since no larger set of vectors in $V$ can be linearly independent, the convex dimension of $V$ is $n$ also.

## Exercises

1. Decide if the following sets of vectors are linearly independent.

   (a) $(1,1,1)^t$, $(-1,1,-1)^t$, $(0,1,0)^t$

   (b) $(1,0,1,0)^t$, $(0,1,1,0)^t$, $(0,0,1,1)^t$

   (c) $\left\{ \begin{bmatrix} 0 & 1 \\ 1 & 0 \end{bmatrix}, \begin{bmatrix} 1 & 0 \\ 0 & 1 \end{bmatrix}, \begin{bmatrix} 1 & 1 \\ 1 & 1 \end{bmatrix} \right\}$

   (d) $1+t$, $1+t^2$, $1-t$

2. Let $V$ be a vector space and $x, y \in V$. Prove that $x, y$ are linearly dependent if and only if one of these vectors is a scalar multiple of the other. (Thus, deciding if two vectors are linearly independent is often a matter of looking.)

3. Let $u, v, w$ be linearly independent vectors in a vector space $V$. Prove that $u$, $u+v$, $u+v+w$ are linearly independent.

4. Prove Lemma 2.2, part b.

5. Prove that every nonempty subset of a linearly independent set is linearly independent.

6. Two parts:

(a) Let $f$ and $g$ be differentiable functions. The Wronskian of $f$ and $g$ is

$$W\left(f\left(t\right), g\left(t\right)\right) = \det \begin{bmatrix} f\left(t\right) & g\left(t\right) \\ f'\left(t\right) & g'\left(t\right) \end{bmatrix}.$$

Prove that if $W\left(f\left(t\right), g\left(t\right)\right) \neq 0$ for some $t$, then $f$ and $g$ are linearly independent.

(b) State and prove the generalization of this result to the $n$ functions.

7. Using Exercise 6, decide which sets of functions are linearly independent.

(a) $e^t, e^{-t}$      (b) $t - 1,\ t + 1,\ t$

(c) $\sin t,\ \cos t$

8. Let $A$ be an $m \times n$ matrix. Prove that $N(A)$ is a subspace.

9. In $R^2$, find the coordinates of $(3, 3)^t$ if the basis is $S = \left\{ (2, 1)^t, (1, 2)^t \right\}$. Draw the axes and the corresponding grid, and geometrically find $(3, 3)^t$ in terms of them.

10. Find the coordinates of $\begin{bmatrix} a & b \\ c & d \end{bmatrix}$ if the basis is

$$\left\{ \begin{bmatrix} 1 & 0 \\ 0 & 0 \end{bmatrix}, \begin{bmatrix} 1 & 1 \\ 0 & 0 \end{bmatrix}, \begin{bmatrix} 1 & 1 \\ 0 & 1 \end{bmatrix}, \begin{bmatrix} 1 & 1 \\ 1 & 1 \end{bmatrix} \right\}.$$

11. Find a basis for each of the following.

(a) span $\left\{ \begin{bmatrix} 1 \\ 1 \\ 1 \end{bmatrix}, \begin{bmatrix} 1 \\ 0 \\ -1 \end{bmatrix}, \begin{bmatrix} 2 \\ 1 \\ 0 \end{bmatrix}, \begin{bmatrix} 3 \\ 1 \\ -1 \end{bmatrix} \right\}$ Give its dimension and draw the shape.

(b) span $\left\{ \begin{bmatrix} 1 & 1 \\ 1 & 0 \end{bmatrix}, \begin{bmatrix} 0 & 1 \\ 1 & 0 \end{bmatrix}, \begin{bmatrix} 1 & 0 \\ 0 & 0 \end{bmatrix}, \begin{bmatrix} 0 & 0 \\ 0 & 1 \end{bmatrix} \right\}$

(c) span $\{t - 1, t + 1, 2t - 1, t - 2\}$

12. Prove that $\begin{bmatrix} a \\ b \end{bmatrix}, \begin{bmatrix} c \\ d \end{bmatrix}, \begin{bmatrix} e \\ f \end{bmatrix}$ are linearly dependent in $R^2$.

13. Prove Lemma 2.3 for a vector space $V$.

14. Find a basis for each of the subspaces given below. Give the dimension of each.

(a) $W = \{A : A \in R^{2 \times 2}$ and $A$ is upper triangular$\}$

(b) $W = \{A : A \in R^{2\times 2}$ and diagonal$\}$

(c) $W = \{A : A \in R^{2\times 2}$ and symmetric$\}$

15. Prove that $C(-\infty, \infty) = \infty$.

 (Hint. Assume that $\dim C(-\infty, \infty) = n$ and consider $1, t, \ldots, t^n$.)

16. The solution set to $y'' + 3y' + 2y = 0$ is a subspace of $C(-\infty, \infty)$. From differential equations, we know that the dimension of this subspace is 2. Thus, if we can guess two linearly independent solutions to this equation, we have a basis for it. Solve the differential equation by guessing.

17. The solution set $S$ to $x(k+2) - 5x(k+1) + 6x(k) = 0$ is a subspace. (See Exercise 6 in Section 1). It is known that $\dim S = 2$. By guessing, find a basis for $S$ and thus $S$ itself. (Hint: Try $x(k) = r^k$ for some scalar $r$. Plug it in and determine which $r$'s work.)

18. Let $V$ be a vector space, $V \neq \{0\}$, and $x_1, \ldots, x_n \in V$. Prove that if span$\{x_1, \ldots, x_n\} = V$, then some subset of $\{x_1, \ldots, x_n\}$ is a basis for $V$.

19. Let $V$ be a vector space and $x_1, x_2, x_3$ in $V$. If $x_1 \neq 0$, show that

 (a) If $x_2 \notin$ span$\{x_1\}$, then $x_1, x_2$ are linearly independent.

 (b) If in addition to (a), $x_3 \notin$ span$\{x_1, x_2\}$, then $x_1, x_2, x_3$ are linearly independent.

20. Let $V$ be a vector space with $\dim V = n$. If $x_1, \ldots, x_n \in V$ and span$\{x_1, \ldots, x_n\} = V$, prove that $x_1, \ldots, x_n$ are linearly independent.

21. (Optional) Find, by making a drawing, the dimensions of the following convex sets in $R^3$.

 (a) A parallelepiped

 (b) A ball, i.e., $\{x : x_1^2 + x_2^2 + x_3^2 \leq 1\}$

 Prove the following are convex and find their dimensions.

 (c) A parallelepiped in $R^n$

 (d) A pyramid in $R^n$

 (e) An affine space in $R^n$

22. (MATLAB) Using MATLAB, rref, Theorem 2.3, and Corollary 2.1,

(a) Decide if $\begin{bmatrix} 1 \\ 0 \\ -1 \\ 0 \end{bmatrix} \begin{bmatrix} 0 \\ 1 \\ 1 \\ -1 \end{bmatrix} \begin{bmatrix} -1 \\ 1 \\ 1 \\ -1 \end{bmatrix} \begin{bmatrix} 1 \\ 0 \\ 0 \\ -1 \end{bmatrix}$ are linearly inde-

pendent.

(b) Find a basis for

$$\text{span} \left\{ \begin{bmatrix} 1 \\ 0 \\ -1 \end{bmatrix}, \begin{bmatrix} 0 \\ 1 \\ 1 \end{bmatrix}, \begin{bmatrix} 1 \\ 0 \\ 0 \end{bmatrix}, \begin{bmatrix} 1 \\ 1 \\ 1 \end{bmatrix} \right\}.$$

(c) Find a basis for span $\left\{ 2t + 1, \ t^2 + t - 1, \ t + 1, \ 4 \right\}$.

(d) Find a basis for span $\left\{ \begin{bmatrix} 1 & 1 \\ 1 & 0 \end{bmatrix}, \begin{bmatrix} 1 & 1 \\ 1 & 1 \end{bmatrix}, \begin{bmatrix} 1 & -1 \\ 0 & 1 \end{bmatrix}, \right.$

$\left. \begin{bmatrix} 0 & 1 \\ 1 & -1 \end{bmatrix} \begin{bmatrix} 1 & -1 \\ 1 & 0 \end{bmatrix}, \begin{bmatrix} 1 & -1 \\ -1 & 1 \end{bmatrix} \right\}.$

(On (c) and (d), use the augmented matrix obtained from the pendent equation.)

## 2.3   Linear Transformations

Functions from vector spaces to vector spaces are called transformations (or maps, operators). As in calculus, they arise in mathematically describing phenomenon.

An example may be helpful.

**Example 2.16** *Let $a \in R^2$ and define $L : R^2 \rightarrow R^2$ by*

$$L(x) = x + a.$$

*This transformation is called a translation. If $a = \begin{bmatrix} a_1 \\ a_2 \end{bmatrix}$, then $L \begin{bmatrix} x_1 \\ x_2 \end{bmatrix} =$*

$\begin{bmatrix} x_1 + a_1 \\ x_2 + a_2 \end{bmatrix}$. *So $L$ shifts $R^2$ $a_1$ units in the direction of the $x_1$-axis and $a_2$*

*units in the direction of the $x_2$-axis. For example, if $a = \begin{bmatrix} 2 \\ 1 \end{bmatrix}$, this shift*

*can be seen in Figure 2.18.*

In this section we give a study of transformations that behave like the derivative and the integral that we saw in calculus.

**Definition 2.7**   *Let $V$ and $W$ be vector spaces. A transformation $L :$ $V \rightarrow W$ is called linear if for all vectors and scalars,*

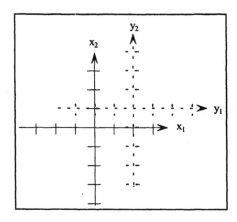

FIGURE 2.18.

*i.* $L(x+y) = L(x) + L(y)$  *(L goes across sums.)*

*ii.* $L(\alpha x) = \alpha L(x)$  *(Scalars can be pulled out.)*

By (ii), $L(0x) = 0L(x)$ so linear transformations also satisfy

$$L(0) = 0.$$

Thus, if $L(0) \neq 0$, then $L$ is not linear. Notice that the transformation in Example 2.16 is not linear.

Putting (i) and (ii) together, we have that

$$L(\alpha x + \beta y) = \alpha L(x) + \beta L(y).$$

So, $L$ maps lines $\alpha x + (1-\alpha)y$ into lines $\alpha L(x) + (1-\alpha)L(y)$ as well as line segments into line segments. (Here we assume $L(x) \neq L(y)$; otherwise $L$ maps the line into a point.)

A matrix map, say $L(x) = Ax$, is a linear transformation. Using this transformation, we demonstrate the line property.

**Example 2.17** *Let* $L(x) = \begin{bmatrix} 1 & 1 \\ 1 & -1 \end{bmatrix} x$ *and* $\ell$ *the line described by* $\alpha \begin{bmatrix} 1 \\ 1 \end{bmatrix} +$
$(1-\alpha)\begin{bmatrix} -1 \\ 2 \end{bmatrix}$. *Then the image of* $\ell$ *is determined by* $\alpha \begin{bmatrix} 2 \\ 0 \end{bmatrix} + (1-\alpha)\begin{bmatrix} 1 \\ -3 \end{bmatrix}$.
*The graphs are shown in Figure 2.19.*

An interesting linear transformation, which we will use later to look at pictures of various sets of matrices, follows.

**Example 2.18** *(Transformation from $R^{2\times 2}$ into $R^4$). Define $L$ by*

$$L\left(\begin{bmatrix} a & b \\ c & d \end{bmatrix}\right) = \begin{bmatrix} a \\ b \\ c \\ d \end{bmatrix}.$$

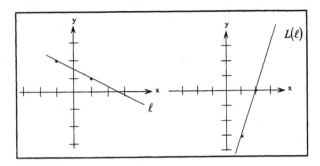

FIGURE 2.19.

*To show that L is linear, we need to verify properties (i) and (ii) of the definition of linear transformation.*

*L goes across sums: Let* $A = \begin{bmatrix} a_{11} & a_{12} \\ a_{21} & a_{22} \end{bmatrix}$ *and* $B = \begin{bmatrix} b_{11} & b_{12} \\ b_{21} & b_{22} \end{bmatrix}$. *Then*

$$L(A+B) = L\left(\begin{bmatrix} a_{11} + b_{11} & a_{12} + b_{12} \\ a_{21} + b_{21} & a_{22} + b_{22} \end{bmatrix}\right)$$

$$= \begin{bmatrix} a_{11} + b_{11} \\ a_{12} + b_{12} \\ a_{21} + b_{21} \\ a_{22} + b_{22} \end{bmatrix} = \begin{bmatrix} a_{11} \\ a_{12} \\ a_{21} \\ a_{22} \end{bmatrix} + \begin{bmatrix} b_{11} \\ b_{12} \\ b_{21} \\ b_{22} \end{bmatrix} = L(A) + L(B).$$

*Scalars can be pulled out: Let* $A = \begin{bmatrix} a_{11} & a_{12} \\ a_{21} & a_{22} \end{bmatrix}$ *and* $\alpha$ *a scalar. Then*

$$L(\alpha A) = L\left(\begin{bmatrix} \alpha a_{11} & \alpha a_{12} \\ \alpha a_{21} & \alpha a_{22} \end{bmatrix}\right) = \begin{bmatrix} \alpha a_{11} \\ \alpha a_{12} \\ \alpha a_{21} \\ \alpha a_{22} \end{bmatrix} = \alpha \begin{bmatrix} a_{11} \\ a_{12} \\ a_{21} \\ a_{22} \end{bmatrix} = \alpha L(A).$$

In this section we will be concerned with matrix maps; however, some of the theorems will be proved for linear transformations in general.

The following theorem lets us see a picture, the grid view, of a linear transformation. These pictures help provide insight into some of the work that follows.

**Theorem 2.5** *Let V and W be vector spaces with V having as a basis* $\{x_1, \ldots, x_n\}$. *Let L be a linear transformation from V to W. Then*

$$L(\alpha_1 x_1 + \cdots + \alpha_n x_n) = \alpha_1 L(x_1) + \ldots + \alpha_n L(x_n) \qquad (2.11)$$

*for all scalars* $\alpha_1, \ldots, \alpha_n$.

**Proof.** This can be seen by sequentially applying the properties of a linear transformation,

$$L\left((\alpha_1 x_1 + \cdots + \alpha_{n-1} x_{n-1}) + \alpha_n x_n\right)$$
$$= L\left(\alpha_1 x_1 + \cdots + \alpha_{n-1} x_{n-1}\right) + L\left(\alpha_n x_n\right)$$

and by continuing,

$$= L\left(\alpha_1 x_1\right) + \cdots + L\left(\alpha_{n-1} x_{n-1}\right) + L\left(\alpha_n x_n\right)$$
$$= \alpha_1 L\left(x_1\right) + \cdots + \alpha_{n-1} L\left(x_{n-1}\right) + \alpha_n L\left(x_n\right).$$

More formally, the proof can be done by induction. ∎

The following example shows how we get a grid view of a linear transformation.

**Example 2.19** *Let*

$$L\left(x\right) = \begin{bmatrix} 2 & 1 \\ 1 & 2 \end{bmatrix} x.$$

*To describe L, note that $\{e_1, e_2\}$ is a basis for $R^2$ and that $L\left(e_1\right) = \begin{bmatrix} 2 \\ 1 \end{bmatrix}$ and $L\left(e_2\right) = \begin{bmatrix} 1 \\ 2 \end{bmatrix}$. And, since $L\left(\alpha_1 e_1 + \alpha_2 e_2\right) = \alpha_1 L\left(e_1\right) + \alpha_2 L\left(e_2\right) = \alpha_1 \begin{bmatrix} 2 \\ 1 \end{bmatrix} + \alpha_2 \begin{bmatrix} 1 \\ 2 \end{bmatrix}$, we see that $\begin{bmatrix} 2 \\ 1 \end{bmatrix}$ and $\begin{bmatrix} 1 \\ 2 \end{bmatrix}$ form axes for a grid in $R^2$ as shown in Figure 2.20.*

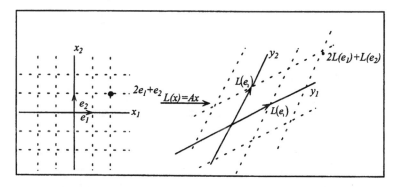

FIGURE 2.20.

*Observe that the image of a square in the grid for $R^2$ is a parallelogram in the image grid. And that the grid view gives a picture of where all points in $R^2$ go in the map.*

*Note that (2.11) also assures that the* range *of L, denoted R(L), is the span of L(x₁),..., L(xₙ), and this is a subspace. (Recall that a basis can be found for R(L) by removing dependent vectors from among L(x₁),..., L(xₙ).)*

Another example may be helpful.

**Example 2.20** *Let*

$$L(x) = \begin{bmatrix} 1 & 0 \\ 1 & 1 \\ 0 & 1 \end{bmatrix} x.$$

*Getting a grid view of L, we plot* $L(e_1) = \begin{bmatrix} 1 \\ 1 \\ 0 \end{bmatrix}$ *and* $L(e_2) = \begin{bmatrix} 0 \\ 1 \\ 1 \end{bmatrix}$ *in R³. Now, drawing the image grid shows where the grid of R² goes under L. (See Figure 2.21.) Note that the image is a subspace spanned by L(e₁),*

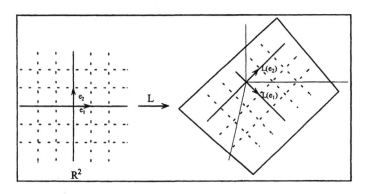

FIGURE 2.21.

$L(e_2)$.

The grid view is often helpful in determining the matrix that does what we want to $R^2$. For example, if we want to skew the plane by moving the points on the $x_2$-axis parallel to the $x_1$-axis so that $\begin{bmatrix} 1 \\ 0 \end{bmatrix}$ ends up at $\begin{bmatrix} 1 \\ 1 \end{bmatrix}$, we would use $A = \begin{bmatrix} 1 & 1 \\ 0 & 1 \end{bmatrix}$ since we want $L(e_1) = e_1$ and $L(e_2) = \begin{bmatrix} 1 \\ 1 \end{bmatrix}$.

And $A = \begin{bmatrix} -1 & 0 \\ 0 & 1 \end{bmatrix}$ reflects $R^2$ about the $x_2$-axis; $A = \begin{bmatrix} \frac{1}{\sqrt{2}} & -\frac{1}{\sqrt{2}} \\ \frac{1}{\sqrt{2}} & \frac{1}{\sqrt{2}} \end{bmatrix}$ rotates the plane $\frac{\pi}{4}$ radians, etc.

An example will show how this can be used in graphics.

**Example 2.21** *Given a sequence of points*

$$\begin{bmatrix} x_1 \\ y_1 \end{bmatrix}, \begin{bmatrix} x_2 \\ y_2 \end{bmatrix}, \ldots, \begin{bmatrix} x_n \\ y_n \end{bmatrix}$$

*in* $R^2$, *we form the matrix* $X = \begin{bmatrix} x_1 & x_2 & \cdots & x_n \\ y_1 & y_2 & \cdots & y_n \end{bmatrix}$. *The command* plot $X$ *will sequentially connect the points with line segments as shown in Figure 2.22. Thus, if* $S = \begin{bmatrix} 0 & 1 & 1 & 0 & 0 \\ 0 & 0 & 1 & 1 & 0 \end{bmatrix}$, *then plot* $S$ *gives a square.*

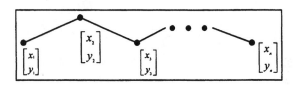

FIGURE 2.22.

*And if* $E = \begin{bmatrix} 2 & 0 & 0 & 1 & 0 & 0 & 2 \\ 0 & 0 & 1 & 1 & 1 & 2 & 2 \end{bmatrix}$, *then plot* $E$ *gives the letter E.*

*Now to rotate* $S$ *by* $\frac{\pi}{4}$ *radians, we let* $A = \begin{bmatrix} \frac{1}{\sqrt{2}} & -\frac{1}{\sqrt{2}} \\ \frac{1}{\sqrt{2}} & \frac{1}{\sqrt{2}} \end{bmatrix}$ *and plot* $AS$.

*And to shear* $E$, *we let* $A = \begin{bmatrix} 1 & 1 \\ 0 & 1 \end{bmatrix}$ *and plot* $AE$. *See Figure 2.23. (If*

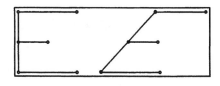

FIGURE 2.23.

*it is not eye appealing, it can be adjusted.)*

We now give a rather easy way of showing that a linear transformation is one-to-one.

**Theorem 2.6** *Let* $V$ *and* $W$ *be vector spaces with* $L : V \longrightarrow W$ *a linear transformation. Then* $L$ *is one-to-one if and only if* $L(x) = 0$ *implies* $x = 0$.

**Proof.** We argue the parts of the biconditional.

Part a. Suppose $L$ is one-to-one and $L(x) = 0$. Since we know that $L(0) = 0$, one-to-one implies that $x = 0$, which is what we need to show.

Part b. Suppose $L(x) = 0$ implies $x = 0$. To show that $L$ is one-to-one, let $x$, $y \in V$ and set

$$L(x) = L(y).$$

Rearranging yields

$$L(x) - L(y) = 0 \qquad \text{or}$$
$$L(x - y) = 0.$$

By hypothesis, this says that $x - y = 0$ and thus $x = y$. Hence $L$ is one-to-one. ∎

An example of how the theorem can be used follows.

**Example 2.22** *Let $S$ be set of $2 \times 2$ symmetric matrices. It can be seen that $S$ is a subspace and $\dim S = 3$. Define $L : S \to R^3$ by*

$$L \begin{bmatrix} a & b \\ b & c \end{bmatrix} = \left( a, \sqrt{2}b, c \right)^t.$$

*The $\sqrt{2}$ occurs since we want to preserve distance. The distance between $\begin{bmatrix} a & b \\ b & c \end{bmatrix}$ and $\begin{bmatrix} \hat{a} & \hat{b} \\ \hat{b} & \hat{c} \end{bmatrix}$ is*

$$\left( (a - \hat{a})^2 + 2 \left( b - \hat{b} \right)^2 + (c - \hat{c})^2 \right)^{\frac{1}{2}}.$$

*The distance between $\left( a, \sqrt{2}b, c \right)^t$ and $\left( \hat{a}, \sqrt{2}\hat{b}, \hat{c} \right)^t$ is*

$$\left( (a - \hat{a})^2 + \left( \sqrt{2}b - \sqrt{2}\hat{b} \right)^2 + (c - \hat{c})^2 \right)^{\frac{1}{2}}.$$

*Note also that $L$ is linear and one-to-one. So $R^3$ gives a model of $S$. The singular matrices in $S$ have determinant $0$, i.e.,*

$$\det \begin{bmatrix} a & b \\ b & c \end{bmatrix} = 0$$

*or*

$$ac - b^2 = 0.$$

*Thus, we can get a view of this set by graphing in $R^3$ those vectors $\left( a, \sqrt{2}b, c \right)^t$ that satisfy $ac - b^2 = 0$. Replacing $b$ by $\sqrt{ac}$, we can then graph $\left( a, \sqrt{2ac}, c \right)^t$. This graph is shown in Figure 2.24.*

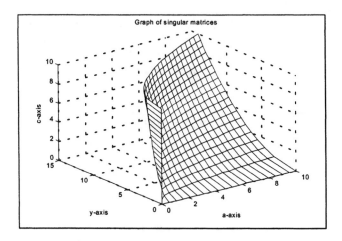

FIGURE 2.24.

It is interesting that special sets of matrices are often not simply infinite sets but actually have some shape.

Not all linear transformations are one-to-one. Some transformations actually collapse the space. To see how collapsing takes place, let

$$N(L) = \{y : y \text{ is a solution to } L(x) = 0\}$$

called the *null space* or *kernel* of $L$. As given in an exercise, $N(L)$ is a subspace.

Let

$$z + N(L) = \{w : w = z + y \text{ for some } y \in N(L)\},$$

called a translate by $z$ of the null space of $L$. Using translates, we can describe how a linear transformation collapses space.

**Theorem 2.7** *Let $V$ and $W$ be vector spaces and $L : V \rightarrow W$, a linear transformation. If $z \in V$ and $L(z) = b$, then the solution to $L(x) = b$ is $z + N(L)$. (Thus $L$ collapses $z + N(L)$ into $b$.)*

**Proof.** We prove two parts. And, we use that

$$S = \{w : w \text{ is a solution to } L(x) = b\}.$$

Part a. We show that $z + N(L) \subseteq S$. For this, let $v \in z + N(L)$. Then $v = z + y$ for some $y \in N(L)$. Thus, $L(v) = L(z + y) = L(z) + L(y) = L(z) = b$. Hence, $v \in S$ and so $z + N(L) \subseteq S$.

Part b. We show that $S \subseteq z + N(L)$. For this, let $w \in S$. Then, set $y = w - z$. Since $L(y) = L(w) - L(z) = 0$, $y \in N(L)$. And since $w = z + y$, $w \in z + N(L)$. Thus, $S \subseteq z + N(L)$. ∎

Since $L$ collapses $z + N(L)$ into $b$ $(b = L(z))$ and the dimension of the affine space $z + N(L)$ is $\dim N(L)$, we see that $L$ collapses $\dim N(L)$ affine spaces in $V$ to each vector in $R(L)$. Thus, intuitively only $\dim V - \dim N(L)$ is left, and we would expect that

$$\dim R(L) = \dim V - \dim N(L)$$

which is correct. However, we will only show this for matrix maps.

**Theorem 2.8** *Let $A$ be an $m \times n$ matrix and $L(x) = Ax$. Then*

$$\dim N(A) + \dim R(A) = n.$$

**Proof.** The pendent equation

$$\alpha_1 a_1 + \cdots + \alpha_n a_n = 0,$$

for the columns of $A$, can be written as

$$Ax = 0$$

where $x = (\alpha_1, \ldots, \alpha_n)^t$. Let $[E \,|\, 0]$ be a row echelon form of $[A \,|\, 0]$.

From (2.10), $\dim N(A) =$ number of free variables determined by $[E \,|\, 0]$. From Corollary 2.1, the columns of $A$ corresponding to the pivot variables are a basis for $R(A)$. Thus,

$$\dim N(A) + \dim R(A) = \text{ total number of variables}$$

$$= n,$$

the desired result. ∎

At the beginning of the chapter we said that we would study spaces which were like $R^2$ and $R^3$. We conclude the chapter by showing that a finite dimensional vector space $V$ is like $E^n$, where $E^n = R^n$ if $V$ is a real vector space and $E^n = \mathcal{C}^n$ if $V$ is a complex vector space.

**Theorem 2.9** *Let $V$ be a vector space with basis $\{x_1, \ldots, x_n\}$. Define $L : V \to E^n$ by $L(\alpha_1 x_1 + \cdots + \alpha_n x_n) = (\alpha_1, \ldots, \alpha_n)^t$. Then $L$ is a one-to-one linear transformation.*

**Proof.** We show that $L$ satisfies the two defining properties of a linear transformation.

$L$ goes across sums: Let $x$, $y \in V$ and write $x = \alpha_1 x_1 + \cdots + \alpha_n x_n$, $y = \beta_1 x_1 + \cdots + \beta_n x_n$, where the $\alpha_i$'s and $\beta_i$'s are scalars. Then $x + y = (\alpha_1 + \beta_1) x_1 + \cdots + (\alpha_n + \beta_n) x_n$. So,

$$L(x + y) = (\alpha_1 + \beta_1, \ldots, \alpha_n + \beta_n)^t$$
$$= (\alpha_1, \ldots, \alpha_n)^t + (\beta_1, \ldots, \beta_n)^t$$
$$= L(x) + L(y).$$

Scalars can be pulled out: left as an exercise.

Finally, $L$ is one-to-one since, if $L(x) = (0, \dots, 0)$,

$$x = 0x_1 + \cdots 0x_n$$
$$= 0$$

This proves the theorem. ∎

A way to view this theorem is that if we express vectors as $x = \alpha_1 x_1 + \cdots + \alpha_n x_n$, then the arithmetic of these vectors is done on the coefficients. So the arithmetic is like that done on vectors $(\alpha_1, \dots, \alpha_n)^t$ in $E^n$.

## 2.3.1  Optional (Graphics of Polygonal Shapes)

Graphics concerns drawing pictures or making movies on a computer screen. In this optional, we want to show how linear transformations play a part in that work.

We look at two problems.

**Example 2.23** *(A house in a strong wind.)  We can start with a house made by drawing a few line segments as in Figure 2.25. If we shear the house*

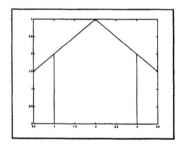

FIGURE 2.25.

*a bit, say by multiplying by*

$$\begin{bmatrix} 1 & .2 \\ 0 & .1 \end{bmatrix},$$

*we have the second picture, Figure 2.26. Of course, if this is a technical drawing, we wouldn't want this roof, since it appears to elongate on the left and contract on the right. We can fix the drawing by shearing the sides of the house and translating the roof by*

$$L(x) = x + \begin{bmatrix} .4 \\ 0 \end{bmatrix}.$$

*And yes, the lengths of the walls in Figure 2.27 (now of length 2.01) are a bit elongated, but that would not be discernible with the eye.*

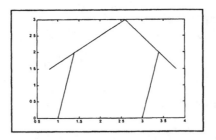

FIGURE 2.26.

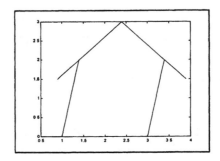

FIGURE 2.27.

**Example 2.24** *(A falling bowling pin.) We find a basic shape of a bowling pin by using a polygonal shape. (See Figure 2.28.) We translate the pin*

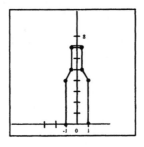

FIGURE 2.28.

*(See Figure 2.29.), so its right lower point (the point at which the rotation for falling takes place) is at the origin by using*

$$L\left(x\right) = x + \begin{bmatrix} -1 \\ 0 \end{bmatrix}.$$

*Now for falling, we show the pin under rotations by various angles. For example, for $-\frac{\pi}{4}$ and $-\frac{\pi}{2}$ we have Figure 2.30.*

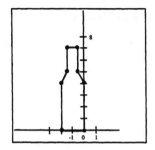

FIGURE 2.29.

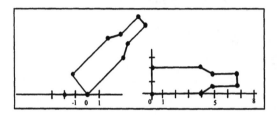

FIGURE 2.30.

## 2.3.2  MATLAB (Codes, including Picture of the Singular Matrices in Matrix Space)

For the polygonal shape obtained by connecting points $\begin{bmatrix} x_1 \\ y_1 \end{bmatrix}, \ldots, \begin{bmatrix} x_n \\ y_n \end{bmatrix}$

sequentially by segments, we define the vector of $x$-values and the vector of corresponding $y$-values:

$x = [x_1 \ldots x_n];$
$y = [y_1 \ldots y_n];$

To connect the points with line segments, we use the command *plot* $(x, y)$.

### 1. Code for Original House

```
wx = [ 1  1  3  3 ];
wy = [ 2  0  0  2 ];    % Determines the points for
                          the walls.
rx = [ .5   2   3.5 ];
ry = [ 1.5  3   1.5 ];   % Determines the points for
                          the roof.
plot (wx wy)
hold                      % Keeps the first plot from
                          being erased.
plot (rx, ry).
```

## 2. Code for Sheared Walls House

Using $wx$, $wy$, $rx$, $ry$ from 1, we add the following.

```
wx1 = wx + .2 * wy;
wy1 = wy;
rx1 = rx + [.2 * ry];
ry1 = ry;
plot (wx1, wy1)
hold
plot (rx1, ry1)
```

## 3. Code for Picture of Singular Matrices

```
a =linspace(0, 10, 20);
c =linspace(0, 10, 20);
[a, c] =meshgrid(a, c) ;
y = (2 * a. * c) . ∧ (.5) ;
mesh(a, y, c)
```

# Exercises

1. Decide which of the following transformations are linear.

   (a) $L(x_1, x_2) = (x_1 + 2x_2, 2x_1 - x_2)^t$

   (b) $L(A) = A + A^t$ for all $k \times k$ matrices $A$

   (c) $L(f(t)) = f'(t) + f(t)$ for all $f \in C_1(-\infty, \infty)$

   (d) $L(x(k)) = x(k+1) + x(k)$ for function $x$ defined on the non-negative integers

2. Which of the following transformations $L : R^{n \times n} \to R$ are linear?

   (a) $L(A) = \text{trace } A$, where trace $A = a_{11} + \cdots + a_{nn}$

   (b) $L(A) = \det A$

3. Draw the grid map of $L(x) = \begin{bmatrix} 3 & 2 \\ 1 & 3 \end{bmatrix} x$, and tell what $L$ does to $R^2$.

4. Let $A = \begin{bmatrix} 1 & 1 & 0 \\ 1 & 1 & 0 \\ 0 & 0 & 1 \end{bmatrix}$. Find the range and the null space in $R^3$ for $L(x) = Ax$. Sketch both.

5. Graph the range of $L$ where $L(x) = \begin{bmatrix} 1 & 1 \\ 1 & 0 \\ 0 & 1 \end{bmatrix} x$. Show the grid of $R^2$ and its image $R^3$.

6. Draw the line $x = \alpha \begin{bmatrix} 1 \\ 1 \end{bmatrix} + (1-\alpha) \begin{bmatrix} 2 \\ 1 \end{bmatrix}$. Compute the image of this line for $L(x) = \begin{bmatrix} 1 & -1 \\ 1 & 1 \end{bmatrix} x$ and draw it.

7. For the given $L_1$ and $L_2$, find $L_1 \circ L_2$ for the following.

(a) $L_1\left((x_1,x_2)^t\right) = (x_1 + x_2, x_1 - x_2)^t$,

  $L_2\left((x_1,x_2)^t\right) = (x_1 - x_2, x_2 + 1)^t$

(b) $L_1\left((x_1,x_2)^t\right) = (2x_1 - 3x_2, x_1 + 2x_2)^t$,

  $L_2\left((x_1,x_2)^t\right) = (x_1 - x_2, x_2)^t$

(c) $L_1\left((x_1,x_2)^t\right) = (x_1, 0)^t$,

  $L_2\left((x_1,x_2)^t\right) = (x_2, x_1)^t + (2, -3)^t$

8. Let $L : V \rightarrow W$ be a linear transformation. Prove that $N(L)$ is a subspace.

9. Find $A$ if $L(x) = Ax$ is such that

(a) $L(e_1) = \begin{bmatrix} 2 \\ 1 \end{bmatrix}$, $L(e_2) = \begin{bmatrix} 1 \\ 2 \end{bmatrix}$ and $A$ is $2 \times 2$.

(b) $L(e_1) = \begin{bmatrix} 1 \\ 1 \\ 0 \end{bmatrix}$, $L(e_2) = \begin{bmatrix} 1 \\ 0 \\ 1 \end{bmatrix}$ and $A$ is $3 \times 2$.

10. Solve the following polygonal graphics problems.

(a) Find the matrix $X$ for the equilateral triangle with base from $(-2,0)^t$ to $(2,0)^t$. Find the matrix $A$ so that $L$ shrinks the $x_1$-axis (and the corresponding space) by $\frac{1}{2}$. Plot $AX$.

(b) Find the matrix $X$ for a tower (4 points should do it) with base from $(-1,0)^t$ to $(1,0)^t$ and height 10. Find the matrix $A$ that leans the tower to the right by $\frac{\pi}{6}$ radians. Plot $AX$.

(c) Move the flag $F = \begin{bmatrix} 0 & 0 & 1 & 1 & 0 & 1 \\ 0 & 1 & 1 & 2 & 2 & 1 \end{bmatrix}$ so its base is at $(1,1)^t$ and it is tilted to the left by $\frac{\pi}{4}$ radian. What transformation $L$ (not linear) is such that plot $LF$ ($L$ applied to each vector in $F$) produces the flag in this position?

11. Let $A$ be an $m \times n$ matrix. Suppose a row echelon form for $Ax = 0$ has $r$ free variables. In writing $x$ in terms of the free variables, we can pull out the free variables so $x$ is a linear combination of vectors which are coefficients the free variables. Explain why those vectors form a basis for the null space of $A$. (Assume that the last $r$ variables are free and that the rest pivot variables. Look at the last $r$ entries in the coefficient vectors.)

12. Find a basis for the following subspaces.

  (a) $W = \{at^2 + bt + a + b : a, b, \in R\}$ (So $a$ and $b$ are free.)

  (b) $W = \left\{ \begin{bmatrix} a+b & a \\ c & b \end{bmatrix} : a, b, c \in R \right\}$

13. Let $L$ be defined on the functions having 2-nd derivatives by

$$L(y) = y'' - 3y' + 2y.$$

Solve $y'' - 3y' + 2y = t$ using Exercise 16 of the previous section, and guessing some $y$ such that the solution set is $y + N(L)$.

14. Using Exercise 17 of the previous section, solve

$$x(k+2) - 5x(k+1) + 6x(k) = 2$$

by guessing a solution.

15. Prove that if $L : V \to W$ is a one-to-one linear transformation and $x_1, \ldots, x_n$ are linearly independent in $V$, then $L(x_1), \ldots, L(x_n)$ are linearly independent in $W$.

16. Let $V$ be a vector space and $V^*$ the set of all linear transformation from $V$ to $V$. With the usual definition of addition and scalar multiplication of functions, show that $V^*$ is a vector space. (Just show closure of addition and scalar multiplication.)

17. Use that the set of all functions of two variables, on which the partial derivatives exist, is a vector space. Then using Exercise 16, decide if $\frac{\delta}{\delta x}$ and $\frac{\delta}{\delta y}$ are linearly independent.

18. Complete the proof of Theorem 2.9.

19. (Optional)  Write the MATLAB code for the third picture of the house with a strong wind sequence.

20. (Optional) Write the MATLAB code that gives all three pictures of the falling pin problem.

21. (MATLAB) Let $L(x) = Ax$ where $A = \begin{bmatrix} 1 & 1 & 0 \\ 1 & 1 & 0 \\ 0 & 0 & 1 \end{bmatrix}$. Using the rref command,

   (a) Find the kernel of $L$.
   (b) Find the range of $L$.

The null and orth commands provide matrices whose columns are (orthonormal) bases for the kernel of $L$ and the range of $L$, respectively. Do (a) and (b) using these commands.

# 3
# Similarity

As shown Figure 3.1, a classical method for solving problems involving a

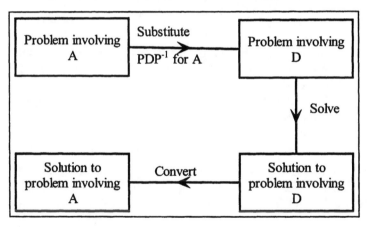

FIGURE 3.1.

matrix, say, $A$, is to first factor the matrix as

$$A = PDP^{-1}$$

where $P$ is a nonsingular matrix and $D$ a diagonal matrix. The expression $PDP^{-1}$ is then substituted into the problem for $A$, thus reducing the problem to one involving $D$. This problem is then solved and its solution converted into the solution of the original problem.

This chapter explains when and how a matrix $A$ can be factored as $A = PDF^{-1}$ and then shows how this factorization is used in problem solving.

## 3.1  Nonsingular Matrices

Nonsingular matrices constitute almost all of the space of $n \times n$ matrices. In this section, we give some of the basic results about nonsingular matrices, describing nonsingular matrices in terms of their rows and columns.

**Theorem 3.1** *Let $A$ be an $n \times n$ matrix. Then $A$ is nonsingular if and only if $A$ has linearly independent columns.*

**Proof.** The biconditional is argued in two parts.

Part a. Suppose $A$ is nonsingular. To show that the columns of $A$ form a linearly independent set, we solve the pendent equation

$$\alpha_1 a_1 + \cdots + \alpha_n a_n = 0$$

where $a_1, \ldots, a_n$ are the columns of $A$. By back multiplication, this can be written as

$$A \begin{bmatrix} \alpha_1 \\ \ldots \\ \alpha_n \end{bmatrix} = 0.$$

Since $A$ is nonsingular, it has an inverse. Multiplying through by this inverse yields $\alpha_1 = \cdots = \alpha_n = 0$. Thus, the columns of $A$ form a linearly independent set.

Part b. Suppose $A$ has linearly independent columns. Then, using that the columns form a basis,

$$Ax = e_i$$

has a solution, say, $b_i$ for each $i$. Set $B = [b_1 \ldots b_n]$. Then by partitioned multiplication, $AB = I$. Now, calculating the determinant of both sides, we have $\det A \det B = 1$. Thus $\det A \neq 0$, and so $A$ is nonsingular. ∎

Since $\det A^t = \det A$, $A$ is nonsingular if and only if $A$ has linearly independent rows. Thus, interestingly, $A$ has linearly independent rows if and only if it has linearly independent columns. We now extend this result.

A word used for the maximum number of linearly independent columns follows. Let $A$ be an $m \times n$ matrix. Define the rank of $A$ as

$$\operatorname{rank} A = \text{largest integer } r \text{ such that } A \text{ has } r$$
$$\text{linearly independent columns.}$$

If $E$ is a row echelon form of $A$, from Corollary 2.1, the columns of $A$ corresponding to pivots in $E$ form a basis for the span of the columns of $A$, and so provide the largest number of linearly independent columns in $A$. Thus,

$$\text{rank } A = \text{ number of pivots in } E.$$

Further, the number of pivots in $E$ is exactly the number of nonzero rows of $E$, e.g., in

$$\begin{bmatrix} \circledast & * & * & * \\ 0 & 0 & \circledast & * \\ 0 & 0 & 0 & 0 \end{bmatrix}$$

both are 2. Thus,

$$\text{rank } A = \text{ number of nonzero rows of } E.$$

**Example 3.1** Let $A = \begin{bmatrix} 1 & 0 & 1 & 1 \\ 1 & 1 & 0 & 1 \\ 2 & 1 & 1 & 2 \end{bmatrix}$. *A row echelon form of $A$ is*

$E = \begin{bmatrix} 1 & 0 & 0 & 1 \\ 0 & 1 & -1 & 0 \\ 0 & 0 & 0 & 0 \end{bmatrix}$. *Thus, $\text{rank } A = 2$. (Columns 1 and 2 of $A$ form a basis for the span of the columns of $A$.)*

The determinant is rarely used in computational work. However, it is a useful tool in developing matrix results. The next theorem, though a bit intricate is worth the effort to learn. It links rank, determinant, and linearly independent rows, linearly independent columns. We show its use in several places in this text.

**Theorem 3.2** Let $A$ be an $m \times n$ matrix, $A \neq 0$. Let $B$ be an $r \times r$ submatrix of $A$ such that $\det B \neq 0$ and such that for any $(r+1) \times (r+1)$ submatrix $C$ containing $B$, $\det C = 0$. Then $\text{rank } A = r$.

**Proof.** We will argue a particular case leaving the general proof as an exercise.

Let $A$ be a $3 \times 4$ matrix and suppose $B$ is the $2 \times 2$ submatrix in the upper left corner of $A$. We use the notation

$$A = \begin{bmatrix} B & \cdot \\ \cdot & \cdot \end{bmatrix} = \left[ \begin{array}{cc|cc} b_1 & b_2 & b_3 & b_4 \\ a_{31} & a_{32} & a_{33} & a_{34} \end{array} \right] = [a_1 a_2 a_3 a_4]$$

where the $b_1, b_2$ and $a_1, a_2, a_3, a_4$ are column vectors of $B$ and $A$, respectively.

Since $\det B \neq 0$, by Theorem 3.1, $b_1, b_2$ are linearly independent and thus $a_1, a_2$ are linearly independent. By deleting dependent vectors, we will show that

$$\text{span}\,\{a_1, a_2\} = \text{span}\,\{a_1, \ldots, a_4\}\,.$$

Thus $\dim \text{span}\,\{a_1, \ldots, a_4\} = 2$ so $A$ has at most 2 linearly independent columns. This assures us that $\text{rank}\, A = 2$.

To prove that $a_3$ is a linear combination of $a_1$ and $a_2$, we proceed as follows. Note that $b_1, b_2$ is a basis, and so we can write

$$b_3 = \alpha_1 b_1 + \alpha_2 b_2$$

for some scalars $\alpha_1, \alpha_2$. Define

$$\ell = a_3 - \alpha_1 a_1 - \alpha_2 a_2 \qquad (3.1)$$

and note that

$$\ell = \begin{bmatrix} 0 \\ 0 \\ \ell_3 \end{bmatrix}.$$

Solving for $a_3$ yields

$$a_3 = \ell + \alpha_1 a_1 + \alpha_2 a_2.$$

By substituting, we have

$$\begin{aligned}
\det [a_1, a_2, a_3] &= \det [a_1, a_2, \ell + \alpha_1 a_1 + \alpha_2 a_2] \\
&= \det [a_1, a_2, \ell] + \det [a_1, a_2, \alpha_1 a_1] \\
&\quad + \det [a_1, a_2, \alpha_2 a_2] \\
&= \det [a_1, a_2, \ell] + 0 + 0 = \det B \cdot \ell_3.
\end{aligned}$$

Since the hypothesis assures $\det [a_1, a_2, a_3] = 0$, it follows that $\ell_3 = 0$. Thus, using (3.1),

$$a_3 = \alpha_1 a_1 + \alpha_2 a_2.$$

Similarly $a_4$ is a linear combination of $a_1, a_2$ and so $\text{rank}\, A = 2$. ∎

An example may be helpful.

**Example 3.2** *Let* $A = \begin{bmatrix} 1 & -1 & 3 & -1 \\ -1 & 1 & 0 & 1 \\ 2 & -2 & 3 & -2 \end{bmatrix}$. *The submatrix* $B = \begin{bmatrix} 1 & 3 \\ -1 & 0 \end{bmatrix}$, *in rows 1, 2 and columns 1, 3 is such that* $\det B = 3$. *All* $3 \times 3$ *submatrices* $C$ *containing* $B$ *are such that* $\det C = 0$. *Thus, by the theorem,* $\text{rank}\, A = 2$.

Three corollaries follow from the theorem.

**Corollary 3.1** *Let $A$ be an $m \times n$ matrix. Then*

*(a)* rank $A = $ rank $A^t$ *and*

*(b)* rank $A = $ rank $A^H$.

**Proof.** We prove Part (a), leaving Part (b) as an exercise.
Suppose rank $A = r$ and let $B$ be an $r \times r$ submatrix of $A$ as described in the theorem. Since det $B^t = $ det $B \neq 0$, $A^t$ contains an $r \times r$ submatrix having a non-zero determinant. Hence, rank $A^t \geq $ rank $A$. Applying the same argument to $A^t$ yields that rank $A \geq $ rank $A^t$. Thus, rank $A = $ rank $A^t$. ∎

Note that this corollary says that the maximum number of linearly independent rows equals precisely the maximum number of linearly independent columns in any matrix.

The next corollary shows that the rank doesn't change when multiplying by nonsingular matrices.

**Corollary 3.2** *Let $A$ be an $m \times n$ matrix, $P$ and $Q$ nonsingular $m \times m$ and $n \times n$ matrices, respectively. Then* rank $PAQ = $ rank $A$.

**Proof.** We outline the proof, leaving the write up as an exercise.
First prove that $a_{i_1}, a_{i_2}, \dots, a_{i_s}$ are linearly independent columns of $A$ if and only if $Pa_{i_1}, Pa_{i_2}, \dots, Pa_{i_s}$ are linearly independent columns of $PA$. (This is a matter of checking the pendent equations.) Thus, rank $PA = $ rank $A$.
Now, set $B = PA$. Then, using the first part of this proof,

$$\text{rank } B = \text{rank } B^t = \text{rank } Q^t B^t = \text{rank } (BQ)^t$$
$$= \text{rank } BQ = \text{rank } PAQ.$$

And putting together,

$$\text{rank } A = \text{rank } B = \text{rank } PAQ,$$

which is what we want. ∎

The last corollary shows how to extend a linearly independent set to a basis.

**Corollary 3.3** *Let $a_1, \dots, a_r$ be linearly independent vectors in Euclidean $n$-space. Then there are vectors $a_{r+1}, \dots, a_n$ such that $a_1, \dots, a_n$ forms a basis for this vector space.*

**Proof.** Let $A = [a_1 \dots a_r]$. Since rank $A = r$, by Theorem 3.2, there is an $r \times r$ submatrix $B$ of $A$ such that det $B \neq 0$.

Suppose $i_1, \ldots, i_{n-r}$ are rows of $A$ which contain no entries of $B$. Suppose further that our indexing is such that $i_1 > \cdots > i_{n-r}$. Then set

$$C = \begin{bmatrix} e_{i_1} \ldots e_{i_{n-r}} A \end{bmatrix}$$

an $n \times n$ matrix. Now, by expanding along the 1-st columns,

$$\det C = (-1)^{i_1+1} \cdots (-1)^{i_{n-r}+1} \det B.$$

Thus, $\det C \neq 0$ and so the columns of $C$ are linearly independent. Setting $a_{r+1} = e_{i_1}, \ldots, a_n = e_{i_{n-r}}$ yields the result. ∎

The following example demonstrates the corollary.

**Example 3.3** *Let $a_1 = (1,1,1,1)^t$ and $a_2 = (1,1,-1,1)^t$. We extend these vectors to a basis. For this let*

$$A = [a_1 a_2] = \begin{bmatrix} 1 & 1 \\ 1 & 1 \\ 1 & -1 \\ 1 & 1 \end{bmatrix}.$$

*Note that the $2 \times 2$ submatrix, in rows 2 and 3 of $A$, is nonsingular. So, we add $e_1$ and $e_4$ to get*

$$C = [e_4 e_1 a_1 a_2]$$

$$= \begin{bmatrix} 0 & 1 & 1 & 1 \\ 0 & 0 & 1 & 1 \\ 0 & 0 & 1 & -1 \\ 1 & 0 & 1 & 1 \end{bmatrix}.$$

*Then, by the proof of the corollary, $C$ is nonsingular and thus $\{a_1, a_2, e_4, e_1\}$ is a basis.*

A useful tool in showing that a matrix is nonsingular follows.

**Lemma 3.1** *Let $A$ be an $n \times n$ matrix. If $Ax = 0$ has only the solution $x = 0$, then $A$ is nonsingular.*

**Proof.** By backward multiplication, write

$$Ax = 0$$

as

$$x_1 a_1 + \cdots + x_n a_n = 0$$

where $a_1, \ldots, a_n$ are the columns of $A$. Since this is a pendence equation, and $x = 0$ is its only solution, the columns of $A$ are linearly independent. Hence, $A$ is nonsingular. ∎

A theorem useful in polynomial interpolation follows.

**Theorem 3.3** *Let* $x_1, \ldots, x_n$ *be* $n$ *distinct real scalars. Then, the Vandermonde matrix*

$$
A = \begin{bmatrix}
1 & x_1 & \cdots & x_1^{n-1} \\
1 & x_2 & \cdots & x_2^{n-1} \\
& & \cdots & \\
1 & x_n & \cdots & x_n^{n-1}
\end{bmatrix}
$$

*is nonsingular.*

**Proof.** By Lemma 3.1, we can show that $A$ is nonsingular by showing that $Ax = 0$ has only the solution $x = 0$. For simplicity we will do this for the case $n = 3$, leaving the general argument as an exercise.

Let $x = \begin{bmatrix} a \\ b \\ c \end{bmatrix}$ be a vector such that

$$
\begin{bmatrix}
1 & x_1 & x_1^2 \\
1 & x_2 & x_2^2 \\
1 & x_3 & x_3^2
\end{bmatrix}
\begin{bmatrix} a \\ b \\ c \end{bmatrix}
=
\begin{bmatrix} 0 \\ 0 \\ 0 \end{bmatrix}. \tag{3.2}
$$

We show that $x = 0$.

Define $p(t) = a + bt + ct^2$. Equation (3.2) can be rewritten as

$$
\begin{aligned}
p(x_1) &= 0 \\
p(x_2) &= 0 \\
p(x_3) &= 0.
\end{aligned}
$$

This means that $p$, a polynomial of at most degree 2, has 3 distinct roots. The Fundamental Theorem of Algebra assures that nonzero polynomials of degree at most 2 cannot have 3 distinct roots. Thus, $p(x)$ must be the zero polynomial and so $a = b = c = 0$. But, this means that $x = 0$ and thus $A$ is nonsingular. ∎

To see where this theorem is useful, suppose we want a polynomial to pass through the data

| $x$ | $x_1$ | $x_2$ | $\cdots$ | $x_n$ |
|---|---|---|---|---|
| $y$ | $y_1$ | $y_2$ | $\cdots$ | $y_n$ |

where $x_1, x_2, \ldots, x_n$ are distinct. We need a polynomial $p$ such that

$$
\begin{aligned}
p(x_1) &= y_1 \tag{3.3} \\
p(x_2) &= y_2 \\
&\cdots \\
p(x_n) &= y_n.
\end{aligned}
$$

To find $p$, we set $p(t) = a_0 + a_1 t + \cdots + a_{n-1} t^{n-1}$ and calculate its coefficients. Note that (3.3) can be written as

$$
\begin{bmatrix}
1 & x_1 & \cdots & x_1^{n-1} \\
1 & x_2 & \cdots & x_2^{n-1} \\
& & \cdots & \\
1 & x_n & \cdots & x_n^{n-1}
\end{bmatrix}
\begin{bmatrix}
a_0 \\
a_1 \\
\cdots \\
a_{n-1}
\end{bmatrix}
=
\begin{bmatrix}
y_1 \\
y_2 \\
\cdots \\
y_n
\end{bmatrix}.
$$

By the theorem, the coefficient matrix is nonsingular, which shows that the system has precisely one solution. This solution determines the coefficients of $p$. Note that we also see that there is precisely one polynomial of degree $n - 1$ or less that passes through these points.

## 3.1.1  Optional (Interpolation and Pictures)

In this optional, we show how to estimate populations and how to use MATLAB to view this work.

Censuses are taken every 10 years, e.g.,

| Year | 1950 | 1960 | 1970 | 1980 |
|---|---|---|---|---|
| Population in millions | 150.7 | 179.3 | 203.2 | 226.5 |

We can get a picture of this data using MATLAB. (See Figure 3.2.)

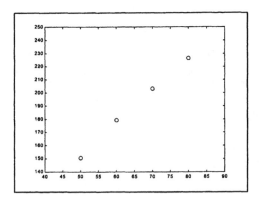

FIGURE 3.2.

Suppose we are preparing a report that requires some estimate of the population in 1965. To get this estimate, we find the polynomial of degree 3 or less that passes through the data points. Using MATLAB, this polynomial is

$$p(x) = 0.0007x^3 - 0.1465\ x^2 + 12.7567x - 206.3000.$$

To get a sense of how well this polynomial will estimate the population in 1965, it is helpful to view the polynomial. The graph of this is shown in Figure 3.3.

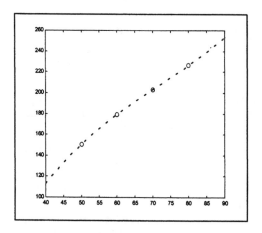

FIGURE 3.3.

Finally, we find the value of $p$ at 65,

$$p(65) = 192.0125$$

computed to 4 decimal places

And, if we want to plot everything, we have the result, which is shown in Figure 3.4

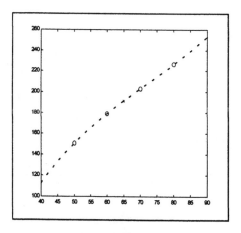

FIGURE 3.4.

Notice that this problem was done with $x = [\ 50\ 60\ 70\ 80\ ]$ rather than $x = [\ 1950\ 1960\ 1970\ 1980\ ]$. Using the latter $x$, MATLAB indicated it was having a problem giving polyfit $(x, y, 3)$. So, we redescribed our problem.

## 3.1.2  MATLAB (Polyfit and Polyval)

Given data
$$x = [x_1\ x_2\ x_3 \ldots x_{n+1}];$$
$$y = [y_1\ y_2\ y_3 \ldots y_{n+1}];$$
the command *polyfit* $(x, y, n)$ finds the coefficients of a polynomial of degree $n$ which passes through the data. And, given a polynomial $p$, *polyval* $(p, xi)$ gives $[p(x_1)\, p(x_2) \ldots p(x_{n+1})]$.

### 1.  Code for Plotting Data

```
x = [50 60 70 80];
y = [150.7 179.3 203.2 226.5];
plot (x, y, 'O')          % Plots data with an O
axis([40 90 100 260])     % Takes data points off edges
                            of the picture by changing
                            the size of the box to [40, 90]
                            by [100, 260]
```

### 2.  Code for Plotting Polynomial

Using lines 1 and 2 from 1, we add the following.
```
p = polyfit (x, y, 3)
ans: 0.0007  −0.1465      % These are the coefficients
                            of the polynomial.
12.7567  −206.3000.
xi = linspace (40, 90, 50);
z = polyval(p, xi);
plot (x, y, 'O', xi, z, ':')   % Plots the data (x, y, 'O')
                                and the 'curve '(xi, z, ':')
```

In the last line above, the symbol 'O' indicates only points are plotted, while ':' indicates points are to be connected by line segments.

### 3.  Code for Plotting Point.

Using all but the last line of 2, we add the following for the last code.
```
polyval (p, 65)
ans = 191.5812
plot (x, y, 'O', xi, z, ' : ', 65, 191.5812, ' × ')
```

For more information, type in *help polyval*, *help polyfit*, and *help plot*.

# Exercises

1. Use Theorem 3.2 to find the rank of

$$A = \begin{bmatrix} 0 & 1 & 2 & -1 \\ 1 & 1 & -1 & 3 \\ 1 & 2 & 1 & -2 \end{bmatrix}.$$

2. Prove that if $A$ is an $n \times n$ singular matrix, then $Ax = 0$ has infinitely many solutions.

3. Extend the given vectors to a basis for $R^3$.

(a) $\begin{bmatrix} 1 \\ 1 \\ 1 \end{bmatrix}$    (b) $\begin{bmatrix} 1 \\ 1 \\ 1 \end{bmatrix}, \begin{bmatrix} 1 \\ 1 \\ 0 \end{bmatrix}$

4. Let $V$ be a vector space. To extend a linearly independent set in $V$ to a basis, we can proceed a follows. Let $u_1, \ldots, u_n$ be a basis for a vector space $V$ and $x_1, \ldots, x_r$ linearly independent vectors in $V$.

   (a) Show that if $r < n$ then $x_1, \ldots, x_r, u_i$ are linearly independent for some $i$. (Some $u_i$ provides a new dimension.)

   (b) If $x_1, \ldots, x_r, u_i$ are linearly independent, set $x_{r+1} = u_i$ and repeat (a) for $r + 1$.

   Prove (a).

5. Apply the algorithm of Exercise 4 to the following.

   (a) $t + 1, t - 1$ and basis $1, t, t^2$

   (b) $\begin{bmatrix} 1 & -1 \\ 1 & 1 \end{bmatrix}, \begin{bmatrix} 1 & 1 \\ -1 & -1 \end{bmatrix}$ and basis
   $\begin{bmatrix} 1 & 0 \\ 0 & 0 \end{bmatrix}, \begin{bmatrix} 0 & 1 \\ 0 & 0 \end{bmatrix}, \begin{bmatrix} 0 & 0 \\ 1 & 0 \end{bmatrix}, \begin{bmatrix} 0 & 0 \\ 0 & 1 \end{bmatrix}$

6. Give a general proof for
   (a) Theorem 3.2    (b) Theorem 3.3.

7. Prove Part (b) of Corollary 3.1.

8. Provide the details for the proof of Corollary 3.2.

9. Is $L(A) = \operatorname{rank} A$ a linear transformation from $R^n \to R$?

10. Find two matrices of rank 2 whose product is rank 1.

11. Prove

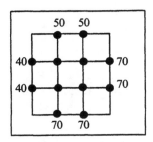

FIGURE 3.5.

(a) That rank $AB \leq$ rank $B$. (Hint: Show $AB$ cannot have more linearly independent columns than $B$.)

(b) That rank $AB \leq$ rank $A$. (Hint: Use the transpose.)

(c) That rank $(AB) =$ rank $B$ if $A$ is nonsingular.

12. Let $A = \begin{bmatrix} 1 & 1 \\ -1 & 1 \end{bmatrix}$. Explain why we can't find a sequence $A_1, A_2, \ldots$ of $2 \times 2$ matrices, having rank 1, which converge to $A$. (Hint: Use that the determinant is continuous.)

13. We will assume that the temperature at an interior point on the plate in Figure 3.5 is the average of the temperatures of the four closest surrounding points. (We are assuming steady-state temperatures and using that each grid point gives an estimate of the temperature there with those estimates getting better when the square has more grid points.)

   (a) Write out the system of linear equations the solution of which gives the unknown temperatures.

   (b) Solve this system. Explain why it is important for there to be precisely one solution.

   (c) Label the points with their temperatures and check to see if it looks right.

   (d) If there were 100 interior points, how many equations would there be?

14. Find a quadratic which passes through the data $(0,0)$, $(1,1)$, $(2,0)$. Graph the quadratic and the data.

15. (Optional) Use the data for 1960, 1970, 1980 to estimate the population in 1977. (This figure has been given in reports as 218.4. But, the figure is actually unknown.)

16. (MATLAB) Let $A = \begin{bmatrix} 1 & 1 & 2 & 1 \\ 0 & 1 & -1 & -2 \\ 1 & 0 & 0 & 1 \\ 2 & 2 & 0 & 0 \end{bmatrix}$.

    (a) Find rank $A$ by using the command rank $(A)$.

    (b) Find rank $A$ by using the command rref $(A)$.

17. (MATLAB) Let $x = \begin{bmatrix} 1 \\ 1 \\ 1 \\ 1 \\ 1 \end{bmatrix}, y = \begin{bmatrix} 1 \\ 1 \\ 0 \\ 0 \\ 0 \end{bmatrix}$.   Extend $x, y$ to a basis by applying rref to $[x\ y\ I]$.

## 3.2   Diagonalization

Let $A$ be an $n \times n$ matrix. If $A$ can be factored as

$$A = PDF^{-1}$$

for some nonsingular matrix $P$ and diagonal matrix D, we say that $A$ is *diagonalizable*. Not all matrices are diagonalizable. However, when they are, we show how this factorization can be done. Before starting this work, we need a few preliminaries.

The function

$$\varphi(\lambda) = \det(A - \lambda I),$$

where $\lambda$ is a scalar, is called the *characteristic polynomial of* $A$ and

$$\varphi(\lambda) = 0$$

its *characteristic equation*. (Some books use $\hat{\varphi}(\lambda) = \det(\lambda I - A)$ as the characteristic polynomial. Note that $\det(\lambda I - A) = (-1)^n \det(A - \lambda I)$, so the solutions to $\varphi(\lambda) = 0$ and $\hat{\varphi}(\lambda) = 0$ are the same.)

The lemma below shows that the characteristic polynomial has, counting multiplicities, $n$ roots.

**Lemma 3.2** *The characteristic polynomial of $A$ is a polynomial of degree $n$.*

**Proof.** Let $B = A - \lambda I$. Expanding $\det B$ along row 1 eliminates row 1 from all submatrices in the cofactors of the expansion. (Recall $c_{ij} = (-1)^{i+j} \det A_{ij}$.) Expanding these minors along row 1 eliminates row 2 of $A$ from all submatrices in the new cofactors. Continuing, we see that

$\det B = (a_{11} - \lambda)(a_{22} - \lambda) \cdots (a_{nn} - \lambda) + p(\lambda)$, where $p(\lambda)$ is a polynomial in $\lambda$ of degree at most $n-1$. Since $(a_{11} - \lambda) \cdots (a_{nn} - \lambda) = (-1)^n \lambda^n + q(\lambda)$, where the degree of $q$ is smaller than $n$, by putting together, the result follows. ■

Using the Fundamental Theorem of Algebra, we can factor

$$\varphi(\lambda) = (\lambda_1 - \lambda)(\lambda_2 - \lambda) \cdots (\lambda_n - \lambda)$$

(or $\hat{\varphi}(\lambda) = (\lambda - \lambda_1)(\lambda - \lambda_2) \cdots (\lambda - \lambda_n)$ if we like). The roots, namely $\lambda_1, \lambda_2, \cdots \lambda_n$, are called the *eigenvalues* (or, sometimes the latent roots or characteristic values) of $A$. We should recall from previous studies of polynomials, that the roots of $\varphi(\lambda)$ are the solutions to $\varphi(\lambda) = 0$. Thus, eigenvalues could be complex numbers even when the entries of $A$ are real numbers. In this case, we must work in $\mathcal{C}$.

An example of finding eigenvalues follows.

**Example 3.4** *To find the eigenvalues of* $A = \begin{bmatrix} 3 & 1 & 1 \\ 0 & 3 & 1 \\ 0 & 0 & 4 \end{bmatrix}$, *we solve*

$$\varphi(\lambda) = 0.$$

*This gives*

$$(3 - \lambda)(3 - \lambda)(4 - \lambda) = 0.$$

*Thus, the eigenvalues are* $\lambda_1 = 3$, $\lambda_2 = 3$, *and* $\lambda_3 = 4$. *Note that the eigenvalue 3 has multiplicity 2.*

We now link the eigenvalues of $A$ to $D$ in any factorization $A = PDP^{-1}$. This requires the following notion: two $n \times n$ matrices $A$ and $B$ are *similar* if there is an $n \times n$ nonsingular matrix $P$ such that

$$A = PBP^{-1}.$$

This equation can be written as $A = S^{-1}BS$ where $S = P^{-1}$. So actually it doesn't matter if the superscript $^{-1}$ is on the first or third factor of $PBP^{-1}$. And, since $P^{-1}AP = B$, $B$ and $A$ are similar so the order of $A$ and $B$ ($A$ and $B$ similar or $B$ and $A$ similar) doesn't matter.

As given below, similar matrices have the same eigenvalues.

**Lemma 3.3** *Let $A$ and $B$ be $n \times n$ matrices. If $A$ and $B$ are similar, their characteristic polynomials are identical. Thus, $A$ and $B$ have precisely the same eigenvalues.*

**Proof.** If $A$ and $B$ are similar, there is a nonsingular matrix $P$ such that

$$A = PBP^{-1}.$$

Thus,

$$\begin{aligned}
\det\left(A - \lambda I\right) &= \det\left(PBF^{-1} - \lambda I\right) \\
&= \det\left[P\left(B - \lambda I\right)P^{-1}\right] \\
&= \det P \det(B - \lambda I)\det P^{-1} \\
&= \det\left(B - \lambda I\right)
\end{aligned}$$

since $\det P \det P^{-1} = \det\left(PF^{-1}\right) = \det I = 1.$ ∎

When $A$ is similar to a diagonal matrix $D$, this lemma assures us that $A$ and $D$ have the same eigenvalues. We now show that the eigenvalues of $D$ are precisely those scalars that are on the main diagonal of $D$. A bit more general result follows.

**Lemma 3.4** *If $T$ is an $n \times n$ triangular matrix, then its characteristic polynomial is*

$$\varphi_T\left(\lambda\right) = \left(t_{11} - \lambda\right)\left(t_{22} - \lambda\right)\cdots\left(t_{nn} - \lambda\right).$$

*Thus, the eigenvalues of $T$ are exactly the main diagonal entries of $T$.*

**Proof.** Suppose $T$ is upper triangular. Then, expanding the determinant along the first column, we have

$$\varphi\left(\lambda\right) = \left(t_{11} - \lambda\right)\det\begin{bmatrix} t_{22} - \lambda & t_{23} & \cdots & t_{2n} \\ 0 & t_{33} - \lambda & \cdots & t_{3n} \\ & & \cdots & \\ 0 & 0 & \cdots & t_{nn} - \lambda \end{bmatrix}.$$

Continuing to expand along the first columns, we have

$$\varphi\left(\lambda\right) = \left(t_{11} - \lambda\right)\left(t_{22} - \lambda\right)\cdots\left(t_{nn} - \lambda\right),$$

the desired result. ∎

Putting the results above together, what we now know is that if $A$ is similar to a diagonal matrix $D$, then the main diagonal entries of $D$ are the eigenvalues of $A$, in some arrangement.

**Example 3.5** *Let $A = \begin{bmatrix} 2 & 1 \\ 1 & 2 \end{bmatrix}$. The eigenvalues of $A$ are $\lambda_1 = 3$ and $\lambda_2 = 1$. So if $A$ is similar to a diagonal matrix $D$, then*

$$D = \begin{bmatrix} 3 & 0 \\ 0 & 1 \end{bmatrix} \text{ or } D = \begin{bmatrix} 1 & 0 \\ 0 & 3 \end{bmatrix}.$$

Thus, when a matrix $A$ is diagonalizable, we can calculate $D$,

$$D = \begin{bmatrix} \lambda_1 & 0 & \cdots & 0 \\ 0 & \lambda_2 & \cdots & 0 \\ & & \cdots & \\ 0 & 0 & \cdots & \lambda_n \end{bmatrix}$$

where $\lambda_1, \ldots, \lambda_n$ are the eigenvalues of $A$ in some order.

We now try to find an $n \times n$ nonsingular matrix $P$ such that

$$A = PDP^{-1}.$$

To find $P$, rearrange this equation to

$$AP = PD.$$

Equating corresponding columns, we have

$$Ap_1 = \lambda_1 p_1$$
$$\cdots$$
$$Ap_n = \lambda_n p_n,$$

(3.4)

where $p_i$ is the $i$-th column of $P$.

For an eigenvalue $\lambda$, any nonzero vector $p$ such that

$$Ap = \lambda p \qquad (3.5)$$

is called an *eigenvector* belonging to $\lambda$. Thus, in (3.4), $p_1$ is an eigenvector belonging to $\lambda_1, \ldots, p_n$ is an eigenvector for $\lambda_n$. And, our problem now is to find linearly independent eigenvectors $p_1, \ldots, p_n$ that satisfy (3.4). That these vectors are linearly independent assures that $P$ is nonsingular.

To find an eigenvector $p$, belonging to eigenvalue $\lambda$, we solve the equation

$$Ap = \lambda p$$

or, by rearranging to a better form

$$Ap - \lambda p = 0$$
$$Ap - \lambda I p = 0$$
$$(A - \lambda I)p = 0,$$

a system of linear equations.

The next lemma shows this equation has a nonzero solution $p$.

**Lemma 3.5** *If $\lambda$ is an eigenvalue of $A$, then there is an eigenvector $p$ belonging to $\lambda$.*

**Proof.** Note that since $\lambda$ is an eigenvalue of $A$,

$$\varphi(\lambda) = 0.$$

Thus,

$$\det(A - \lambda I) = 0$$

and so $A - \lambda I$ is singular. Hence, from Lemma 3.1, we see that there is a nonzero vector $p$ such that

$$(A - \lambda I)p = 0.$$

Thus, $Ap = \lambda p$ and $p$ is an eigenvector belonging to $\lambda$. ∎

We call the null space of $A - \lambda I$, the *eigenspace* for the eigenvalue $\lambda$. Note this is a subspace, whose dimension is positive, and that all vectors in $N(A - \lambda I)$, except 0, are eigenvectors belonging to $\lambda$. So there are lots of eigenvectors belonging to any eigenvalue.

To find $P$, we only need to find a linearly independent set of eigenvectors, say, $p_1, p_2, \ldots, p_n$ belonging to $\lambda_1, \ldots, \lambda_n$, respectively. Then $P = [p_1 \ldots p_n]$ is nonsingular and $A = PDP^{-1}$. Note also from (3.4), that the order of the eigenvalues in $D$ is determined by the order of eigenvectors in $P$ or vice versa.

**Example 3.6** *Let* $A = \begin{bmatrix} 2 & 1 & 1 \\ 1 & 2 & 1 \\ 1 & 1 & 2 \end{bmatrix}$. *We find $D$ and $P$ such that $A = PDP^{-1}$.*

*(a) Computing D: We solve*

$$\det(A - \lambda I) = 0,$$

*which is*

$$(\lambda - 1)^2(\lambda - 4) = 0.$$

*Thus, $\lambda_1 = 1, \lambda_2 = 1$, and $\lambda_3 = 4$, are the eigenvalues of $A$. This yields*

$$D = \begin{bmatrix} 1 & 0 & 0 \\ 0 & 1 & 0 \\ 0 & 0 & 3 \end{bmatrix}.$$

*(b) Computing P: We find corresponding eigenvectors.*

i. *Eigenvectors for $\lambda_1 = \lambda_2 = 1$. Here we solve*

$$(A - \lambda_1 I) x = 0$$

*to find $p_1$ and $p_2$. Solving* $\begin{bmatrix} 1 & 1 & 1 \\ 1 & 1 & 1 \\ 1 & 1 & 1 \end{bmatrix} x = 0$ *by Gaussian elimination, we get the row echelon form*

$$\begin{array}{ccc} x_1 & x_2 & x_3 \\ \left[\begin{array}{ccc|c} 1 & 1 & 1 & 0 \\ 0 & 0 & 0 & 0 \\ 0 & 0 & 0 & 0 \end{array}\right] \end{array}.$$

*Note $x_2$, $x_3$ are free so,*

$$x_3 = \alpha$$
$$x_2 = \beta$$

*where $\alpha, \beta$ are arbitrarily chosen. Then*

$$x_1 = -\alpha - \beta.$$

*Thus*

$$x = \begin{bmatrix} x_1 \\ x_2 \\ x_3 \end{bmatrix} = \begin{bmatrix} -\alpha - \beta \\ \beta \\ \alpha \end{bmatrix}$$

$$= \alpha \begin{bmatrix} -1 \\ 0 \\ 1 \end{bmatrix} + \beta \begin{bmatrix} -1 \\ 1 \\ 0 \end{bmatrix}.$$

*Since $\lambda_1 = \lambda_2 = 1$, we need two solutions $p_1$ and $p_2$, which form a linearly independent set. We take $\alpha = 1$, $\beta = 0$ for*

$$p_1 = \begin{bmatrix} -1 \\ 0 \\ 1 \end{bmatrix} \text{ and } \alpha = 0, \ \beta = 1 \text{ for } p_2 = \begin{bmatrix} -1 \\ 1 \\ 0 \end{bmatrix}. \ (Different$$

*choices for $\alpha$ and $\beta$ could have been made.) Observe, by looking at the last 2 entries of each vector, that $p_1$ and $p_2$ are linearly independent.*

ii. *Eigenvector for $\lambda_3 = 4$. Solving*

$$(A - \lambda_3 I) x = 0 \text{ we get}$$

$$x = \alpha \begin{bmatrix} 1 \\ 1 \\ 1 \end{bmatrix}, \text{ where } \alpha \text{ is arbitrary.}$$

*Let $\alpha = 1$ and so $p_3 = \begin{bmatrix} 1 \\ 1 \\ 1 \end{bmatrix}.$*

It can be shown that $p_1$, $p_2$, and $p_3$ are linearly independent, and so

$$P = [p_1 \ p_2 \ p_3] = \begin{bmatrix} -1 & -1 & 1 \\ 0 & 1 & 1 \\ 1 & 0 & 1 \end{bmatrix}.$$

*(To check, we can always compute $PDF^{-1}$ to see if we actually get A.)*

It might also be interesting to see the eigenspaces. Observe in Figure 3.6 that they intersect only at the origin and that $p_1$, $p_2$, and $p_3$ are linearly independent.

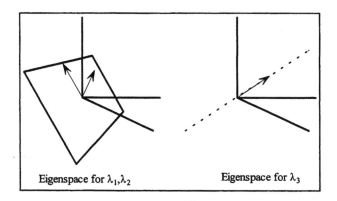

Eigenspace for $\lambda_1, \lambda_2$      Eigenspace for $\lambda_3$

FIGURE 3.6.

## 3.2.1 Optional (Buckling Beam)

A uniform column of length $l$ compressed by a load (force) $F$ at its top is shown in Figure 3.7. We let $y(x)$ be the deflection of the column at $x$, as given in Figure 3.8. The mathematical equations for this deflection are

$$\frac{d^2y}{dx} = -ky,$$
$$y(0) = 0, \ y(\ell) = 0$$

where $k$ is a positive constant depending on the force and the composition of the column. ($k = \frac{F}{EI}$ where $E$ is the modulus of elasticity of the beam and $I$ is the moment of inertia of the cross-sectional area by the column.) Actually, this differential equation is not difficult to solve directly. However, we will use it to show how differential equations can be solved numerically.

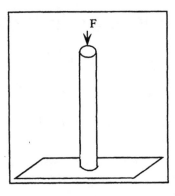

FIGURE 3.7.

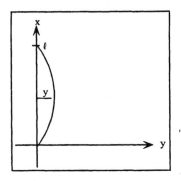

FIGURE 3.8.

This differential equation can be converted into a system of linear equations by approximating $\frac{d^2y}{dx^2}$ at equally spaced points $x_0, x_1, \ldots, x_n$ in $[0, \ell]$. We use the usual approximation

$$\frac{d^2y\,(x_i)}{dx^2} \approx \frac{y_{i+1} - 2y_i + y_{i-1}}{h^2}$$

where $h = \frac{\ell}{n}$ and $y_i$ $(y_i = y\,(x_i))$ the deflection shown in Figure 3.9 of the column at $x_i$.

Then we have, for five points (Actually, more points would lead to a better approximation and a better description of the deflection of the column.)

$$\frac{y_2 - 2y_1 + y_0}{\left(\frac{\ell}{4}\right)^2} = -ky_1$$

$$\frac{y_3 - 2y_2 + y_1}{\left(\frac{\ell}{4}\right)^2} = -ky_2.$$

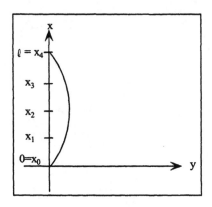

FIGURE 3.9.

To keep the problem small, we will assume the column's deflection is symmetric about $x_2$, the center of the beam, and so $y_3 = y_1$. Thus we have, using that $y_0 = 0$,

$$2y_1 - y_2 = \frac{kl^2}{16}y_1$$
$$-2y_1 + 2y_2 = \frac{kl^2}{16}y_2$$

or

$$\begin{bmatrix} 2 & -1 \\ -2 & 2 \end{bmatrix} \begin{bmatrix} y_1 \\ y_2 \end{bmatrix} = \lambda \begin{bmatrix} y_1 \\ y_2 \end{bmatrix} \qquad (3.6)$$

where $\lambda = \frac{kl^2}{16}$.

Observe that if $\begin{bmatrix} y_1 \\ y_2 \end{bmatrix} \neq \begin{bmatrix} 0 \\ 0 \end{bmatrix}$, by (3.6), $\lambda$ must be an eigenvalue of $\begin{bmatrix} 2 & -1 \\ -2 & 2 \end{bmatrix}$ and so $\lambda = 2 - \sqrt{2}$ or $\lambda = 2 + \sqrt{2}$.

Plugging $\lambda_1 = 2 - \sqrt{2}$ (the smallest eigenvalue) into (3.6), and rearranging, we have

$$\begin{bmatrix} \sqrt{2} & -1 \\ -2 & \sqrt{2} \end{bmatrix} \begin{bmatrix} y_1 \\ y_2 \end{bmatrix} = \begin{bmatrix} 0 \\ 0 \end{bmatrix}.$$

Thus $\begin{bmatrix} y_1 \\ y_2 \end{bmatrix} = \alpha \begin{bmatrix} 1 \\ \sqrt{2} \end{bmatrix}$, where $\alpha$ is free. So $y_1 = \alpha$, $y_2 = \sqrt{2}\alpha$, and consequently the column could appear in any of the buckling shapes in Figure 3.10.

Buckling theory indicates that if the force is small, so that $\lambda < 2 - \sqrt{2}$ (approximately, since our equations approximate the solution to the

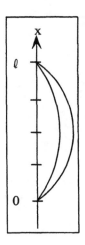

FIGURE 3.10.

differential equation), then $\begin{bmatrix} y_1 \\ y_2 \end{bmatrix} = \begin{bmatrix} 0 \\ 0 \end{bmatrix}$ and if small deflections occur, the column returns to this position. When $\lambda = 2 - \sqrt{2}$, the column can buckle a bit and stand as the one of the shapes described. When $\lambda > 2 - \sqrt{2}$, and not at other eigenvalues, slight deflections can collapse the column. At the remaining eigenvalue, at least in theory, buckling can occur with shapes different from that given.

### 3.2.2   MATLAB (Eig and [P, D])

The matrix function eig($A$) computes the eigenvalues and corresponding eigenvectors of the matrix $A$. As a single command, *eig* will provide a list of these eigenvalues. If we want corresponding eigenvectors, in particular $P$ and $D$ so that $A = PDP^{-1}$, we must ask for those matrices using the command $[P, D] = eig\,(A)$. For example,

$A = [1\ 2;\ 3\ 0]$;
$eig\,(A)$

ans $= \begin{bmatrix} 3 \\ -2 \end{bmatrix}$

$[P, D] = eig\,(A)$

ans: $P = \begin{bmatrix} 0.7071 & -0.5547 \\ 0.7071 & 0.8321 \end{bmatrix}$

ans: $D = \begin{bmatrix} 3 & 0 \\ 0 & -2 \end{bmatrix}$

For more, type in *help eig*.

# Exercises

1. Diagonalize (find $P$ and $D$ such that $A = PDF^{-1}$) each of the following matrices.

   (a) $A = \begin{bmatrix} 0 & 1 \\ 1 & 0 \end{bmatrix}$
   
   (b) $A = \begin{bmatrix} 3 & -1 \\ -1 & 3 \end{bmatrix}$

   (c) $A = \begin{bmatrix} 4 & 5 \\ 1 & 0 \end{bmatrix}$
   
   (d) $A = \begin{bmatrix} 0 & 1 \\ -1 & 0 \end{bmatrix}$

   (e) $A = \begin{bmatrix} 1 & 1 & 1 \\ 1 & 1 & 1 \\ 1 & 1 & 1 \end{bmatrix}$
   
   (f) $A = \begin{bmatrix} 1 & -1 & 0 \\ -1 & 1 & 0 \\ 0 & 0 & 2 \end{bmatrix}$

2. Draw the eigenspaces for each of (a), (b), (c), (e), and (f) of Exercise 1.

3. Find a matrix with eigenvalues $\lambda_1 = 1$ and $\lambda_2 = 2$ and eigenvectors $p_1 = \begin{bmatrix} 1 \\ 1 \end{bmatrix}$ and $p_2 = \begin{bmatrix} 1 \\ -1 \end{bmatrix}$, respectively.

4. Let $A = \begin{bmatrix} 1 & 1 \\ 1 & 1 \end{bmatrix}$. Then $A$ is similar to $D = \begin{bmatrix} 2 & 0 \\ 0 & 0 \end{bmatrix}$. Find two matrices for $P$ such that $A = PDF^{-1}$.

5. Give an example of a $2 \times 2$ matrix which is not similar to a diagonal matrix. Draw the eigenspaces for the example. (Hint: Look at triangular matrices with multiple eigenvalues.)

6. Find $2 \times 2$ matrices $A$ and $B$ such that the eigenvalues of $A + B$ are not, in any order, the sums of the eigenvalues of $A$ and $B$.

7. Let $A$ be a $3 \times 3$ diagonalizable matrix with $\lambda$ an eigenvalue of multiplicity 2. Prove that rank$(A - \lambda I) = 1$.

8. Let $A = \begin{bmatrix} \cos\theta & -\sin\theta \\ \sin\theta & \cos\theta \end{bmatrix}$ where $\theta \neq 0$ or $\pi$. Explain, using $Ap = \lambda p$, why $A$ has no real eigenvalues.

9. If $A$ is similar to a diagonal matrix, and $\alpha$ a scalar, what are the eigenvalues of $A - \alpha I$ in terms of $\alpha$ and the eigenvalues of $A$?

10. Suppose $A = PDF^{-1}$ where $D = diag(\lambda_1, \ldots, \lambda_n)$. Prove

    (a) If $A$ is $2 \times 2$, then $A^2 = P \begin{bmatrix} \lambda_1^2 & 0 \\ 0 & \lambda_2^2 \end{bmatrix} P^{-1}$.

    (b) $A^k = P\, diag\left(\lambda_1^k, \ldots, \lambda_n^k\right) P^{-1}$ for any positive integer $k$.

11. Let $A$ be an $3 \times 3$ matrix with linearly independent eigenvectors $p_1$, $p_2$, and $p_3$. Let $P = [p_1 p_2 p_3]$. What is $P^{-1}AF$?

12. Two parts: Let $A$ be an $n \times n$ real matrix.

    (a) Show that if $\lambda$ is an eigenvalue of $A$, then $\bar{\lambda}$ is an eigenvalue of $A$. (Complex eigenvalues come in conjugate pairs.)

    (b) Show that if $p$ is an eigenvector of $A$ belonging to $\lambda$, then $\bar{p}$ is an eigenvector for $A$ belonging to $\bar{\lambda}$.

13. Prove that

    (a) $A$ is similar to $A$.

    (b) If $A$ is similar to $B$ and $B$ is similar to $C$, then $A$ is similar to $C$.

14. Give an example of two matrices that have the same eigenvalues but are not similar.

15. Find (using, say, CRC Standard Math Tables) formulas for the solution to quadratic, cubic, and quartic equations. (For polynomial equations of degree 5 or more, no such formulas exist. Thus, for $k \times k$ matrices with $k \geq 5$, approximation techniques are used to find eigenvalues.)

16. (MATLAB) Let $p(t) = t^n - a_{n-1}t^{n-1} - \cdots - a_0 1$. Set

$$C = \begin{bmatrix} 0 & 1 & 0 & \cdots & 0 & 0 \\ 0 & 0 & 1 & \cdots & 0 & 0 \\ & & & \cdots & & \\ 0 & 0 & 0 & \cdots & 0 & 1 \\ a_0 & a_1 & a_2 & \cdots & a_{n-2} & a_{n-1} \end{bmatrix}.$$

    (a) Show that $\varphi(t) = (-1)^n p(t)$ so $\varphi(t)$ and $p(t)$ have the same roots ($\varphi(t)$ is the characteristic polynomial of $C$).

    (b) Use MATLAB and (a), to solve $t^4 - 3t^3 + 2t^2 - 3t + 1 = 0$.

17. (MATLAB) Find $P$ and $D$ for each of the matrices in Exercise 1. Use rank to check $P$ to see if it is nonsingular.

18. (Optional) Repeat the Optional work using $m = 6$.

19. (Optional) The boundary valve problem

$$y'' + y' + y = x^2 + 4x + 6$$
$$y(0) = 2, \ \ y(1) = 5$$

has solution $y = x^2 + 2x + 2$. To approximate the solution by finite difference methods, we set $x_0 = 0, x_1 = \frac{1}{n}, x_2 = \frac{2}{n}, \ldots, x_n = 1$ and use the approximations

$$y''(x_i) = \frac{y(x_{i+1}) - 2y(x_i) + y(x_{i-1})}{h^2}$$

$$y'(x_i) = \frac{y(x_{i+1}) - y(x_{i-1})}{2h}$$

where $h = \frac{1}{n}$ (the step size for this problem).

Use the approximation above to convert the boundary value problem into a system of linear equations and solve that system. Using MATLAB, plot $y = x^2 + 2x + 2$ and the approximations. Use

(a) $n = 4$

(b) $n = 8$


# 3.3    Conditions for Diagonalization

In this section, we describe when there are $n$ linearly independent eigenvectors $p_1, \ldots, p_n$ for the $n$ eigenvalues $\lambda_1, \ldots, \lambda_n$ of an $n \times n$ matrix $A$. Of course, in this case, since $P$ can be constructed from these eigenvectors, $P$ is nonsingular and $A$ is similar to a diagonal matrix.

We need a lemma.

**Lemma 3.6**  *Let $A$ be an $n \times n$ matrix with distinct eigenvalues $\lambda_1, \ldots, \lambda_r$. Then any corresponding eigenvectors $p_1, \ldots, p_r$ to these eigenvalues, respectively, form a linearly independent set.*

**Proof.** Consider the pendent equation

$$\alpha_1 p_1 + \cdots + \alpha_r p_r = 0.$$

Multiplying both sides of this equation by $A$, then $A^2, \ldots$ yields

$$\alpha_1 \lambda_1 p_1 + \cdots + \alpha_r \lambda_r p_r = 0$$
$$\cdots$$
$$\alpha_1 \lambda_1^{r-1} p_1 + \cdots + \alpha_r \lambda_r^{r-1} p_r = 0.$$

Writing these equations in matrix form yields, by backward multiplication,

$$[\alpha_1 p_1 \ldots \alpha_r p_r] \begin{bmatrix} 1 & \lambda_1 & \cdots & \lambda_1^{r-1} \\ & & \cdots & \\ 1 & \lambda_r & \cdots & \lambda_r^{r-1} \end{bmatrix} = 0.$$

Since the Vandermonde matrix is nonsingular, we can multiply though by its inverse to get

$$[\alpha_1 p_1 \ldots \alpha_r p_r] = 0.$$

So

$$\alpha_1 p_1 = 0$$
$$\cdots$$
$$\alpha_r p_r = 0$$

and thus $\alpha_1 = \cdots = \alpha_r = 0$. Hence we see that the eigenvectors form a linearly independent set. ∎

As a consequence of this theorem, we have one of the most important results in matrix theory.

**Corollary 3.4** *Let $A$ be an $n \times n$ matrix with $n$ distinct eigenvalues. Then $A$ is similar to a diagonal matrix.*

**Proof.** By the lemma, $A$ has, corresponding to the $n$ distinct eigenvalues, a set of $n$ linearly independent eigenvectors. These eigenvectors form a nonsingular matrix $P$ such that $A = PDP^{-1}$, $D$ the diagonal matrix made up of eigenvalues corresponding to the eigenvectors in $P$. ∎

Of course, matrices don't always have distinct eigenvalues. In those cases, to diagonalize, we need some further information about eigenvectors.

**Lemma 3.7** *Let $A$ be any $n \times n$ matrix. If $\lambda$ is an eigenvalue of $A$ of multiplicity $m$, then $A$ cannot have more than $m$ linearly independent eigenvectors belong to $\lambda$. (Thus, $\dim N(A - \lambda I) \leq m$.)*

**Proof.** We will argue a special case of the lemma, using proof by contradiction, leaving the general case as an exercise.

Let $A$ be a $3 \times 3$ matrix and suppose $m = 1$ and $x, y$ linearly independent eigenvectors for $\lambda$. Extend $x, y$ to $x, y, z$ a basis for Euclidean 3-space and set $P = [x\ y\ z]$. Then

$$AP = [\lambda x\ \lambda y\ w]$$

where $w = Az$. Factoring yields

$$AP = P \begin{bmatrix} \lambda & 0 & \alpha \\ 0 & \lambda & \beta \\ 0 & 0 & \gamma \end{bmatrix}$$

where $\alpha$, $\beta$, and $\gamma$ are chosen to satisfy $w = \alpha x + \beta y + \gamma z$. Now

$$A = P \begin{bmatrix} \lambda & 0 & \alpha \\ 0 & \lambda & \beta \\ 0 & 0 & \gamma \end{bmatrix} P^{-1}$$

and so the eigenvalues of $A$ are $\lambda, \lambda, \gamma$. This yields the contradiction. ∎

This lemma assures that if an eigenvalue of $A$ has fewer linearly independent eigenvectors than its multiplicity, then we simply cannot get enough linearly independent eigenvectors to form $P$. For example, if the eigenvalues are $\lambda_1 = \lambda_2 = \lambda_3 = 2$, $\lambda_4 = \lambda_5 = 3$ and $\dim(A - \lambda_1 I) = 2$, then we cannot get three linearly independent eigenvectors for the eigenvalue 2. And, thus $A$ is not diagonalizable. An example of such a matrix follows.

**Example 3.7** *Let* $A = \begin{bmatrix} 0 & 1 \\ 0 & 0 \end{bmatrix}$. *Then* $\lambda_1 = \lambda_2 = 0$. *Computing the corresponding eigenspace, we solve, in augmented matrix form*

$$\left[ \begin{array}{cc|c} 0 & 1 & 0 \\ 0 & 0 & 0 \end{array} \right].$$

*Thus,*

$$x_1 = \alpha, \ \alpha \text{ is arbitrary}$$
$$x_2 = 0$$

*and so*

$$x = \alpha \begin{bmatrix} 1 \\ 0 \end{bmatrix}.$$

*(The eigenspace is the $x_1$-axis.) Thus, there are not two linearly independent vectors belonging to the eigenvalue 0, and so $A$ is not diagonalizable. Note also, that if $A$ were diagonalizable then*

$$A = P \begin{bmatrix} 0 & 0 \\ 0 & 0 \end{bmatrix} P^{-1}$$

*since both eigenvalues of $A$ are 0. But this means that $A = 0$, not the given $A$.*

The following theorem gives necessary and sufficient conditions for a matrix to be diagonalizable.

**Theorem 3.4** *Let $A$ be an $n \times n$ matrix with distinct eigenvalues $\lambda_1, \ldots, \lambda_r$ having multiplicities $m_1, \ldots, m_r$, respectively. Then $A$ is similar to a diagonal matrix if and only if each $\lambda_i$ has a linearly independent set of $m_i$ eigenvectors (i.e., the dimension of its eigenspace is $m_i$).*

**Proof.** We prove this result for a $3 \times 3$ matrix $A$ with eigenvalues $\lambda_1 = \lambda_2, \lambda_3$ and corresponding eigenvectors $p_1, p_2, p_3$ where $p_1$ and $p_2$ are linearly independent. We need to show that $p_1, p_2, p_3$ are linearly independent.

Arguing now by contradiction, we suppose $(\beta_1, \beta_2, \beta_3)$ is a nontrivial solution to the pendent equation for $p_1, p_2, p_3$. Thus,

$$\beta_1 p_1 + \beta_2 p_2 + \beta_3 p_3 = 0. \tag{3.7}$$

We suppose that $\beta_1 \neq 0$. (The same line of reasoning that follows applies to any choice of $\beta_i \neq 0$.) Then $\beta_1 p_1 + \beta_2 p_2$ is an eigenvector belonging to $\lambda_1$. And $\beta_3 p_3$ is either an eigenvector belonging to $\lambda_3$ or it is 0. Regardless, by rearranging (3.7), we have

$$1 \left( \beta_1 p_1 + \beta_2 p_2 \right) + 1 \left( \beta_3 p_3 \right) = 0.$$

But this says that we have some eigenvectors, belonging to distinct eigenvalues, which are linearly dependent. However, this remark contradicts Lemma 3.6. ∎

It is interesting that an alternate approach, using row vectors, could have been taken to diagonalize a matrix. To see this, let $\lambda$ be an eigenvalue for an $n \times n$ matrix $A$. Then $\det(A - \lambda I) = 0$ and so, taking the transpose of the matrix $(A - \lambda I)$, $\det(A^t - \lambda I) = 0$. Thus, there is a nonzero row vector $y$ such that

$$\left( A^t - \lambda I \right) y^t = 0$$

or, taking the transpose

$$y \left( A - \lambda I \right) = 0$$

and so

$$yA = \lambda y.$$

Such a nonzero row vector is called a *left eigenvector*, belonging to $\lambda$, for $A$. When emphasis is desired, we call an eigenvector, as previously defined, a *right eigenvector* for $A$.

Using left eigenvectors, we could have formed a matrix $R$, whose rows are left eigenvectors. Then

$$RA = DR$$

where $D = diag(\lambda_1, \dots, \lambda_n)$ where $\lambda_1, \dots, \lambda_n$ are the èigenvalues of $A$. If $R$ is nonsingular, $A = R^{-1} DR$.

There is a useful relationship, called the Principle of Biorthogonality, between left and right eigenvectors, which we give below.

**Theorem 3.5** *Let $A$ be an $n \times n$ matrix with distinct eigenvalues $\lambda_1, \dots, \lambda_n$ and corresponding left and right eigenvectors $y_1, \dots, y_n$ and $x_1, \dots, x_n$, respectively.*

(a) If $i \neq j$, then $y_i x_j = 0$.

(b) Otherwise, $y_i x_i \neq 0$.

**Proof.** There are two parts.
Part a. Since

$$y_i A = \lambda_i y_i \text{ and } A x_j = \lambda_j x_j,$$

we have

$$y_i A x_j = \lambda_i y_i x_j \text{ and } y_i A x_j = \lambda_j y_i x_j.$$

Thus,

$$\lambda_i y_i x_j = \lambda_j y_i x_j$$

or

$$(\lambda_i - \lambda_j) y_i x_j = 0.$$

Since $\lambda_i \neq \lambda_j$, it follows that $y_i x_j = 0$.
Part b. Since $\lambda_1, \dots, \lambda_n$ are distinct, $x_1, \dots, x_n$ forms a basis for Euclidean $n$-space. Thus,

$$y_i^H = \alpha_1 x_1 + \cdots + \alpha_n x_n$$

for some scalars $\alpha_1, \dots, \alpha_n$. Multiplying through by $y_i$ yields, using that $y_i x_j = 0$ for $i \neq j$,

$$\|y_i\|_2^2 = \alpha_i y_i x_i.$$

Since $\|y_i\|_2^2 \neq 0$, $y_i x_i \neq 0$. ∎

The following example numerically demonstrates the property.

**Example 3.8** Let $A = \begin{bmatrix} 1 & 3 \\ 2 & 2 \end{bmatrix}$. The eigenvalues of $A$ are $\lambda_1 = 4$, $\lambda_2 = -1$ with corresponding right and left eigenvectors $x_1 = \begin{bmatrix} 1 \\ 1 \end{bmatrix}$, $x_2 = \begin{bmatrix} 3 \\ -2 \end{bmatrix}$, and $y_1 = (2, 3)$, $y_2 = (1, -1)$. Note that $y_1 x_1 = 5$, $y_1 x_2 = 0$, $y_2 x_1 = 0$, and $y_2 x_2 = 5$.

To conclude this section, we show how to use diagonalization in helping understand linear transformation geometrically. To do this, we let $A$ be an $n \times n$ diagonalizable matrix, with real eigenvalues. So, we can factor $A = PDP^{-1}$ using real numbers.

The columns of $P$ form a basis, say, $Y = \{p_1, \ldots, p_n\}$. Thus given any $x$,

$$x = y_1 p_1 + \cdots + y_n p_n$$

for some scalars $y_1, \ldots, y_n$. And $y = (y_1, \ldots, y_n)^t$ gives the $Y$-coordinates of $x$. Note that

$$x = Py$$

so $P$ converts $Y$-coordinates of $x$ into the vector $x$.

Now, we describe the linear transformation $L(x) = Ax$ in $Y$-coordinates. We do this in two steps.

1. Converting $L(x)$ into $Y$-coordinates, we have

$$P^{-1}L(x) = P^{-1}Ax.$$

2. Converting $x$ into $Y$-coordinates, we have

$$P^{-1}L(Py) = P^{-1}APy.$$

Thus, if we set

$$L_Y(y) = P^{-1}APy$$
$$= Dy,$$

we have described the transformation $L(x) = Ax$ in terms of $Y$-coordinates. And, with respect to these coordinates, $L$ stretches, shrinks, reflects the axes, etc. (and thus the corresponding space) in the $Y$-coordinate system.

We show a particular example.

**Example 3.9** *For* $A = \begin{bmatrix} 2 & 1 \\ 1 & 2 \end{bmatrix}$, $D = \begin{bmatrix} 3 & 0 \\ 0 & 1 \end{bmatrix}$, *and* $P = \begin{bmatrix} \frac{1}{\sqrt{2}} & -\frac{1}{\sqrt{2}} \\ \frac{1}{\sqrt{2}} & \frac{1}{\sqrt{2}} \end{bmatrix}$.

*Thus,*

$$Y = \left\{ \begin{bmatrix} \frac{1}{\sqrt{2}} \\ \frac{1}{\sqrt{2}} \end{bmatrix}, \begin{bmatrix} -\frac{1}{\sqrt{2}} \\ \frac{1}{\sqrt{2}} \end{bmatrix} \right\}$$

*and* $L(x) = Ax$ *is described by* $L_Y(y) = Dy$ *in the $Y$-coordinate system.*

*Looking at $L$ through the $Y$-coordinates, we see in Figure 3.11 that $L$ leaves the $y_2$-axis alone but stretched the $y_1$-axis (and corresponding space) by 3.*

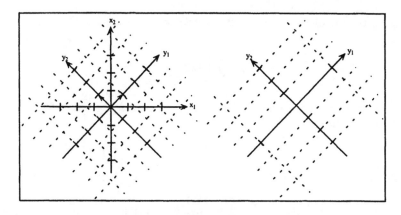

FIGURE 3.11.

### 3.3.1 Optional (Picture of Multiple Eigenvalue Matrices in Matrix Space)

Matrices with distinct eigenvalues can be diagonalized. These matrices make up most all of matrix space. To view this, we give a picture.

Let

$$A = \begin{bmatrix} a & b \\ c & d \end{bmatrix}.$$

The characteristic equation for $A$ is

$$\lambda^2 - (a+d)\lambda + ad - bc = 0$$

so

$$\lambda = \frac{(a+d) \pm \sqrt{(a+d)^2 - 4(ad-bc)}}{2}.$$

Thus, $A$ has a multiple eigenvalue if and only if $(a+d)^2 - 4(ad-bc) = 0$. Expanding and rearranging yields $(a-d)^2 + 4bc = 0$. If $b$ and $c$ are 0, then $a = d$, and we have a diagonal matrix. Thus, we will suppose $c \neq 0$.

Now

$$b = \frac{-(a-d)^2}{4c}.$$

Thus,

$$A = \begin{bmatrix} a & \frac{-(a-d)^2}{4c} \\ c & d \end{bmatrix}$$

has multiple eigenvalues for all $c \neq 0$.

To get some picture of this set, we set $c = d$. Define

$$L\left(\begin{bmatrix} a & b \\ d & d \end{bmatrix}\right) = \begin{bmatrix} a \\ b \\ \sqrt{2}d \end{bmatrix}$$

a linear map which preserves distances as shown in Optional, Chapter 2, Section 3. Since

$$L\left(\begin{bmatrix} a & \frac{-(a-d)^2}{4d} \\ d & d \end{bmatrix}\right) = \begin{bmatrix} a \\ \frac{-(a-d)^2}{4d} \\ \sqrt{2}d \end{bmatrix},$$

we have a picture of the multiple eigenvalue matrices in a piece of the space of $2 \times 2$ matrices. To draw this picture, we graph $\left(a, \frac{-(a-d)^2}{4d}, \sqrt{2}d\right)^t$. Notice in Figure 3.12 that this set of matrices has shape. And, notice that

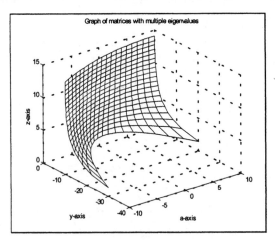

FIGURE 3.12.

it does not take up much of the matrix space.

### 3.3.2   MATLAB (Code for Picture)

#### Code for Picture of Multiple Eigenvalue Matrices

```
a = linspace (-10, 10, 20);
d = linspace (1, 10, 20);
[a, d] = meshgrid (a, d);
y = - ((a - d) . ^ 2)./ (4 * d);
z = sqrt (2) * d;
mesh (a, y, z)
```

# Exercises

1. If possible, diagonalize (find $D$ and $P$) the given matrix. If not, draw the eigenspaces and explain why the matrix cannot be diagonalized.

   (a) $\begin{bmatrix} 2 & 1 \\ 0 & 2 \end{bmatrix}$    (b) $\begin{bmatrix} 1 & 1 \\ 0 & 2 \end{bmatrix}$

   (c) $\begin{bmatrix} 1 & 1 & 0 \\ 1 & 1 & 0 \\ 0 & 0 & 2 \end{bmatrix}$    (d) $\begin{bmatrix} 1 & 1 & 0 \\ 1 & 1 & 0 \\ 1 & 1 & 2 \end{bmatrix}$

2. Prove

   (a) Lemma 3.7.    (b) Theorem 3.4.

3. Find left and right eigenvectors for each eigenvalue of the matrices below.

   (a) $A = \begin{bmatrix} 1 & 4 \\ 1 & 1 \end{bmatrix}$    (b) $A = \begin{bmatrix} 1 & 1 & 0 \\ 0 & 1 & 1 \\ 0 & 1 & 1 \end{bmatrix}$

4. If $A = PDF^{-1}$, how can we find $n$ linearly independent left eigenvectors for $A$, by using $P$?

5. Let $A$ be an $n \times n$ matrix and $E$ a row echelon form of $A$. Are the eigenvalues of $A$ on the main diagonal of $E$?

6. Let $L(x) = Ax$. As in Example 3.9, describe $L$ in the $Y$-coordinate system for the matrices given below.

   (a) $A = \begin{bmatrix} 0 & 1 \\ 1 & 0 \end{bmatrix}$    (b) $a = \begin{bmatrix} 1 & 2 \\ 1 & 0 \end{bmatrix}$

7. Prove that if $A$ is an $n \times n$ matrix, then $A$ and $A^t$ have the same eigenvalues.

8. Two parts.

   (a) Prove $\det \begin{bmatrix} A & 0 \\ B & C \end{bmatrix} = (\det A)(\det C)$, where $A$ and $C$ are square matrices. (Use induction on the number of rows of $A$.)

   (b) Tell how to find the eigenvalues of $\begin{bmatrix} A & 0 \\ B & C \end{bmatrix}$ in terms of $A$ and $C$.

9. If $Ax = \lambda x$, $x \neq 0$, and $B = PAF^{-1}$, show how to find an eigenvector for $B$ belonging to $\lambda$ by using $x$ and $P$.

10. Prove that $A$ is singular if and only if 0 is an eigenvalue of $A$.

11. Let $A$ be an $m \times n$ matrix and $B$ an $n \times m$ matrix. Prove that the $m \times m$ matrix $AB$ and the $n \times n$ matrix $BA$ have the same nonzero eigenvalues.

(Hint: $\begin{bmatrix} AB & 0 \\ B & 0 \end{bmatrix} \begin{bmatrix} I & A \\ 0 & I \end{bmatrix} = \begin{bmatrix} I & A \\ 0 & I \end{bmatrix} \begin{bmatrix} 0 & 0 \\ B & BA \end{bmatrix}$.)

12. Two parts:

   (a) Prove that if $A$ is diagonalizable, then so is $A^2$.

   (b) Find a matrix $A$ which is not diagonalizable, but $A^2$ is diagonalizable.

13. Let $A$ be an $n \times n$ matrix with linearly independent right eigenvectors $x_1, \ldots, x_n$ and left eigenvectors $y_1, \ldots, y_n$. If

$$x = \alpha_1 x_1 + \cdots + \alpha_n x_n,$$

prove that $\alpha_i = \frac{y_i x}{y_i x_i}$ for all $i$.

14. (Optional)  View the $2 \times 2$ matrices, with multiple eigenvalues, in $R^3$ by setting $d = 0$ and graphing $\left( a, \frac{-a^2}{4c}, c \right)^t$ over $1 \le a \le 10$, $1 \le c \le 10$.

15. (MATLAB)  Let $S = \begin{bmatrix} 1 & -1 & -1 & 1 & 1 \\ 1 & 1 & -1 & -1 & 1 \end{bmatrix}$, the matrix for a square. Find the linear transformation $L(x) = Ax$ that stretches the square (and corresponding space), along the line $y = x$, by 2. Plot $S$ and plot $(AS)$. (Hint: Find $L_Y(y) = Dy$ that does this and then contruct $L(x) = Ax$ from it.)

## 3.4  Jordan Forms

As we saw in the last section, not all matrices are diagonalizable. Those matrices which are not diagonalizable are often called *defective*. In this section, we describe another $n \times n$ matrix $J$, quite close to a diagonal matrix, except the superdiagonal entries $j_{12}, j_{23}, \ldots, j_{n-1,n}$ need not be 0. This matrix is called a Jordan form. It can be shown that every square matrix, diagonalizable or defective, is similar to a Jordan form.

The proof of the Jordan form result is much more intricate than what we saw in Sections 2 and 3. However, the use of the Jordan form is not much beyond that for diagonal matrices, and so there is no reason not to use it.

The Jordan form is described below.

**Definition 3.1** *Let $J$ be an upper triangular matrix with a super diagonal of 0's and 1's and all entries above the superdiagonal 0's ($j_{rs} = 0$ for $s > r+1$). If a 1 appearing on the superdiagonal implies the diagonal entries in its row and column are identical ($j_{r,r+1} = 1 \Rightarrow j_{rr} = j_{r+1,r+1}$), then $J$ is called a Jordan form.*

Thus, Jordan forms appear as

$$
\begin{bmatrix} \lambda_1 & 0 \\ 0 & \lambda_2 \end{bmatrix},
\begin{bmatrix} \lambda_1 & 1 \\ 0 & \lambda_1 \end{bmatrix},
\begin{bmatrix} \lambda_1 & 0 & 0 \\ 0 & \lambda_2 & 0 \\ 0 & 0 & \lambda_3 \end{bmatrix},
\begin{bmatrix} \lambda_1 & 1 & 0 \\ 0 & \lambda_1 & 0 \\ 0 & 0 & \lambda_3 \end{bmatrix}
$$

$$
\begin{bmatrix} \lambda_1 & 0 & 0 \\ 0 & \lambda_2 & 1 \\ 0 & 0 & \lambda_2 \end{bmatrix},
\begin{bmatrix} \lambda_1 & 1 & 0 \\ 0 & \lambda_1 & 1 \\ 0 & 0 & \lambda_1 \end{bmatrix}, \cdots
$$

From these remarks we can state Jordan's theorem.

**Theorem 3.6** *Any $n \times n$ matrix $A$ is similar to a Jordan form.*

Notice that since $J$ is upper triangular, by Lemma 3.4, the eigenvalues of $A$ are the main diagonal entries of $J$. And we can partition $J$ so that all main diagonal blocks are square. Each main diagonal block contains the same eigenvalue and has a super diagonal of 1's. All other blocks have entries 0. For example,

$$
\left[\begin{array}{cc|c} 2 & 1 & 0 \\ 0 & 2 & 0 \\ \hline 0 & 0 & 3 \end{array}\right]
\quad \text{or} \quad
\left[\begin{array}{ccc|c|cc} 3 & 1 & 0 & 0 & 0 & 0 \\ 0 & 3 & 1 & 0 & 0 & 0 \\ 0 & 0 & 3 & 0 & 0 & 0 \\ \hline 0 & 0 & 0 & 2 & 0 & 0 \\ \hline 0 & 0 & 0 & 0 & 2 & 1 \\ 0 & 0 & 0 & 0 & 0 & 2 \end{array}\right].
$$

These main diagonal blocks are called Jordan blocks. If the Jordan blocks are $J_1, \ldots, J_r$, we can write this as

$$
J = \operatorname{diag}(J_1, \ldots, J_r).
$$

Except for the arrangement of the Jordan blocks, it is known that the Jordan form is unique. A general method to find $J$, as well as the nonsingular matrix $P$, can be found in the theory books in the Bibliography. In this text, for $2 \times 2$ and $3 \times 3$ matrices, we will simply try to solve $AP = PJ$ for both $P$ and $J$. We show how in the example below.

**Example 3.10** *Let $A = \begin{bmatrix} 2 & 0 & 0 \\ 3 & 2 & 0 \\ 0 & 0 & 4 \end{bmatrix}$. Here, the eigenvalues of $A$ are given by $\lambda_1 = 2$, $\lambda_2 = 2$, and $\lambda_3 = 4$. It is clear that the Jordan block for*

$\lambda_3 = 4$ is $1 \times 1$. *We need to decide, however, if there are two $1 \times 1$ Jordan blocks for the eigenvalues $\lambda_1$ and $\lambda_2$ or only one $2 \times 2$ Jordan block.*
   *We solve*

$$A = PJF^{-1}$$

*or, obtaining a better form,*

$$AF = PJ$$
$$A\,[p_1 p_2 p_3] = [p_1 p_2 p_3]\,J$$

*where $p_1$, $p_2$, $p_3$ are the columns of $P$.*
   *Placing $\lambda_3$ as the last Jordan block in $J$, we have*

$$A\,[p_1 p_2 p_3] = [p_1 p_2 p_3] \begin{bmatrix} \hat{J} & 0 \\ 0 & 4 \end{bmatrix}.$$

*So*

$$Ap_3 = 4p_3,$$

*(We use backward multiplication to get $4p_3$.) an eigenvector problem which we solve to get*

$$p_3 = \begin{bmatrix} 0 \\ 0 \\ 1 \end{bmatrix}.$$

*(Other choices could have been made for $p_3$.)*
   *Now*

$$A\,[p_1 p_2] = [p_1 p_2]\,\hat{J} \tag{3.8}$$

*where $\hat{J}$ has two Jordan blocks of size $1 \times 1$ or one $2 \times 2$ Jordan block. Thus,*

$$\hat{J} = \begin{bmatrix} 2 & \beta \\ 0 & 2 \end{bmatrix}$$

*where $\beta = 0$ or $1$.*
   *Using (3.8), we know that*

$$Ap_1 = 2p_1.$$

*So we solve*

$$(A - 2I)\,p_1 = 0.$$

*This yields*

$$p_1 = \alpha \begin{bmatrix} 0 \\ 1 \\ 0 \end{bmatrix}, \quad \alpha \text{ arbitrary.}$$

*We let $\alpha = 1$, so*

$$p_1 = \begin{bmatrix} 0 \\ 1 \\ 0 \end{bmatrix}.$$

*Thus, the eigenspace for the eigenvalue 2 has dimension 1. This assures us that $\hat{J}$ is not a diagonal matrix. So*

$$\hat{J} = \begin{bmatrix} 2 & 1 \\ 0 & 2 \end{bmatrix}.$$

*Now by (3.8) and backward multiplication, we need to solve*

$$A p_2 = p_1 + 2 p_2$$

*or*

$$(A - 2I) p_2 = p_1$$

*for $p_2$. Since $p_1 = \begin{bmatrix} 0 \\ 1 \\ 0 \end{bmatrix}$, we solve*

$$(A - 2I) p_2 = \begin{bmatrix} 0 \\ 1 \\ 0 \end{bmatrix}.$$

*By Gaussian elimination and choosing one solution, we get*

$$p_2 = \begin{bmatrix} \frac{1}{3} \\ 0 \\ 0 \end{bmatrix}.$$

*Thus, $P = [p_1 p_2 p_3] = \begin{bmatrix} 0 & \frac{1}{3} & 0 \\ 1 & 0 & 0 \\ 0 & 0 & 1 \end{bmatrix}$ and $J = \begin{bmatrix} 2 & 1 & 0 \\ 0 & 2 & 0 \\ 0 & 0 & 4 \end{bmatrix}$.*

The 1's on the super diagonal of $J$ are there by choice. Other numbers could also have been chosen as shown below.

**Theorem 3.7** *Let $A$ be an $n \times n$ matrix. Then $A$ is similar to a Jordan form with the 1's on the superdiagonal replaced by any $\epsilon \neq 0$.*

**Proof.** Let $A = PJF^{-1}$ where $J$ is a Jordan form of $A$. Let $D = \text{diag}\left(\epsilon^{-1}, \epsilon^{-2}, \ldots, \epsilon^{-n}\right)$. Then $DJD^{-1}$ is $J$ with the 1's on the superdiagonal replaced by $\epsilon$'s. This occurs since premultiplying $J$ by $D$ multiplies the $i$-th row of $J$ by $\epsilon^{-i}$ and postmultiplying by $D^{-1}$ multiplies the $j$-th column by $\epsilon^j$. For example,

$$DJD^{-1} = \begin{bmatrix} \epsilon^{-1} & 0 & 0 \\ 0 & \epsilon^{-2} & 0 \\ 0 & 0 & \epsilon^{-3} \end{bmatrix} \begin{bmatrix} 3 & 1 & 0 \\ 0 & 3 & 1 \\ 0 & 0 & 3 \end{bmatrix} \begin{bmatrix} \epsilon & 0 & 0 \\ 0 & \epsilon^2 & 0 \\ 0 & 0 & \epsilon^3 \end{bmatrix}$$

$$= \begin{bmatrix} 3 & \epsilon & 0 \\ 0 & 3 & \epsilon \\ 0 & 0 & 3 \end{bmatrix}.$$

Now, if we set $R = PD^{-1}$, then

$$A = PJF^{-1}$$
$$= \left(PD^{-1}\right) DJD^{-1} \left(DF^{-1}\right)$$
$$= RJ_\epsilon R^{-1}$$

where $J_\epsilon$ is $J$ with all 1's on the superdiagonal of $J$ replaced by $\epsilon$. This yields the result. ∎

Mostly, the Jordan form (for the defective case) is used for theoretical purposes. However, it is important to have some kind of diagonal-like form for any matrix. The Jordan form is such a form. We conclude by showing some uses of the form.

**Theorem 3.8** *Let $A$ be an $n \times n$ matrix having eigenvalues $\lambda_1, \ldots, \lambda_n$. Then*

*(a)* $\det A = \lambda_1 \cdots \lambda_n$.

*(b)* $\text{trace } A = \lambda_1 + \cdots + \lambda_n$.

*(c)* $\alpha A$ *has eigenvalues* $\alpha\lambda_1, \ldots, \alpha\lambda_n$ *for any scalar $\alpha$.*

*(d)* $A^k$ *has eigenvalues* $\lambda_1^k, \ldots, \lambda_n^k$ *for any natural number $k$.*

**Proof.** We argue parts (a) and (c), leaving part (b) and (d) as exercises. In both parts we let $A = PJP^{-1}$ where $J$ is a Jordan form of $A$.
    Part a. Since $A = PJP^{-1}$,

$$\det A = \det P \det J \det P^{-1}$$
$$= \det P \det P^{-1} \det J$$
$$= \det J = \lambda_1 \cdots \lambda_n.$$

Part c.  Since $A = PJF^{-1}$,

$$\alpha A = P\left[\alpha J\right]P^{-1}.$$

Thus, $\alpha A$ and $\alpha J$ are similar and thus have the same eigenvalues. Since $\alpha J$ is upper triangular, its eigenvalues are the entries on its main diagonal. These eigenvalues are $\alpha\lambda_1, \dots, \alpha\lambda_n$. ∎

Other uses of the Jordan form will be seen in the remainder of the text.

### 3.4.1   Optional (Numerical Problems in Finding the Jordan Form)

Let

$$A_\epsilon = \begin{bmatrix} \frac{7+\epsilon}{2} & \frac{-1+\epsilon}{2} \\ \frac{1+\epsilon}{2} & \frac{5+\epsilon}{2} \end{bmatrix}.$$

where $\epsilon$ is a scalar.  We can factor

$$A_\epsilon = \begin{bmatrix} \frac{1}{\sqrt{2}} & \frac{1}{\sqrt{2}} \\ \frac{1}{\sqrt{2}} & \frac{-1}{\sqrt{2}} \end{bmatrix} \begin{bmatrix} 3+\epsilon & 1 \\ 0 & 3 \end{bmatrix} \begin{bmatrix} \frac{1}{\sqrt{2}} & \frac{1}{\sqrt{2}} \\ \frac{1}{\sqrt{2}} & \frac{-1}{\sqrt{2}} \end{bmatrix}^{-1}.$$

Thus, the eigenvalues of $A_\epsilon$ are $\lambda_1 = 3 + \epsilon$ and $\lambda_2 = 3$. Note that the Jordan form of $A_0$ is

$$J_0 = \begin{bmatrix} 3 & 1 \\ 0 & 3 \end{bmatrix}.$$

But, if $\epsilon \neq 0$, then $A_\epsilon$ has distinct eigenvalues and so it has Jordan form

$$J_\epsilon = \begin{bmatrix} 3+\epsilon & 0 \\ 0 & 3 \end{bmatrix}.$$

If $\epsilon \to 0$, $A_\epsilon \to A_0$; however, $J_\epsilon \to \begin{bmatrix} 3 & 0 \\ 0 & 3 \end{bmatrix}$ which isn't the Jordan form of $A_0$.

In numerical computations involving a matrix, the answer obtained is not necessarily accurate. However, it usually can be proved that it is correct for a matrix $\hat{A}$ which is close to $A$. ($\hat{A}$ is obtained from the numerical calculation on $A$ by adjusting for round off.) Thus, loosely speaking, if close matrices have close exact answers, then a numerical calculation provides a good approximation to the desired answer.

However, note that this isn't the case for Jordan forms. Round off errors on defective matrices can produce close matrices which are diagonalizable.

## 3.4.2  MATLAB ([P, D] and Defective A)

MATLAB does not calculate Jordan forms for defective matrices. If there are close or multiple eigenvalues, there may be a problem in computing $P$. Type in *help eig* and carefully read any information about this. (We will look at this problem mathematically in Chapter 9.)

When $A$ is defective, instead of using the Jordan form, it is sometimes possible to use the Schur form in its place. This form can be calculated numerically. (MATLAB does it.) We will cover this in Chapter 6.

## Exercises

1. If the eigenvalues of $A$ are given by

    (a) $\lambda_1 = 2$, $\lambda_2 = 2$, and $\lambda_3 = 3$; what are the possible Jordan forms?

    (b) Do the same for $\lambda_1 = \lambda_2 = \lambda_3 = 4$.

    (c) Do the same for $\lambda_1 = 12$, $\lambda_2 = 2$.

2. Find $P$ and $J$ for the following $A$'s.

    (a) $A = \begin{bmatrix} 3 & 0 \\ 2 & 3 \end{bmatrix}$    (b) $A = \begin{bmatrix} 1 & -1 \\ 1 & -1 \end{bmatrix}$

    (c) $A = \begin{bmatrix} 2 & 1 \\ 1 & 2 \end{bmatrix}$    (d) $A = \begin{bmatrix} 1 & -1 & 0 \\ -1 & 1 & 0 \\ 1 & -1 & 2 \end{bmatrix}$

    (e) $A = \begin{bmatrix} 4 & 0 & 0 \\ 1 & 2 & 1 \\ 2 & 0 & 4 \end{bmatrix}$

3. If the Jordan form for $A$ is

$$\begin{bmatrix} \lambda & 1 & 0 \\ 0 & \lambda & 0 \\ 0 & 0 & \lambda \end{bmatrix},$$

    what is the dimension of the eigenspace for $\lambda$?

4. Let $A = \begin{bmatrix} 5 & -1 \\ 1 & 3 \end{bmatrix}$. Find $P$ such that $P^{-1}AF = \begin{bmatrix} 4 & 0 \\ 1 & 4 \end{bmatrix}$.

5. Is the set of all diagonalizable matrices a subspace?

6. Let $A = \begin{bmatrix} 4 & 0 & 0 \\ 0 & 4 & 0 \\ 0 & 3 & 4 \end{bmatrix}$. Then $\lambda_1 = \lambda_2 = \lambda_3 = 4$. The eigenspace for 4 is dimension 2. Find two linearly independent eigenvectors $p_1, p_2$ for 4. Now, $p_3$ must satisfy $(A - 4I)p_3 = p$ for some eigenvector $p$ belonging to 4. Try $p_1$ and then $p_2$ for $p$. Is there always a solution? (So, $P$ can be a bit difficult to find.)

7. Let $P$ be an $n \times n$ nonsingular matrix. Prove that $L : R^{n \times n} \to R^{n \times n}$ defined by $L(A) = P^{-1}AF$ is linear.

8. Suppose

$$Ap_1 = \lambda p_1$$
$$Ap_2 = \lambda p_2 + p_1.$$

Prove that $p_2 \in N\left((A - \lambda I)^2\right)$.

9. Let $A = \begin{bmatrix} 2 & 0 \\ 3 & 2 \end{bmatrix}$. Find $P$ such that $A = PJ_\epsilon P^{-1}$ where $J = \begin{bmatrix} 2 & \epsilon \\ 0 & 2 \end{bmatrix}$, $\epsilon > 0$.

10. Let $A$ be a $3 \times 3$ matrix which has $\begin{bmatrix} \lambda & 1 & 0 \\ 0 & \lambda & 1 \\ 0 & 0 & \lambda \end{bmatrix}$ as a Jordan form.

Prove that $A$ is similar to $\begin{bmatrix} \lambda & a & 0 \\ 0 & \lambda & b \\ 0 & 0 & \lambda \end{bmatrix}$ where $a \neq 0$ and $b \neq 0$.

11. Prove that if $A$ and $B$ are $n \times n$ matrices then trace $AB = $ trace $BA$.

12. For Theorem 3.8, prove (b) and (d). (Hint: On (b), use Exercise 11.)

13. Let $A$ be a nonsingular matrix. If the eigenvalues of $A$ are $\lambda_1, \dots, \lambda_n$ prove that the eigenvalues of $A^{-1}$ are $\lambda_1^{-1}, \dots, \lambda_n^{-1}$.

14. (Optional) For the given matrices, using MATLAB, decide which of the following matrices are diagonalizable.

(a) $A = \begin{bmatrix} 1 & -1 & 0 \\ -1 & 1 & 0 \\ 1 & 1 & 2 \end{bmatrix}$     (b) $A = \begin{bmatrix} 1 & 1 & 1 \\ 1 & 1 & 1 \\ 1 & 1 & 1 \end{bmatrix}$

(c) $A = \begin{bmatrix} 1 & -1 & 0 & 1 \\ -1 & 1 & 0 & 1 \\ 1 & 1 & 2 & 1 \\ 0 & 0 & 0 & 3 \end{bmatrix}$

# 4

# Matrix Calculus

In previous courses we studied calculus for functions of one variable and calculus for functions of several variables. In this chapter we extend these studies to a calculus for matrices.

## 4.1 Calculus of Matrices

To develop a calculus for matrices, we need a way to measure distance between matrices. We use the standard definition of Euclidean distance.

**Definition 4.1** *Let $A$ and $B$ be $m \times n$ matrices. Define*

$$d_E(A, B) = \left( \sum_{i=1}^{m} \sum_{j=1}^{n} |a_{ij} - b_{ij}|^2 \right)^{\frac{1}{2}}$$

*which we call the Euclidean distance between $A$ and $B$. (Thus, if $A$ and $B$ are close, then all of their corresponding entries are close and vice versa.)*

The calculus for matrices extends, a bit, the calculus for functions of several variables. The work will be familiar (even duplicative), and thus we need only give a sampling of the results. We begin with sequences.

**Definition 4.2** *Let $A_1, A_2, \ldots$ be a sequence of $m \times n$ matrices. If there is an $m \times n$ matrix $A$ such that*

$$\lim_{k \to \infty} d_E(A_k, A) = 0$$

*we say that the sequence converges to $A_0$ and write*

$$\lim_{k \to \infty} A_k = A.$$

We now show that limits can be calculated entrywise, thus allowing us to use the results from calculus in our work. In doing this we use the notation $A_k = \left[ a_{ij}^{(k)} \right]$.

**Theorem 4.1** *In the set of $m \times n$ matrices, $\lim_{k \to \infty} A_k = A$ if and only if*

$$\lim_{k \to \infty} a_{ij}^{(k)} = a_{ij} \text{ for all } i, j.$$

**Proof.** We argue the two parts of the biconditional.

Part a. Suppose $\lim_{k \to \infty} A_k = A$. Since for any $i, j$

$$0 \le \left| a_{ij}^{(k)} - a_{ij} \right| \le d_E \left( A_k, A \right)$$

it follows from the Squeeze Theorem in calculus that $\lim_{k \to \infty} a_{ij}^{(k)} = a_{ij}$ for any $i, j$.

Part b. Suppose $\lim_{k \to \infty} a_{ij}^{(k)} = a_{ij}$ for each $i, j$. Using that the square root and the absolute value functions are continuous and properties of the limit from calculus,

$$\lim_{k \to \infty} \left( \sum_{i,j} \left| a_{ij}^{(k)} - a_{ij} \right|^2 \right)^{\frac{1}{2}} = 0.$$

So

$$\lim_{k \to \infty} d_E \left( A_k, A \right) = 0$$

or $\lim_{k \to \infty} A_k = A$. ∎

As an example, we have the following.

**Example 4.1** *Let $A_k = \begin{bmatrix} 1 & \frac{1}{k} \\ -\frac{1}{k^2} & -1 \end{bmatrix}$ for $k = 1, 2, \ldots$. Then, using the theorem,*

$$\lim_{k \to \infty} A_k = \begin{bmatrix} \lim_{k \to \infty} 1 & \lim_{k \to \infty} \frac{1}{k} \\ \lim_{k \to \infty} -\frac{1}{k^2} & \lim_{k \to \infty} -1 \end{bmatrix} = \begin{bmatrix} 1 & 0 \\ 0 & -1 \end{bmatrix}.$$

Although there are many results about sequences of matrices, we will consider only two. This is enough to show how these kinds of results are developed.

**Theorem 4.2** *Let $A_1, A_2, \ldots$ and $B_1, B_2, \ldots$ be sequences of $m \times s$ matrices and $s \times n$ matrices, which converge to $A$ and $B$, respectively. Then*

$$\lim_{k \to \infty} A_k B_k = AB.$$

**Proof.** By Theorem 4.1, we can prove this result by an entrywise argument. For this, note that the ij-th entry of the k-th term, $A_k B_k$, of the sequence, is $\sum_{r=1}^{s} a_{ir}^{(k)} b_{rj}^{(k)}$. By the sum and product rule in calculus,

$$\lim_{k \to \infty} \sum_{r=1}^{s} a_{ir}^{(k)} b_{rj}^{(k)} = \sum_{r=1}^{s} a_{ir} b_{rj}$$

the $ij$-th entry in $AB$. ∎

The following corollary shows a bit more about how we obtain these limit results.

**Corollary 4.1** *Let $A_1, A_2, \ldots$ be a sequence of $m \times n$ matrices that converge to $A$. Let $P$ and $Q$ be $p \times m$ and $n \times q$ matrices, respectively. Then*

$$\lim_{k \to \infty} P A_k Q = PAQ.$$

**Proof.** Consider the sequences $P, P, \ldots$ and $A_1, A_2, \ldots$. Then by the Theorem 4.2

$$\lim_{k \to \infty} P A_k = PA.$$

Now consider the sequences $PA_1, PA_2, \ldots$ and $Q, Q, \ldots$. Again using the theorem,

$$\lim_{k \to \infty} P A_k Q = (PA)Q$$

which yields the result. ∎

Series of matrices are defined by sequences as shown in the following.

**Definition 4.3** *Let $A_1, A_2, \ldots$ be a sequence of $m \times n$ matrices. Construct the sequence of partial sums $A_1, A_1 + A_2, A_1 + A_2 + A_3, \ldots$. If this sequence of matrices converges to $A$, we write*

$$\sum_{k=1}^{\infty} A_k = A$$

*and say that the series $\sum_{k=1}^{\infty} A_k$ converges to $A$.*

We show two basic series results. Other such results are derived in the same way.

**Theorem 4.3**  *Let $P$ and $Q$ be $p \times m$ and $n \times q$ matrices, respectively. Let $\sum_{k=1}^{\infty} A_k$ be a series of $m \times n$ matrices that converges to $A$. Then*

$$\sum_{k=1}^{\infty} P A_k Q = PAQ.$$

**Proof.** To keep our notation simple, we first show that $\sum_{k=1}^{\infty} P A_k = PA$. We can do this by an entrywise argument.

The $ij$-th entry in the $t$-th partial sum, $\sum_{k=1}^{t} P A_k$, is

$$\sum_{k=1}^{t} \left( \sum_{s=1}^{m} p_{is} a_{sj}^{(k)} \right).$$

And, by the sum and product rule of calculus

$$\lim_{t \to \infty} \sum_{k=1}^{t} \left( \sum_{s=1}^{m} p_{is} a_{sj}^{(k)} \right) =$$

$$\lim_{t \to \infty} \left[ \left( \sum_{k=1}^{t} p_{i1} a_{1j}^{(k)} \right) + \cdots + \left( \sum_{k=1}^{t} p_{im} a_{mj}^{(k)} \right) \right] =$$

$$\lim_{t \to \infty} \left( p_{i1} \sum_{k=1}^{t} a_{1j}^{(k)} \right) + \cdots + \lim_{t \to \infty} \left( p_{im} \sum_{k=1}^{t} a_{mj}^{(k)} \right) =$$

$$p_{i1} a_{1j} + \cdots + p_{im} a_{mj} = \sum_{k=1}^{m} p_{ik} a_{kj}$$

the $ij$-th entry of $PA$.

Similarly, setting $B_k = P A_k$ for $k = 1, 2, \ldots$ and noting that $\sum_{k=1}^{\infty} B_k = PA$, we can show that $\sum_{k=1}^{\infty} B_k Q = (PA) Q$. Thus,

$$\sum_{k=1}^{\infty} P A_k Q = PAQ,$$

the intended result. ∎

For the second result, we give Neumann's formula for the sum of a particular series.

**Theorem 4.4** *Let $A$ be an $n \times n$ matrix such that $\lim_{k \to \infty} A^k = 0$. Then*

$$I + A + A^2 + \cdots = (I - A)^{-1}.$$

**Proof.** We prove this in two parts.

Part a. We show $I - A$ is nonsingular. Arguing by contradiction, suppose that $I - A$ is singular. Then there is a nonzero solution to

$$(I - A)\, x = 0,$$

say, $\hat{x}$. Thus,

$$A\hat{x} = \hat{x}.$$

Multiplying through by $A$ yields

$$A^2 \hat{x} = A\,(A\hat{x}) = A\hat{x} = \hat{x}$$

and in general

$$A^k \hat{x} = \hat{x}.$$

Thus, taking the limit as $k \to \infty$, we have

$$0 = \hat{x}$$

a contradiction from which Part a follows.

Part b. We show $I + A + A^2 + \cdots = (I - A)^{-1}$. To do this, note that

$$(I - A)\,(I + A + \cdots + A^{k-1}) = I - A^k.$$

So

$$I + A + \cdots + A^{k-1} = (I - A)^{-1}\,(I - A^k).$$

Taking the limit on $k \to \infty$, we see that the partial sums converge, and

$$I + A + A^2 + \cdots = (I - A)^{-1},$$

the desired result.    ∎

Let $f$ be a function from a set of $m \times n$ matrices to a set of $p \times q$ matrices. An example may help.

**Example 4.2** *Define $S = \left\{ \begin{bmatrix} a & 1 \\ 1 & b \end{bmatrix} : a, b \in R \right\}$. Define*

$$f\left(\begin{bmatrix} a & 1 \\ 1 & b \end{bmatrix}\right) = \det \begin{bmatrix} a & 1 \\ 1 & b \end{bmatrix} = ab - 1.$$

*This is a function of a subset of $R^{2 \times 2}$ into $R^{1 \times 1}$. (Recall that the braces about a matrix are cosmetic, so $R^{1 \times 1} = R$.)*
   *Identifying*

$$\begin{bmatrix} a & 1 \\ 1 & b \end{bmatrix} \leftrightarrow (a, b)^t,$$

*we graph $f$ by graphing $(a, b, ab - 1)$. This is done in Figure 4.1.*

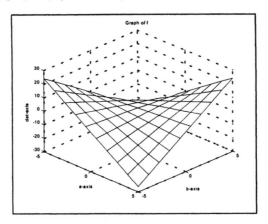

FIGURE 4.1.

Let $B$ and $L$ be $m \times n$ matrices. As in calculus,

$$\lim_{A \to B} f(A) = L \text{ means } \lim_{A \to B} d_E(f(A), L) = 0.$$

(In terms of $\epsilon - \delta$, given $\epsilon > 0$, there is a $\delta > 0$ such that if $d_E(A, B) < \delta$, then $d_E(f(A), L) < \epsilon$.)
   Writing

$$f(A) = [f_{ij}(A)],$$

where $f_{ij}(A)$ is the $ij$-th entry of $f(A)$, we can show that

$$\lim_{A \to B} f(A) = L \text{ if and only if } \left[\lim_{A \to B} f_{ij}(A)\right] = L.$$

The limit properties can be derived as well.
   The function $f$ is *continuous* at $B$ means $\lim\limits_{A \to B} f(A) = f(B)$, and $f$ is continuous on a set means $f$ is continuous at each matrix of the set. And, $f$ is continuous at $B$ or on a set if and only if this is also true for each $f_{ij}$.

   From calculus we know that $f\left(\begin{bmatrix} a & b \\ c & d \end{bmatrix}\right) = ad - bc$ is continuous ($ad - bc$ is a polynomial in $a$, $b$, $c$, $d$), and more generally, $f(A) = \det A$ is continuous on the set of all $n \times n$ matrices.

As seen in Chapter 1, there are determinantal formulas for the entries in $A^{-1}$ and $x = A^{-1}b$. For example, if

$$A = \begin{bmatrix} a & b \\ c & d \end{bmatrix}$$

then

$$A^{-1} = \begin{bmatrix} \frac{d}{ad-bc} & \frac{-b}{ad-bc} \\ \frac{-c}{ad-bc} & \frac{a}{ad-bc} \end{bmatrix}.$$

Note that the entries are rational functions of $a$, $b$, $c$, and $d$, which are continuous where the denominator is not 0. From this, we see that, for nonsingular matrices,

$$f(A) = A^{-1}$$

and

$$f(A, b) = A^{-1}b \quad \text{(the solution to } Ax = b\text{)}$$

are continuous functions, as well.

Now we look at matrices whose entries are functions of a real variable $t$, say,

$$A(t) = [a_{ij}(t)].$$

Then

$$\lim_{t \to t_0} A(t) = \left[ \lim_{t \to t_0} a_{ij}(t) \right],$$

assuming that the limits of the entries exist. (Note that $a_{ij}(t)$ is a scalar function, and we know the calculus results for such functions.) The limit properties are as those in calculus.

Continuing, $A(t)$ is continuous at $t_0$, or on an interval for $t$, if and only if the same is true for all entries $a_{ij}(t)$ of $A(t)$.

For the derivative, if $A(t)$ is such that its $ij$-th entry $a_{ij}(t)$ is differentiable for all $i, j$, then

$$\frac{d}{dt} A(t) = \left[ \frac{d}{dt} a_{ij}(t) \right].$$

The following theorem shows a result about the derivative of matrix products. Recall, for this work, matrices don't commute.

**Theorem 4.5** *Let $A(t)$ be $m \times s$ and let $B(t)$ be $s \times n$, both matrices with entries differentiable on $(a, b)$. Then*

$$\frac{d}{dt}(A(t)B(t)) = \left( \frac{d}{dt} A(t) \right) B(t) + A(t) \left( \frac{d}{dt} B(t) \right).$$

**Proof.** Note that

$$\frac{d}{dt}\left(A\left(t\right)B(t)\right) = \left[\frac{d}{dt}\left(\sum_{k=1}^{s} a_{ik}\left(t\right)b_{kj}\left(t\right)\right)\right]$$

$$= \left[\sum_{k=1}^{s}\left(a'_{ik}\left(t\right)b_{kj}\left(t\right) + a_{ik}\left(t\right)b'_{kj}\left(t\right)\right)\right]$$

$$= \left[\sum_{k=1}^{s} a'_{ik}\left(t\right)b_{kj}\left(t\right)\right] + \left[\sum_{k=1}^{s} a_{ik}\left(t\right)b'_{kj}\left(t\right)\right]$$

$$= \left(\frac{d}{dt}A\left(t\right)\right)B\left(t\right) + A\left(t\right)\left(\frac{d}{dt}B\left(t\right)\right),$$

as desired. ∎

Finally, for the integral, we define

$$\int_a^b A\left(t\right)dt = \left[\int_a^b a_{ij}\left(t\right)dt\right]$$

provided the integrals of the entries are defined. Properties of the integral can also be proved by entrywise arguments.

### 4.1.1   Optional (Modeling Spring-Mass Problems)

We give an example showing how the calculus just described can be used in mathematical modeling.

Two particles of masses $m_1$ and $m_2$ are attached to springs in the configuration shown in Figure 4.2. The particles move on a frictionless floor in a horizontal line. If the spring constants are $k_1$ and $k_2$, respectively, we want to find the mathematical model that describes the motion of the particles.

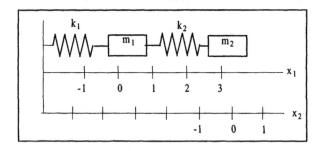

FIGURE 4.2.

When the masses are not in motion (equilibrium position), we associate an $x_1$-axis and an $x_2$-axis so that their origins are at the positions of particle

1 and particle 2, respectively. Now, if the particles have been put in motion, let

$$x_1(t) = \text{position of particle 1 at time t on the } x_1\text{-axis and}$$
$$x_2(t) = \text{position of particle 2 at time t on the } x_2\text{-axis.}$$

Hooke's law implies that the restoring force on a particle due to a spring is the product of the spring constant and the displacement of the particle from the equilibrium position. Using Hooke's law and applying Newton's law, that mass times acceleration is equal to force (See Figure 4.3.), we have

$$m_1 \frac{d^2}{dt^2} x_1(t) = \text{force on particle 1 due to the springs.}$$

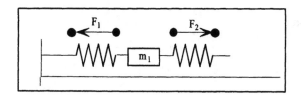

FIGURE 4.3.

There are two forces on particle 1, namely

$$F_1 = -k_1 x_1(t),$$

and

$$F_2 = k_2 [x_2(t) - x_1(t)].$$

Thus

$$m_1 \frac{d^2}{dt^2} x_1(t) = -k_1 x_1(t) + k_2 [x_2(t) - x_1(t)]$$
$$= -(k_1 + k_2) x_1(t) + k_2 x_2(t).$$

For particle 2 we have

$$m_2 \frac{d^2}{dt^2} x_2(t) = -k_2 [x_2(t) - x_1(t)].$$

Putting these into a matrix equation we have

$$\begin{bmatrix} m_1 & 0 \\ 0 & m_2 \end{bmatrix} \frac{d^2}{dt^2} \begin{bmatrix} x_1(t) \\ x_2(t) \end{bmatrix} + \begin{bmatrix} (k_1 + k_2) & -k_2 \\ -k_2 & k_2 \end{bmatrix} \begin{bmatrix} x_1(t) \\ x_2(t) \end{bmatrix} = 0.$$

Solving this equation for $x_1(t)$ and $x_2(t)$ gives the positions of the particles at any time $t$.

## 4.1.2   MATLAB (Code for Graph of Function)

**Code for Figure 4.1**

```
a =linspace(−5, 5, 10);
b = a;
[a, b] =meshgrid(a, b);
mesh(a, b, a. * b − 1)
view([1, −1, 1])            %  View with (1, −1, 1)ᵗ
                               pointing toward us.
```

# Exercises

1. Find $d_E(A, B)$ for the given $A$ and $B$.

   (a) $A = \begin{bmatrix} 1 & 0 \\ 2 & -1 \end{bmatrix}$, $B = \begin{bmatrix} 2 & -1 \\ 1 & 0 \end{bmatrix}$

   (b) $A = \begin{bmatrix} i & 0 \\ 1-i & 2+i \end{bmatrix}$, $B = \begin{bmatrix} 0 & 1-i \\ i & 1+i \end{bmatrix}$

2. Compute

   (a) $\lim\limits_{k \to \infty} \begin{bmatrix} \frac{1}{k} & \frac{k+1}{k} \\ \frac{-k}{k+1} & e^{-k} \end{bmatrix}$.

   (b) $\lim\limits_{t \to 0} \begin{bmatrix} \frac{t}{t+1} & \sin t \\ e^t & t \end{bmatrix}$.

3. Prove that if $A_k$ and $B_k$ are $m \times n$ matrices for all $k$ and $\lim\limits_{k \to \infty} A_k = A$, $\lim\limits_{k \to \infty} B_k = B$ then $\lim\limits_{k \to \infty} (A_k + B_k) = A + B$.

4. Let $\alpha(t)$ is a real valued function and $A(t)$ a matrix of functions. If $\lim\limits_{t \to a} \alpha(t) = \alpha_0$ and $\lim\limits_{t \to a} A(t) = A$, prove the result that $\lim\limits_{t \to a} \alpha_0(t) A(t) = \alpha_0 A$.

5. Let $A_1, A_2, \ldots$ be a sequence of matrices that converge to $A$. If $A$ is nonsingular, show that $A_1^{-1}, A_2^{-1}, \ldots$ converge to $A^{-1}$.

6. Let $A_k = \begin{bmatrix} \frac{1}{2^k} & \frac{1}{3^k} \\ 0 & \frac{1}{4^k} \end{bmatrix}$ for $k = 1, 2, \ldots$.    Find $A_1 + A_2 + \cdots$.  (Recall that $1 + r + r^2 + \cdots = \frac{1}{1-r}$ for any $r$, $|r| < 1$.)

7. Let $A(t) = \begin{bmatrix} t & 1 \\ 1 & t \end{bmatrix}$.  Calculate and graph each of the following.

   (a) $\det A(t)$

(b) The 1,2-entry of $A\left(t\right)^{-1}$

(c) The first entry of $A\left(t\right)^{-1}b$, where $b = \begin{bmatrix} 1 \\ 1 \end{bmatrix}$

8. Let $f : R^{2\times 2} \rightarrow R^2$. Set $f\left(A\right) = \begin{bmatrix} f_1\left(A\right) \\ f_2\left(A\right) \end{bmatrix}$ where $f_i\left(A\right)$ is the i-th entry of $f\left(A\right)$. Prove that $f$ is continuous if and only if both $f_1$ and $f_2$ are continuous.

9. Let $A\left(t\right) = \begin{bmatrix} 2t-1 & e^t \\ \frac{t}{t-1} & 0 \end{bmatrix}$. Show that

(a) $A\left(t\right)$ is continuous at $t = 0$.

(b) $A\left(t\right)$ is not continuous at $t = 1$.

10. Let $A = \begin{bmatrix} \cos t & \sin t \\ t+2 & 0 \end{bmatrix}$. Find

(a) $\frac{d}{dt}A\left(t\right)$.

(b) $\int_0^\pi A\left(t\right)dt..$

11. Suppose $A\left(t\right)$, $B\left(t\right)$ are differentiable $n \times n$ matrices. Prove that

$$\frac{d}{dt}\left[A\left(t\right)+B\left(t\right)\right] = \frac{d}{dt}A\left(t\right) + \frac{d}{dt}B\left(t\right).$$

12. Suppose $P$ and $A\left(t\right)$ are $n \times n$ matrices with $A\left(t\right)$ differentiable. Prove that

$$\frac{d}{dt}PA\left(t\right) = P\frac{d}{dt}A\left(t\right).$$

13. Suppose $A(t)$ and $B(t)$ are integrable $n \times n$ matrices. Prove that $\int_a^b \left(A(t)+B(t)\right)dt = \int_a^b A(t)dt + \int_a^b B(t)dt.$

14. (Optional) Attach a third spring to $m_2$ and to a wall as diagramed in Figure 4.4. Find the mathematical model for this system.

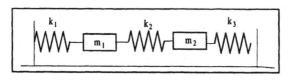

FIGURE 4.4.

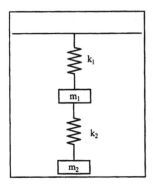

FIGURE 4.5.

15. (Optional) Derive the mathematical model for the spring-mass system shown in Figure 4.5.

16. (MATLAB) Let $A = \begin{bmatrix} a & b \\ b & a \end{bmatrix}$ and $c = \begin{bmatrix} 1 \\ 2 \end{bmatrix}$. In the square $[0,4] \times [5,9]$ graph

    (a) The 1,1-entry and the 1,2-entry of $A^{-1}$.
    (b) Both $x_1$ and $x_2$ of $x = A^{-1}c$.

## 4.2   Difference Equations

In this section, we show how to solve systems of difference equations, as well as show that eigenvalues determine the solution's behavior. We demonstrate the technique to solve systems with a small example. Extensions of the technique should be clear.

Let $x_1(k)$ and $x_2(k)$ be functions defined on the nonnegative integers that satisfy

$$x_1(k+1) = a_{11}x_1(k) + a_{12}x_2(k)$$
$$x_2(k+1) = a_{21}x_1(k) + a_{22}x_2(k)$$

where $a_{11}$, $a_{12}$, $a_{21}$, and $a_{22}$ are scalars. We can write these equations as a matrix equation

$$x(k+1) = Ax(k) \tag{4.1}$$

where $x(k) = \begin{bmatrix} x_1(k) \\ x_2(k) \end{bmatrix}$ and $A = \begin{bmatrix} a_{11} & a_{12} \\ a_{21} & a_{21} \end{bmatrix}$. (The equation in (4.1) is called a difference equation.)

If $x(0)$ is a given vector then (4.1) determines a sequence

$$x(0), x(1), x(2), \ldots$$

We intend to find a formula for $x(k)$ in terms of the eigenvalues and eigenvectors of $A$. To get this, note that

$$x(1) = Ax(0), \tag{4.2}$$
$$x(2) = Ax(1) = A^2 x(0)$$
$$\cdots$$
$$x(k) = A^k x(0)$$
$$\cdots$$

Observe that if $A$ and $x(0)$ are real, so is $x(k)$ for all $k$.

We now assume that $A$ is diagonalizable, say,

$$A = PDF^{-1}$$

where $P = [p_1 \ p_2]$ and $D = \begin{bmatrix} \lambda_1 & 0 \\ 0 & \lambda_2 \end{bmatrix}$. We substitute $PDF^{-1}$ for $A$ in (4.2) to obtain

$$x(k) = PD^k P^{-1} x(0).$$

Set

$$P^{-1} x(0) = \begin{bmatrix} \alpha_1 \\ \alpha_2 \end{bmatrix}. \tag{4.3}$$

(Note that $\begin{bmatrix} \alpha_1 \\ \alpha_2 \end{bmatrix}$ can be computed by solving $P \begin{bmatrix} \alpha_1 \\ \alpha_2 \end{bmatrix} = x(0)$ or $\alpha_1 p_1 + \alpha_2 p_2 = x(0)$. So $P^{-1}$ need not be calculated.) Thus, we have

$$x(k) = [p_1 \ p_2] \begin{bmatrix} \lambda_1^k & 0 \\ 0 & \lambda_2^k \end{bmatrix} \begin{bmatrix} \alpha_1 \\ \alpha_2 \end{bmatrix} \tag{4.4}$$
$$= \begin{bmatrix} \lambda_1^k p_1 & \lambda_2^k p_2 \end{bmatrix} \begin{bmatrix} \alpha_1 \\ \alpha_2 \end{bmatrix}$$
$$= \alpha_1 \lambda_1^k p_1 + \alpha_2 \lambda_2^k p_2$$

the desired formula involving eigenvalues and eigenvectors. More generally, if $A$ is $n \times n$ and diagonalizable, we would get

$$x(k) = \alpha_1 \lambda_1^k p_1 + \cdots + \alpha_n \lambda_n^k p_n.$$

An example showing how to use the formula to solve a difference equation follows.

**Example 4.3** *Let* $x(0) = \begin{bmatrix} 6 \\ -2 \end{bmatrix}$. *We solve*

$$x(k+1) = \begin{bmatrix} 2 & 1 \\ 1 & 2 \end{bmatrix} x(k).$$

*Here* $\lambda_1 = 3$, $\lambda_2 = 1$ *with corresponding eigenvectors* $p_1 = \begin{bmatrix} 1 \\ 1 \end{bmatrix}$, $p_2 = \begin{bmatrix} -1 \\ 1 \end{bmatrix}$, *respectively. Thus, using our formula,*

$$x(k) = \alpha_1 \lambda_1^k p_1 + \alpha_2 \lambda_2^k p_2 \qquad (4.5)$$
$$= \alpha_1 3^k \begin{bmatrix} 1 \\ 1 \end{bmatrix} + \alpha_2 \begin{bmatrix} -1 \\ 1 \end{bmatrix}.$$

*Now, since* $x(0) = \begin{bmatrix} 6 \\ -2 \end{bmatrix}$, *using (4.5) and plugging in $k = 0$, we have*

$$\begin{bmatrix} 6 \\ -2 \end{bmatrix} = \alpha_1 \begin{bmatrix} 1 \\ 1 \end{bmatrix} + \alpha_2 \begin{bmatrix} -1 \\ 1 \end{bmatrix}.$$

*Solving for $\alpha_1$ and $\alpha_2$ yields $\alpha_1 = 2$ and $\alpha_2 = -4$. Thus, our solution is*

$$x(k) = \alpha_1 3^k \begin{bmatrix} 1 \\ 1 \end{bmatrix} + \alpha_2 \begin{bmatrix} -1 \\ 1 \end{bmatrix}$$
$$= 2 \cdot 3^k \begin{bmatrix} 1 \\ 1 \end{bmatrix} - 4 \begin{bmatrix} -1 \\ 1 \end{bmatrix}.$$

*Note that as $k \to \infty$, the entries in $x(k) \to \infty$ tend to $\infty$.*

We now extend our work a bit to defective matrices. Observe that if $A = PJP^{-1}$, where $J = \mathrm{diag}(J_1, \ldots, J_r)$ is a Jordan form of $A$,

$$\lim_{k\to\infty} A^k = \lim_{k\to\infty} PJ^kP^{-1} \qquad (4.6)$$
$$= P\left(\lim_{k\to\infty} J^k\right) P^{-1}$$
$$= P \begin{bmatrix} \lim_{k\to\infty} J_1^k & 0 & \cdots & 0 \\ 0 & \lim_{k\to\infty} J_2^k & \cdots & 0 \\ & & \cdots & \\ 0 & 0 & \cdots & \lim_{k\to\infty} J_r^k \end{bmatrix} P^{-1}.$$

Thus, convergence of $A, A^2, \ldots$ depends on the Jordan blocks of $A$. Formulas for their powers follows.

If $J_i = \begin{bmatrix} \lambda & 1 & 0 & \cdots & 0 \\ 0 & \lambda & 1 & \cdots & 0 \\ & & \cdot & & \\ & & \cdot & & \\ 0 & 0 & 0 & \cdots & \lambda \end{bmatrix}$, an $s \times s$ Jordan block, then

$$J_i^2 = \begin{bmatrix} \lambda^2 & 2\lambda & 1 & \cdots & 0 \\ 0 & \lambda^2 & 2\lambda & \cdots & 0 \\ & & \cdot & & \\ 0 & 0 & 0 & \cdots & \lambda^2 \end{bmatrix},$$

$$J_i^3 = \begin{bmatrix} \lambda^3 & 3\lambda^2 & 3\lambda & 1 & \cdots & 0 \\ 0 & \lambda^3 & 3\lambda^2 & 3\lambda & \cdots & 0 \\ & & \cdot & \cdot & & \\ 0 & 0 & 0 & & \cdots & \lambda^3 \end{bmatrix}$$

and in general (We leave it as an exercise.),

$$J_i^k = \begin{bmatrix} \lambda^k & k\lambda^{k-1} & \binom{k}{2}\lambda^{k-2} & \cdots & \binom{k}{s-1}\lambda^{k-s+1} \\ 0 & \lambda^k & k\lambda^{k-1} & \cdots & \binom{k}{s-2}\lambda^{k-s} \\ & & \cdot & \cdot & \\ 0 & 0 & 0 & \cdots & \lambda^k \end{bmatrix} \qquad (4.7)$$

where $\binom{k}{r} = 0$ if $k < r$ and $\binom{k}{r} = \frac{k!}{(k-r)!r!}$, otherwise.

Using these formulas, we can solve difference equations even when $A$ is defective.

**Example 4.4** *Solve*

$$x(k+1) = \begin{bmatrix} 1 & .5 \\ -.5 & 0 \end{bmatrix} x(k).$$

*Factoring, we have*

$$\begin{bmatrix} 1 & .5 \\ -.5 & 0 \end{bmatrix} = \begin{bmatrix} 1 & 1 \\ -1 & 1 \end{bmatrix} \begin{bmatrix} .5 & 1 \\ 0 & .5 \end{bmatrix} \begin{bmatrix} .5 & -.5 \\ .5 & .5 \end{bmatrix}.$$

*By direct calculation,*

$$x(k) = PJ^k P^{-1} x(0)$$

$$= \begin{bmatrix} 1 & 1 \\ -1 & 1 \end{bmatrix} \begin{bmatrix} .5^k & k(.5)^{k-1} \\ 0 & .5^k \end{bmatrix} \begin{bmatrix} \alpha_1 \\ \alpha_2 \end{bmatrix}$$

*where* $\begin{bmatrix} \alpha_1 \\ \alpha_2 \end{bmatrix} = P^{-1} x(0)$. *And by backward multiplication, we have*

$$x(k) = \left( (.5)^k \begin{bmatrix} 1 \\ -1 \end{bmatrix}, k(.5)^{k-1} \begin{bmatrix} 1 \\ -1 \end{bmatrix} + .5^k \begin{bmatrix} 1 \\ 1 \end{bmatrix} \right) \begin{bmatrix} \alpha_1 \\ \alpha_2 \end{bmatrix}$$

$$= \alpha_1 (.5)^k \begin{bmatrix} 1 \\ -1 \end{bmatrix} + \alpha_2 \left( k(.5)^{k-1} \begin{bmatrix} 1 \\ -1 \end{bmatrix} + .5^k \begin{bmatrix} 1 \\ 1 \end{bmatrix} \right).$$

*Note that, as a consequence of the eigenvalues,*

$$\lim_{k \to \infty} x(k) = 0.$$

By observing (4.6) and the formulas for the Jordan blocks in (4.7), we have the following.

**Theorem 4.6** *Let A be an $n \times n$ matrix. Then*

(a) $\lim_{k \to \infty} A^k$ *converges if each eigenvalue $\lambda$ of A is such that $|\lambda| < 1$ or if $|\lambda| = 1$, then $\lambda = 1$ and it is on $1 \times 1$ Jordan blocks.*

(b) *And for all other cases, $\lim_{k \to \infty} A^k$ doesn't exist.*

An example demonstrating the theorem follows.

**Example 4.5** *Let $A = \begin{bmatrix} \frac{2}{3} & \frac{1}{3} \\ \frac{1}{3} & \frac{2}{3} \end{bmatrix}$. Then A is diagonalizable with $P = \begin{bmatrix} 1 & -1 \\ 1 & 1 \end{bmatrix}$ and $D = \begin{bmatrix} 1 & 0 \\ 0 & \frac{1}{3} \end{bmatrix}$. Thus*

$$\lim_{k \to \infty} A^k = P \left( \lim_{k \to \infty} \begin{bmatrix} 1^k & 0 \\ 0 & (\frac{1}{3})^k \end{bmatrix} \right) P^{-1}$$

$$= P \begin{bmatrix} 1 & 0 \\ 0 & 0 \end{bmatrix} P^{-1} = \begin{bmatrix} \frac{1}{2} & \frac{1}{2} \\ \frac{1}{2} & \frac{1}{2} \end{bmatrix}.$$

If $\lim_{k \to \infty} A^k$ doesn't exist, often we can still say something about the behavior of $x(k)$. A small example can show this.

**Example 4.6** *If we solve*

$$x(k+1) = \begin{bmatrix} .3 & .9 \\ .9 & 0 \end{bmatrix} x(k),$$

*we get (using 5 digits in our answers)*

$$x(k) = \alpha_1 (1.0624)^k \begin{bmatrix} 0.7630 \\ 0.6464 \end{bmatrix} \qquad (4.8)$$

$$+ \alpha_2 (-0.7624)^k \begin{bmatrix} 0.6464 \\ -0.7630 \end{bmatrix}.$$

*We find the dominant term (the term having the largest eigenvalue, in absolute value, in $x(k)$). This is $\alpha_1 (1.0624)^k \begin{bmatrix} 0.7630 \\ 0.6464 \end{bmatrix}$. Note that by factoring out this coefficient in (4.8), we have (assuming $\alpha_1 \neq 0$)*

$$x(k) = \alpha_1 (1.0624)^k \left( \begin{bmatrix} 0.7630 \\ 0.6464 \end{bmatrix} + \frac{\alpha_2 (-0.7624)^k}{\alpha_1 (1.0624)^k} \begin{bmatrix} 0.6464 \\ -0.7630 \end{bmatrix} \right).$$

*Now, since the second term, within the parentheses, approaches 0 as $k \to 0$, we see that the contribution of*

$$\alpha_2 \left(-0.7624\right)^k \begin{bmatrix} 0.6464 \\ -0.7630 \end{bmatrix}$$

*to the size of $x(k)$ is small compared to that of*

$$\alpha_1 \left(1.0624\right)^k \begin{bmatrix} 0.7630 \\ 0.6464 \end{bmatrix}.$$

*We indicate this by writing*

$$x(k) \sim \alpha_1 \left(1.0624\right)^k \begin{bmatrix} 0.7630 \\ 0.6464 \end{bmatrix}$$

*and say that $x(k)$ has dominant term $\alpha_1 \left(1.0624\right)^k \begin{bmatrix} 0.7630 \\ 0.6464 \end{bmatrix}$.*

*For $x(0) = \begin{bmatrix} 1 \\ 1 \end{bmatrix}$, in Figure 4.6, we can see a picture of the iterates from $k = 0$ to $k = 20$. In this picture 0 indicates the initial vector and each * a following vector, $x(1), x(2), \ldots, x(20)$. The polygonal line indicates the order of occurrence of these vectors. Notice that the vectors look like the dominant term as $k$ increases.*
*Since*

$$x(k) \sim \alpha_1 \left(1.0624\right)^k \begin{bmatrix} 0.7630 \\ 0.6464 \end{bmatrix},$$

*we see that $x(k)$ increases by about 6% on each iteration.*

As a final consequence of Theorem 4.6, we consider the nonhomogeneous difference equation

$$x(k+1) = Ax(k) + b$$

where $A$ is an $n \times n$ matrix and $b$ an $n \times 1$ vector.
Writing out a few iterates, we have

$$\begin{aligned} x(1) &= Ax(0) + b \hspace{4em} (4.9) \\ x(2) &= Ax(1) + b \\ &= A^2 x(0) + Ab + b \\ &\cdots \\ x(k+1) &= A^k x(0) + A^{k-1}b + A^{k-2}b + \cdots + b \\ &= A^k x(0) + \left(A^{k-1} + A^{k-2} + \cdots + I\right) b. \end{aligned}$$

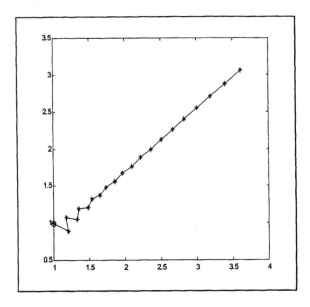

FIGURE 4.6.

Now, if each eigenvalue, say, $\lambda$, of $A$ satisfies $|\lambda| < 1$, then by Theorem 4.6, $\lim_{k \to \infty} A^k = 0$. And by Neumann's formula,

$$I + A + A^2 + \cdots = (I - A)^{-1}.$$

Thus, calculating the limit in (4.9), we have

$$\lim_{k \to \infty} x(k+1) = (I - A)^{-1} b. \qquad (4.10)$$

($x(k+1)$ here can be replaced by $x(k)$ since we are talking about the convergence of a sequence.)

We can see some use of the result in the following example of a production process.

**Example 4.7** *We consider a two-grade school (7th and 8th grades). Each year, 1000 new students enter the 7th grade. Of those currently in the school, 80% of the students are promoted, 10% retained for another year and 10% of each class drops out. A diagram of the situation follows in Figure 4.7.*

*Let $x_1(k)$ and $x_2(k)$ denote the number of students in the 7th and 8th grades in the k-th year, respectively. Then*

$$x_1(k+1) = .1x_1(k) + 1000$$
$$x_2(k+1) = .8x_1(k) + .1x_2(k)$$

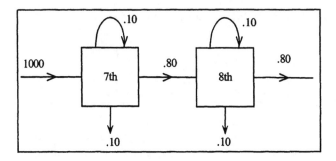

FIGURE 4.7.

*or*

$$x(k+1) = Ax(k) + b$$

*where* $x(k) = \begin{bmatrix} x_1(k) \\ x_2(k) \end{bmatrix}$, $A = \begin{bmatrix} .1 & 0 \\ .8 & .1 \end{bmatrix}$, *and* $b = \begin{bmatrix} 1000 \\ 0 \end{bmatrix}$.

*Since the eigenvalues of the matrix are .1 and .1, we have by (4.10)*

$$\lim_{k \to \infty} x(k) = (I - A)^{-1} b$$

$$= \begin{bmatrix} 1.111 & 0 \\ 0.988 & 1.111 \end{bmatrix} \begin{bmatrix} 1000 \\ 0 \end{bmatrix}$$

$$= \begin{bmatrix} 1111. \\ 988 \end{bmatrix}.$$

*Thus, as k increases, we expect to see about 1111 students in the 7th grade and 988 students in the 8th grade.*

## 4.2.1  Optional (Long-Run Prediction)

Being able to see what is going to happen if trends continue is important in many areas. We look at one such problem.

An important social science (demographic) problem is to predict the population of a region or country in future years. Such information is used in planning (roads, water, schools, food, etc.) for that area.

To describe the technique in general, suppose that some population is divided into three age groups: young, adult, and older, where the number of years in each group, called the period, is the same. Survival rates (% of those in one group that live to be in the next) are computed. These rates, $s_1$ for young to adult, $s_2$ for adult to older, can be obtained from official records. Birth rates (number of offsprings per member in each group per period) say, $b_1$, $b_2$, $b_3$, for the groups, respectively, can be obtained in the same way. Using this data we form the matrix

$$P = \begin{bmatrix} b_1 & b_2 & b_3 \\ s_1 & 0 & 0 \\ 0 & s_2 & 0 \end{bmatrix},$$

which is called a Leslie (or population) matrix.

Suppose we compute populations in each group every period. Let $x_1(k)$, $x_2(k)$, $x_3(k)$ denote the number of people in groups 1, 2, and 3, respectively, at period $k$. Then at period $k+1$ we have

$$x_1(k+1) = b_1 x_1(k) + b_2 x_2(k) + b_3 x_3(k)$$
$$x_2(k+1) = s_1 x_1(k)$$
$$x_3(k+1) = s_2 x_2(k)$$

or in matrix form

$$x(k+1) = Px(k)$$

where

$$x(k) = \begin{bmatrix} x_1(k) \\ x_2(k) \\ x_3(k) \end{bmatrix}.$$

The Leslie matrix given below was obtained (taking some liberties with the data) from a third world country. The age groupings were 0–4, 5–9, 10–14, ..., 45–49.

$$A = \begin{bmatrix} 0 & 0 & 0 & .83 & .83 & .50 & .50 & .11 & .11 & 0 \\ .94 & 0 & 0 & 0 & 0 & 0 & 0 & 0 & 0 & 0 \\ 0 & .98 & 0 & 0 & 0 & 0 & 0 & 0 & 0 & 0 \\ 0 & 0 & .98 & 0 & 0 & 0 & 0 & 0 & 0 & 0 \\ 0 & 0 & 0 & .98 & 0 & 0 & 0 & 0 & 0 & 0 \\ 0 & 0 & 0 & 0 & .98 & 0 & 0 & 0 & 0 & 0 \\ 0 & 0 & 0 & 0 & 0 & .98 & 0 & 0 & 0 & 0 \\ 0 & 0 & 0 & 0 & 0 & 0 & .98 & 0 & 0 & 0 \\ 0 & 0 & 0 & 0 & 0 & 0 & 0 & .97 & 0 & 0 \\ 0 & 0 & 0 & 0 & 0 & 0 & 0 & 0 & .97 & 0 \end{bmatrix}$$

The largest eigenvalue for $A$ is $\lambda_1 = 1.1903$, with corresponding eigenvector

$$p_1 = \begin{bmatrix} 0.5897 \\ 0.4657 \\ 0.3834 \\ 0.3157 \\ 0.2599 \\ 0.2140 \\ 0.1762 \\ 0.1451 \\ 0.1182 \\ 0.0964 \end{bmatrix}.$$

Since $x(k) \sim \alpha_1 \lambda_1^k p_1$, we see that the population here is increasing and tends to $\infty$ as $k \to \infty$. For large $k$, if the population is $x(k)$, in 5 years, it is about 1.190 times greater, so we are seeing a growth of about 19% every 5 years.

Another interesting observation is that, if we sum the entries in $p_1$, obtaining 2.7643, and divide $p_1$ by that number, getting

$$\hat{p} = \begin{bmatrix} 0.2133 \\ 0.1685 \\ 0.1387 \\ 0.1142 \\ 0.0940 \\ 0.0774 \\ 0.0637 \\ 0.0525 \\ 0.0428 \\ 0.0349 \end{bmatrix}$$

then, for $k$ large, the entries in $\hat{p}$ indicate the percentage of population in each category. For example, 0.2133 indicates that about 21% of the population will be 0–4 years old, etc. Note that the majority of the population here is very young.

### 4.2.2 MATLAB (Code for Viewing Solution to Difference Equations; Handling Large Matrices)

There are some useful commands when working with population matrices. The command $A = zeros(n, n)$ provides an $n \times n$ matrix, all entries of which are 0's. Now to obtain a population matrix, we can change some entries in $A$, using say, $A(1, 4) = .83$, which changes the $1, 4$-th entry in $A$ to .83.

Also, if the command $[V, D] = eig(A)$ is used, the columns of $V$ are eigenvectors. To get an individual column of $V$, say, the second, use $V(:, 2)$. Of course, for a row, the companion command is $V(2, :)$. If we want to sum the entries in say, $V(:,2)$ and divide $V(:,2)$ by that sum, we can use $e = ones(1,10)$ which provides a $1 \times 10$ vector having 1's as its entries. Then we use $(e * V(:,2)) \wedge (-1) * V(:,2)$. (Exponents are done before multiplication. Type in *help precedence* for more.) An exercise will be provided on which these commands can be helpful.

## 1.  Code for Computing Limits

```
A = [.2 .8; .6 .4];
[P, D] = eig (A)
D (1, 1) = 0;              %  The 1,1-entry of D was −.4.
                              We set it to 0.

limit = P * D * inv (P)
```

## 2.  Code for Computing Limits

```
A = [.3 .7; .4 .6];
L = zero (2, 2);
while norm (A − L, 'fro')    %  Tests to see if the distance
            > 10 ∧ (−7)         between A and L ≤ 10⁻⁷.
                                This condition can change
                                with different problems.

    L = A;
    A = A * A;
end, A                     %  If distance ≤ 10⁻⁷, prints
                              out A.
```

## 3.  Code for Viewing Solution to Difference Equation

```
x = [1; 1];
A = [.3 .9; .9 0];
for k = 1 : 21
    p (k) = x (1);      %  Generates x-values [p(1)...p(21)]
    q (k) = x (2);          and y-values [q(1)...q(21)]
                            for the (p(k), q(k)) to be plotted.

    x = A * x;          %  Gets to the next point in
                            the iteration.
end
plot (1, 1, 'O')        %  Plots starting point with O.
hold
plot (p, q, ' * ')      %  Plots points (p(k), q(k)) with *.
plot (p, q)             %  Draws 'curve' through points.
```

In iterations like this, it is helpful to include a stopping criteria so that the iteration won't run forever. For example, insert $c = 1$ between the 2nd and 3rd lines, and

$$c = c + 1$$
if $c > 1000$
  break
end
between the 6th and 7th lines.

## Exercises

1. Compute $\lim_{k \to \infty} A^k$ for the given $A$, if possible. If not possible, explain why the sequence does not converge.

   (a) $A = \begin{bmatrix} .2 & .4 \\ .3 & .3 \end{bmatrix}$     (b) $A = \begin{bmatrix} \frac{1}{4} & \frac{3}{4} \\ \frac{1}{2} & \frac{1}{2} \end{bmatrix}$

   (c) $A = \begin{bmatrix} 0 & 1 \\ 1 & 0 \end{bmatrix}$     (d) $A = \begin{bmatrix} 3 & 5 \\ 4 & 4 \end{bmatrix}$

   (e) $A = \begin{bmatrix} 1 & 1 & 1 \\ 0 & .2 & .3 \\ 0 & .3 & .2 \end{bmatrix}$     (f) $A = \begin{bmatrix} \frac{1}{3} & \frac{1}{3} & \frac{1}{3} \\ \frac{1}{3} & \frac{1}{3} & \frac{1}{3} \\ \frac{1}{3} & \frac{1}{3} & \frac{1}{3} \end{bmatrix}$

2. Solve

   (a) $\begin{aligned} x_1(k+1) &= 3x_1(k) & - x_2(k) \\ x_2(k+1) &= -x_1(k) & +3x_2(k) \end{aligned}$

   (b) $\begin{aligned} x_1(k+1) &= 3x_1(k) & +2x_2(k) \\ x_2(k+1) &= x_1(k) & +2x_2(k) \\ x_1(0) &= 9 \\ x_2(0) &= 3 \end{aligned}$

   (c) $\begin{aligned} x_1(k+1) &= 3x_1(k) \\ x_2(k+1) &= x_1(k) & +3x_2(k) \end{aligned}$

3. For the given matrices, compute $J^2$, $J^3$, $J^7$, and $J^k$.

   (a) $J = \begin{bmatrix} 2 & 1 \\ 0 & 2 \end{bmatrix}$     (b) $J = \begin{bmatrix} 3 & 1 & 0 \\ 0 & 3 & 1 \\ 0 & 0 & 3 \end{bmatrix}$

4. Find the dominant term in (a) and (b) of the solutions of Exercise 2.

5. Let

$$x(k+1) = \begin{bmatrix} 1 & 0 \\ 2 & 3 \end{bmatrix} x(k)$$

$$x(0) = c.$$

Find a vector $c$ such that $x(k)$ is constant for all $k$.

6. The solution to $y(k+1) = \begin{bmatrix} 1 & -1 \\ 1 & 1 \end{bmatrix} y(k)$, $y(0) = \begin{bmatrix} 2 \\ 0 \end{bmatrix}$ is $y(k) =$

$(1+i)^k \begin{bmatrix} 1 \\ -i \end{bmatrix} + (1-i)^k \begin{bmatrix} 1 \\ i \end{bmatrix}$.

(a) Show that $\overline{y(k)} = y(k)$ so $y(k)$ is real. (The imaginary part is 0.)

(b) Find $y(k)$ as an expression involving real numbers. (Hint: Use $(a+ib)^k = r^k (\cos k\theta + i \sin k\theta)$ where $a+ib = r(\cos\theta + i\sin\theta)$.)

7. Let $A$ be an $n \times n$ matrix. Prove that if $\lim_{k\to\infty} A^k$ exists, then so does $\lim_{k\to\infty} J^k$. (Hint: Write $J^k = P^{-1}A^k P$ and compute the limit.)

8. To solve the scalar difference equation

$$x(k+2) - 3x(k+1) + 2x(k) = 0$$

set

$$y_1(k) = x(k)$$
$$y_2(k) = x(k+1).$$

Then, using the three equations above, we have

$$y_1(k+1) = y_2(k)$$
$$y_2(k+1) = 3y_2(k) - 2y_1(k).$$

Solve this system for $y_1$, to find $x$.

9. Company $A$ has machines that periodically break down. When a machine does break, it costs about \$1,000 to fix it. (We will assume at most one machine breaks per month.) In monthly intervals, the probability that if a machine broke the previous month, one will break this month is .1, while if no machine broke the previous month, the probably that one will break this month is .15. (See Figure 4.8.) Let

$$y_1(k) = \text{probability that a machine breaks in month } k,$$
$$y_2(k) = \text{probability that a machine doesn't break}$$
$$\text{in month } k.$$

Then

$$y_1(k+1) = .1y_1(k) + .15y_2(k)$$
$$y_2(k+1) = .9y_1(k) + .85y_2(k)$$

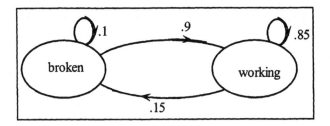

FIGURE 4.8.

or

$$y(k+1) = \begin{bmatrix} .1 & .15 \\ .9 & .85 \end{bmatrix} y(k)$$

where $y(k) = \begin{bmatrix} y_1(k) \\ y_2(k) \end{bmatrix}$.

(a) Compute $y(k)$ and $\lim_{k\to\infty} y(k)$. (Use $y(0) = \begin{bmatrix} 0 \\ 1 \end{bmatrix}$.)

(b) We estimate the average cost of fixing the machine as follows: the first entry in $\lim_{k\to\infty} y(k)$ is the long-run probability of the machine breaking in any month. So that number times $1,000 will give us an estimate of the monthly cost of fixing machines. Compute this number.

10. For the 3-class school diagramed in Figure 4.9, find $\lim_{k\to\infty} y(k)$ and interpret this vector.

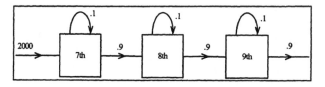

FIGURE 4.9.

11. Let $A$ be a $2 \times 2$ diagonalizable matrix and $b$ a $2 \times 1$ vector. Find, by using the eigenvalue-eigenvector approach, a formula for the solution to

$$x(k+1) = Ax(k) + b.$$

12. Let $A$ be an $n \times n$ matrix. Prove that if $x(k+1) = Ax(k)$ for all nonnegative integers $k$, then $x(0), x(1), \ldots$ converges, for all $x(0)$, if and only if $A, A^2, \ldots$ converges. (Hint: Use $x(0) = e_1, e_2, \ldots, e_n$.)

13. (MATLAB) By computing eigenvalues, eigenvectors, and solving a system of linear equations for the scalars, solve

$$x_1(k+1) = 2x_1(k) + x_2(k)$$
$$x_2(k+1) = x_1(k) + 2x_2(k)$$
$$x_3(k+1) = x_2(k) + 2x_3(k)$$
$$x_1(0) = 1$$
$$x_2(0) = 2$$
$$x_3(0) = 4.$$

14. (MATLAB) Find $\lim_{k\to\infty} A^k$ for $A = \begin{bmatrix} 1 & 0 & 0 & 0 \\ 0 & 1 & 0 & 0 \\ 0 & .3 & .7 & 0 \\ .2 & 0 & 0 & .8 \end{bmatrix}.$

    (a) Use the diagonalization approach.

    (b) Use the iteration approach.

15. (MATLAB) As in Example 4.6, graph the solution to Example 4.7, where $x(0) = \begin{bmatrix} 400 \\ 500 \end{bmatrix}.$

16. (Optional) The following population matrix, with age groups as in Optional, is for a small county.

$$\begin{bmatrix} 0 & 0 & 0 & 0 & .14 & .34 & .26 & .14 & .08 & .06 \\ .94 & 0 & 0 & 0 & 0 & 0 & 0 & 0 & 0 & 0 \\ 0 & .98 & 0 & 0 & 0 & 0 & 0 & 0 & 0 & 0 \\ 0 & 0 & .99 & 0 & 0 & 0 & 0 & 0 & 0 & 0 \\ 0 & 0 & 0 & .99 & 0 & 0 & 0 & 0 & 0 & 0 \\ 0 & 0 & 0 & 0 & .99 & 0 & 0 & 0 & 0 & 0 \\ 0 & 0 & 0 & 0 & 0 & .99 & 0 & 0 & 0 & 0 \\ 0 & 0 & 0 & 0 & 0 & 0 & .98 & 0 & 0 & 0 \\ 0 & 0 & 0 & 0 & 0 & 0 & 0 & .97 & 0 & 0 \\ 0 & 0 & 0 & 0 & 0 & 0 & 0 & 0 & .97 & 0 \end{bmatrix}$$

Analyze this matrix as was done in Optional. Use commands described in MATLAB..

## 4.3  Differential Equations

In this section, we show how to solve systems of differential equations. We start with a small example which can be generalized.

Let $x_1(t)$ and $x_2(t)$ be differentiable functions that satisfy

$$x_1'(t) = a_{11}x_1(t) + a_{12}x_2(t)$$
$$x_2'(t) = a_{21}x_1(t) + a_{22}x_2(t).$$

where $a_{11}, a_{12}, a_{21}$, and $a_{22}$ are scalars. Putting these equations into a matrix equation yields

$$\frac{d}{dt}x(t) = Ax(t) \tag{4.11}$$

where $x = \begin{bmatrix} x_1(t) \\ x_2(t) \end{bmatrix}$ and $A = \begin{bmatrix} a_{11} & a_{12} \\ a_{21} & a_{22} \end{bmatrix}$. We assume that $A$ is diagonalizable and that $A = PDP^{-1}$, where $D = \begin{bmatrix} \lambda_1 & 0 \\ 0 & \lambda_2 \end{bmatrix}$.

To find the function $x(t)$, we substitute $PDP^{-1}$ for $A$ in (4.11), obtaining

$$\frac{d}{dt}x(t) = PDP^{-1}x(t).$$

Rearrangement yields

$$P^{-1}\frac{d}{dt}x(t) = DP^{-1}x(t)$$

or

$$\frac{d}{dt}\left(P^{-1}x(t)\right) = D\left(P^{-1}x(t)\right). \tag{4.12}$$

Now define

$$y(t) = P^{-1}x(t) \tag{4.13}$$

and substitute this expression into (4.12) to obtain

$$\frac{d}{dt}y(t) = Dy(t).$$

In terms of entries, we now have

$$\frac{d}{dt}y_1(t) = \lambda_1 y_1(t) \tag{4.14}$$
$$\frac{d}{dt}y_2(t) = \lambda_2 y_2(t).$$

Since, in general, the scalar differential equation

$$\frac{d}{dt}z(t) = \lambda z(t)$$

has solution $z(t) = ae^{\lambda t}$, where $a$ is an arbitrary constant, the solution to (4.14) is

$$y_1(t) = \alpha_1 e^{\lambda_1 t}$$
$$y_2(t) = \alpha_2 e^{\lambda_2 t}$$

where $\alpha_1$ and $\alpha_2$ are arbitrary scalars. Thus

$$y(t) = \begin{bmatrix} \alpha_1 e^{\lambda_1 t} \\ \alpha_2 e^{\lambda_2 t} \end{bmatrix}$$

and so by (4.13), and backward multiplication,

$$x(t) = Py(t) \qquad\qquad (4.15)$$
$$= \alpha_1 e^{\lambda_1 t} p_1 + \alpha_2 e^{\lambda_2 t} p_2.$$

If $A$ is $n \times n$ and diagonalizable, this extends to

$$x(t) = \alpha_1 e^{\lambda_1 t} p_1 + \cdots + \alpha_n e^{\lambda_n t} p_n.$$

Thus to solve (4.11), we need only find the eigenvalues $\lambda_1, \lambda_2$ and corresponding eigenvectors $p_1, p_2$ of $A$, respectively, and write out the solution using them.

The following example shows how to use the formula to solve systems of differential equations.

**Example 4.8** *Solve*

$$\begin{aligned} x_1'(t) &= -3x_1(t) + x_2(t) \\ x_2'(t) &= x_1(t) - 3x_2(t). \end{aligned}$$

*Here* $A = \begin{bmatrix} -3 & 1 \\ 1 & -3 \end{bmatrix}$. *The eigenvalues of $A$ are $\lambda_1 = -2$ and $\lambda_2 = -4$ with corresponding eigenvectors $p_1 = \begin{bmatrix} 1 \\ 1 \end{bmatrix}$ and $p_2 = \begin{bmatrix} -1 \\ 1 \end{bmatrix}$, respectively. Thus,*

$$x(t) = \alpha_1 e^{\lambda_1 t} p_1 + \alpha_2 e^{\lambda_2 t} p_2$$
$$= \alpha_1 e^{-2t} \begin{bmatrix} 1 \\ 1 \end{bmatrix} + \alpha_2 e^{-4t} \begin{bmatrix} -1 \\ 1 \end{bmatrix}$$

*is the solution.*

*Note that because the eigenvalues of $A$ are negative,* $\lim_{t \to \infty} x(t) = 0$.

Similarly, if $A$ is diagonalizable and has positive eigenvalues, we can show that

$$\frac{d^2}{dt^2}x(t) + Ax(t) = 0$$

has solution

$$x(t) = \left(\alpha_1 \sin\left(\sqrt{\lambda_1}\, t\right) + \beta_1 \cos\left(\sqrt{\lambda_1}\, t\right)\right)p_1 \qquad (4.16)$$
$$+ \left(\alpha_2 \sin\left(\sqrt{\lambda_2}\, t\right) + \beta_2 \cos\left(\sqrt{\lambda_2}\, t\right)\right)p_2$$

where the $\alpha_i's$ and $\beta_i's$ are arbitrary constants. (The extension to $n \times n$ diagonalizable matrices should be clear.)

Another way to solve differential equations is by using functions. Functions will give neat, compact expressions for solutions which don't depend on the Jordan form. However, they can be difficult to compute.

To see how to develop functions of matrices, let $f(\tau)$ be a scalar function with Maclaurin series

$$f(\tau) = a_0 + a_1\tau + a_2\tau^2 + \cdots \qquad (4.17)$$

where $\tau$ is a variable and $a_0, a_1, \ldots$ constants. We assume that this series converges for all $\tau$ and thus converges absolutely for all $\tau$.

For an $n \times n$ matrix $A$, define correspondingly

$$f(A) = a_0 I + a_1 A + a_2 A^2 + \cdots \qquad (4.18)$$

As given in the exercises, if $m = \max_{i,j}|a_{ij}|$ (the largest entry, in absolute value, in $A$), then

$$|a_{ij}| \le m, \ \left|a_{ij}^{(2)}\right| \le nm^2, \le \left|a_{ij}^{(3)}\right| \le n^2 m^3, \ldots$$

So since (4.17) converges absolutely, using $\tau = nm$,

$$|a_0| + |a_1|\, nm + |a_2|\,(nm)^2 + \cdots$$

converges. Thus by the comparison test, using $\delta_{ij}$ as the Kronecker $\delta$, $|a_0|\,\delta_{ij} + |a_1|\,|a_{ij}| + |a_2|\,\left|a_{ij}^{(2)}\right| + \cdots$ converges and so $a_0\delta_{ij} + a_1 a_{ij} + a_2 a_{ij}^{(2)} + \cdots$ converges. So the series in (4.18) converges. (Thus, $f(A)$ can be computed without knowing the Jordan form.)

An example may be helpful.

**Example 4.9** Let $A = \begin{bmatrix} 1 & 1 \\ 0 & 1 \end{bmatrix}$. Then $A^k = \begin{bmatrix} 1 & k \\ 0 & 1 \end{bmatrix}$ for all $k$. Now let $f(\tau) = e^\tau$. Since

$$e^\tau = 1 + \frac{\tau}{1!} + \frac{\tau^2}{2!} + \cdots,$$
$$e^A = I + \frac{1}{1!}A + \frac{1}{2!}A^2 + \cdots$$

$$= \begin{bmatrix} 1+\frac{1}{1!}+\frac{1^2}{2!}+\cdots & \frac{1}{1!}+\frac{2}{2!}+\frac{3}{3!}+\cdots \\ 0 & 1+\frac{1}{1!}+\frac{1^2}{2!}+\cdots \end{bmatrix}$$

$$= \begin{bmatrix} e & e \\ 0 & e \end{bmatrix}.$$

*It should be mentioned that this matrix was chosen so that the series that occurred could be summed. In general, we can't find $f(A)$ so easily.*

We now obtain formulas for $f(A)$ in terms of the Jordan form of $A$. Note that if

$$A = PJP^{-1}$$

where $J$ is the Jordan form of $A$, then by substitution,

$$f(A) = a_0 I + a_1 A + a_2 A^2 + \cdots$$
$$= a_0 PP^{-1} + a_1 PJP^{-1} + a_2 PJ^2 P^{-1} +$$
$$= P\left(a_0 I + a_1 J + a_2 J^2 + \cdots\right) P^{-1}$$

which yields

$$f(A) = Pf(J)P^{-1}. \tag{4.19}$$

And, if $J = \mathrm{diag}\,(J_1,\ldots,J_r)$, where each $J_k$ is a Jordan block,

$$f(J) = a_0 I_1 + a_1 \,\mathrm{diag}\,(J_1,\ldots,J_r) + a_2\,\mathrm{diag}\,(J_1^2,\ldots,J_r^2) + \cdots$$
$$= \mathrm{diag}(a_0 I_1 + a_1 J_1 + a_2 J_1^2 + \cdots,\ldots, a_0 I + a_1 J_r + \cdots)$$

where $I = \mathrm{diag}\,(I_1,\ldots,I_r)$ is partitioned as is $J$. So

$$f(J) = \mathrm{diag}\,(f(J_1),\ldots,f(J_r)).$$

Thus, to compute $f(A)$, we need only find a formula for $f(J_i)$, where $J_i$ is some Jordan block.

**Lemma 4.1** *If $J_i$ is an $n \times n$ Jordan block, say,*

$$J_i = \begin{bmatrix} \lambda & 1 & 0 & \cdots & 0 \\ 0 & \lambda & 1 & \cdots & 0 \\ & & \cdots & & \\ 0 & 0 & 0 & \cdots & \lambda \end{bmatrix}$$

*then*

$$f(J_i) = \begin{bmatrix} f(\lambda) & \frac{f^{(1)}(\lambda)}{1!} & \frac{f^{(2)}(\lambda)}{2!} & \cdots & \frac{f^{(n-1)}(\lambda)}{(n-1)!} \\ 0 & f(\lambda) & \frac{f^{(1)}(\lambda)}{1!} & \cdots & \frac{f^{(n-2)}(\lambda)}{(n-2)!} \\ & & \cdots & & \\ 0 & 0 & 0 & \cdots & f(\lambda) \end{bmatrix} \tag{4.20}$$

where $f^{(k)}$ denotes the k-th derivative of f. (Recall here that $f^{(k)}(\lambda)$ means that $f(\tau)$ is differentiated k times and then $\tau$ is replaced by $\lambda$.)

**Proof.** We sum the series $a_0 I + a_1 J_i + a_2 J_i^2 + \cdots$. This yields, as the 1, 1-entry of the sum,

$$a_0 + a_1 \lambda + a_2 \lambda^2 + \cdots = f(\lambda).$$

For the 1, r + 1 entry, we have

$$a_r \begin{pmatrix} r \\ r \end{pmatrix} + a_{r+1} \begin{pmatrix} r+1 \\ r \end{pmatrix} \lambda + a_{r+2} \begin{pmatrix} r+2 \\ r \end{pmatrix} \lambda^2 + \cdots$$

$$= \frac{r!}{r!} a_r + \frac{(r+1)!}{r!1!} a_{r+1} \lambda + \frac{(r+2)!}{r!2!} a_{r+2} \lambda^2 + \cdots$$

$$= \frac{1}{r!} \left( r! a_r + \frac{(r+1)!}{1!} a_{r+1} \lambda + \frac{(r+2)!}{2!} a_{r+2} \lambda^2 + \cdots \right)$$

$$= \frac{1}{r!} f^{(r)}(\lambda).$$

These expressions yield the entries of $f(J_i)$ that appear in the formula of the lemma. ∎

An example follows.

**Example 4.10** Let $f(\tau) = \sin \tau$ and $A = \begin{bmatrix} \frac{\pi}{6} & 1 & 0 \\ 0 & \frac{\pi}{6} & 1 \\ 0 & 0 & \frac{\pi}{6} \end{bmatrix}$. Then by using

(4.20), we have

$$\sin A = \begin{bmatrix} f(\lambda) & \frac{f^{(1)}(\lambda)}{1!} & \frac{f^{(2)}(\lambda)}{2!} \\ 0 & f(\lambda) & \frac{f^{(1)}(\lambda)}{1!} \\ 0 & 0 & f(\lambda) \end{bmatrix}$$

$$= \begin{bmatrix} \sin \frac{\pi}{6} & \frac{\cos \frac{\pi}{6}}{1!} & \frac{-\sin \frac{\pi}{6}}{2!} \\ 0 & \sin \frac{\pi}{6} & \frac{\cos \frac{\pi}{6}}{1!} \\ 0 & 0 & \sin \frac{\pi}{6} \end{bmatrix}$$

$$= \begin{bmatrix} \frac{1}{2} & \frac{\sqrt{3}}{2} & -\frac{1}{4} \\ 0 & \frac{1}{2} & \frac{\sqrt{3}}{2} \\ 0 & 0 & \frac{1}{2} \end{bmatrix}.$$

More generally, we need to look at an example of the type $f(At)$ where t is a real variable. (Scalars usually precede matrices; however, in this setting, by tradition, the roles are reversed.)

As shown previously for $A$, we can show that

$$f(At) = Pf(Jt)P^{-1}.$$

However, we cannot use the formulas for $f(Jt)$ since, for example, $Jt = \begin{bmatrix} \lambda t & t \\ 0 & \lambda t \end{bmatrix}$, the super diagonal is composed of 0's and $t$'s, not 0's and 1's. Simply put, $Jt$ is not a Jordan form. This, however, is easily remedied. For example, we can write

$$Jt = \begin{bmatrix} \lambda t & t & 0 \\ 0 & \lambda t & t \\ 0 & 0 & \lambda t \end{bmatrix}$$

$$= \begin{bmatrix} 1 & 0 & 0 \\ 0 & t^{-1} & 0 \\ 0 & 0 & t^{-2} \end{bmatrix} \begin{bmatrix} \lambda t & 1 & 0 \\ 0 & \lambda t & 1 \\ 0 & 0 & \lambda t \end{bmatrix} \begin{bmatrix} 1 & 0 & 0 \\ 0 & t & 0 \\ 0 & 0 & t^2 \end{bmatrix}.$$

And since $\begin{bmatrix} \lambda t & 1 & 0 \\ 0 & \lambda t & 1 \\ 0 & 0 & \lambda t \end{bmatrix}$ is a Jordan block, we have

$$e^{Jt} = \begin{bmatrix} 1 & 0 & 0 \\ 0 & t^{-1} & 0 \\ 0 & 0 & t^{-2} \end{bmatrix} \begin{bmatrix} e^{\lambda t} & \frac{e^{\lambda t}}{1!} & \frac{e^{\lambda t}}{2!} \\ 0 & e^{\lambda t} & \frac{e^{\lambda t}}{1!} \\ 0 & 0 & e^{\lambda t} \end{bmatrix} \begin{bmatrix} 1 & 0 & 0 \\ 0 & t & 0 \\ 0 & 0 & t^2 \end{bmatrix}$$

$$= \begin{bmatrix} e^{\lambda t} & \frac{te^{\lambda t}}{1!} & \frac{t^2 e^{\lambda t}}{2!} \\ 0 & e^{\lambda t} & \frac{te^{\lambda t}}{1!} \\ 0 & 0 & e^{\lambda t} \end{bmatrix}.$$

More generally, for an $n \times n$ Jordan block $J$, we have

$$e^{Jt} = \begin{bmatrix} e^{\lambda t} & \frac{te^{\lambda t}}{1!} & \frac{t^2 e^{\lambda t}}{2!} & \cdots & \frac{t^{n-1}e^{\lambda t}}{(n-1)!} \\ 0 & e^{\lambda t} & \frac{te^{\lambda t}}{1!} & \cdots & \frac{t^{n-2}e^{\lambda t}}{(n-2)!} \\ & & & \cdots & \\ 0 & 0 & 0 & \cdots & e^{\lambda t} \end{bmatrix}. \tag{4.21}$$

In the example below, we show how to compute $e^{At}$.

**Example 4.11** Let $A = \begin{bmatrix} 2 & 0 \\ 1 & 2 \end{bmatrix}$. Then $J = \begin{bmatrix} 2 & 1 \\ 0 & 2 \end{bmatrix}$ and $P = \begin{bmatrix} 0 & 1 \\ 1 & 0 \end{bmatrix}$. Thus, using (4.21),

$$e^{At} = Pe^{Jt}P^{-1} = P \begin{bmatrix} e^{2t} & te^{2t} \\ 0 & e^{2t} \end{bmatrix} P^{-1}$$

$$= \begin{bmatrix} e^{2t} & 0 \\ te^{2t} & e^{2t} \end{bmatrix}.$$

The remainder of the section concerns $e^{At}$ and $\lim_{t \to \infty} e^{At}$.    We use that if $\lambda = a + ib$, then $e^{\lambda t} = e^{at} e^{ibt}$ where $e^{ibt} = \cos(bt) + i \sin(bt)$.    Thus, $\left| e^{ibt} \right| = 1$ for all $t$.    In addition, we use that

$$\lim_{t \to \infty} e^{at} = \begin{cases} 0 \text{ if } a < 0 \\ 1 \text{ if } a = 0 \\ \infty \text{ if } a > 0 \end{cases}.$$

We now describe when $\lim_{t \to \infty} e^{At}$ exists.

**Theorem 4.7** *Let $A$ be an $n \times n$ matrix.  Then*

(a) $\lim_{t \to \infty} e^{At} = 0$ *if and only if all eigenvalues $\lambda = a + bi$ of $A$ satisfy $a < 0$.*

(b) $\lim_{t \to \infty} e^{At}$ *exists if all eigenvalues $\lambda = a + bi$ of $A$ satisfy $a \leq 0$, and when $a = 0$, then $\lambda = 0$ and its corresponding Jordan blocks are $1 \times 1$.*

**Proof.** This follows by using L'Hospital's Rule on the entries of $e^{Jt}$. ∎

We now solve the differential equation, with initial condition

$$y'(t) = Ay(t) \tag{4.22}$$
$$y(0) = c,$$

by using functions of matrices.  We know that the scalar differential equation

$$x'(t) = ax(t)$$
$$x(0) = x_0$$

has solution

$$x(t) = e^{at} x_0.$$

Using functions of matrices, we mimic this solution.
    Note that, as in the scalar case,

$$\frac{d}{dt} e^{At} = A + \frac{2A^2 t}{2!} + \frac{3A^3 t^2}{3!} + \cdots$$
$$= A \left( I + At + \frac{A^2 t^2}{2!} + \cdots \right)$$
$$= A e^{At}.$$

Hence by direct computation, we can show that

$$y(t) = e^{At} c$$

is the solution to (4.22).  Thus, if $A$ is real, since $e^{At}$ is a series in $At$, it is real and so is $y(t)$, provided $c$ is real.
    An example follows.

**Example 4.12** *Solve*

$$\frac{d}{dt}x(t) = \begin{bmatrix} 2 & 0 \\ 1 & 2 \end{bmatrix} x(t)$$

$$x(0) = \begin{bmatrix} 1 \\ 1 \end{bmatrix}.$$

*Using the data for the previous example,*

$$x(t) = e^{\begin{bmatrix} 2 & 0 \\ 1 & 2 \end{bmatrix}t} \begin{bmatrix} 1 \\ 1 \end{bmatrix}$$

$$= \begin{bmatrix} e^{2t} & 0 \\ te^{2t} & e^{2t} \end{bmatrix} \begin{bmatrix} 1 \\ 1 \end{bmatrix}$$

$$= \begin{bmatrix} e^{2t} \\ te^{2t} + e^{2t} \end{bmatrix}.$$

*We can get a view of the solution $x(t)$ by graphing the vector $x(t)$, or by using the exponential. The latter method does not require our knowing the Jordan form, so we will demonstrate this technique. We graph*

$$x(t) = e^{At} \begin{bmatrix} 1 \\ 1 \end{bmatrix}$$

*on $[0,2]$ in increments of .1. So, we plot*

$$e^0 \begin{bmatrix} 1 \\ 1 \end{bmatrix}, e^{A(.1)} \begin{bmatrix} 1 \\ 1 \end{bmatrix}, e^{A(.2)} \begin{bmatrix} 1 \\ 1 \end{bmatrix}, \ldots, e^{A(2)} \begin{bmatrix} 1 \\ 1 \end{bmatrix}$$

*to achieve the $*$'s in Figure 4.10. Again, 0 indicates the position of the initial*

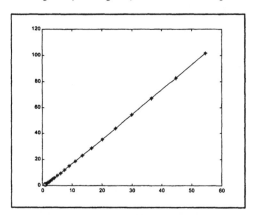

FIGURE 4.10.

*vector and the segments connecting $*$'s indicate the order of occurrence of the $x(t)$'s. Observe that as $t$ increases, the $x(t)$'s cover more distance so there is some acceleration.*

## 4.3.1    Optional (Modeling Motions of a Building)

The two walls of a building, sketched in Figure 4.11, provide a restoring force on the floor above them. This force is equal to the stiffness constant

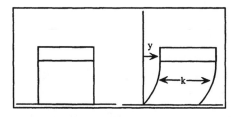

FIGURE 4.11.

$k$ of the walls times the displacement of the floor from equilibrium.

We now model the two story building in Figure 4.12 with floor masses $m_1$, $m_2$ and stiffness constants $k_1$, $k_2$.

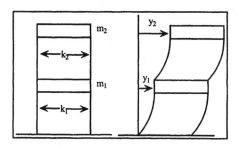

FIGURE 4.12.

Let

$$y_1(t) = \text{ displacement of floor 1 from}$$
$$\text{equilibrium at time } t, \text{ and}$$
$$y_2(t) = \text{displacement of floor 2 from}$$
$$\text{equilibrium at time } t.$$

(Positive values indicate the building is to the right of equilibrium.)

The restoring force on floor 2 is $-k_2(y_2(t) - y_1(t))$ and thus, by Newton's law,

$$m_2 y_2''(t) = -k_2(y_2(t) - y_1(t)).$$

The restoring force on floor 1 is $-k_1 y_1(t) + k_2(y_2(t) - y_1(t))$, so we have

$$m_1 y_1''(t) = -k_1 y_1(t) + k_2(y_2(t) - y_1(t)).$$

Or, in matrix form,

$$\begin{bmatrix} m_1 & 0 \\ 0 & m_2 \end{bmatrix} y''(t) + \begin{bmatrix} k_1 + k_2 & -k_2 \\ -k_2 & k_2 \end{bmatrix} y(t) = 0$$

where $y(t) = \begin{bmatrix} y_1(t) \\ y_2(t) \end{bmatrix}$.

## 4.3.2   MATLAB (Code for Viewing Solutions of Differential Equations Using expm)

MATLAB does not provide a command for $e^{At}$; however, we can use $expm\,(At_k)$, for values $t_1, t_2, \ldots$ instead. We demonstrated this in Example 4.12.
  For more information, type in *help expm*.

**Code for Viewing Solution of Differential Equations**

```
x = [1; 1]; y = x;
A = [2 0; 2 1];
t = .1;
for k = 1 : 21
        p(k) = y(1);            %  Generates e^{t_k A} x for
        q(k) = y(2);                t = 0, .1, . . . , 2.
        y = expm(t * A) * x;
        t = t + .1;
end
plot (1, 1, 'O')               %  Plots starting point with O.
hold
plot (p, q, ' * ')             %  Plots points [p(k), p(k)]
                                   for k = 1, . . . , 20 with *.
plot (p, q)                    %  Plots 'curve.'
```

# Exercises

1. Solve, using the eigenvalue-eigenvector formula (4.15).

   (a) $x_1'(t) = -2x_1(t) + x_2(t)$
   $\phantom{(a)}\ x_2'(t) = \phantom{-2}x_1(t) - 2x_2(t)$

   (b) $x_1'(t) = 2x_1(t) + x_2(t)$
   $\phantom{(b)}\ x_2'(t) = 2x_1(t) + x_2(t)$
   $\phantom{(b)}\ x_1(0) = -1$
   $\phantom{(b)}\ x_2(0) = 5$

   (c) $x_1'(t) = x_1(t) + x_2(t) + x_3(t)$
   $\phantom{(c)}\ x_2'(t) = x_1(t) + x_2(t) + x_3(t)$

$$x_3'(t) = x_1(t) + x_2(t) + x_3(t)$$
$$x_1(0) = 3$$
$$x_2(0) = 0$$
$$x_3(0) = 0.$$

2. Solve, using the eigenvalue-eigenvector formula (4.16).

   (a) $x_1''(t) + 2x_1(t) + x_2(t) = 0$
   $x_2''(t) + x_1(t) + 2x_2(t) = 0$
   $x_1(0) = 2$
   $x_2(0) = 0$
   $x_1'(0) = -4$
   $x_2'(0) = -2$

   (b) $x_1''(t) + 3x_1(t) + 1x_2(t) = 0$
   $x_2''(t) + 2x_1(t) + 2x_2(t) = 0$
   $x_1(0) = 0$
   $x_2(0) = 3$
   $x_1'(0) = 3$
   $x_2'(0) = 2$

3. Solve the spring-mass problem in Optional of Section 1, for $m_1 = m_2 = 1$, $k_1 = 3$, $k_2 = 2$.

4. Solve the two-floor building problem for $m_1 = m_2 = 1$, $k_1 = 3$, $k_2 = 4$ in Optional. Also use $y_1(0) = 1$, $y_2(0) = 2$, $y_1'(0) = y_2'(0) = 0$.

5. Let $A = \begin{bmatrix} 3 & -1 \\ 1 & 1 \end{bmatrix}$. Compute each of the following, using formulas (4.20)

   (a) $e^A$    (b) $\sin A$
   (c) $e^{At}$    (d) $\sin At$

6. Let $f(t) = a_0 + a_1 t + a_2 t^2 + \cdots$. Assume that the series converges absolutely. Calculate each of the following.

   (a) $f'(t)$    (b) $f''(t)$
   (c) $f^{(k)}(t)$

7. Let $A = \begin{bmatrix} 2 & 1 \\ 0 & 2 \end{bmatrix}$.

   (a) Find $e^A$ by summing the series.
   (b) Find $e^{At}$.

8. Solve $y'(t) = \begin{bmatrix} -2 & 1 \\ 1 & -2 \end{bmatrix} y$, by using (4.21), where $y(0) = \begin{bmatrix} 1 \\ 1 \end{bmatrix}$.

    Compute $\lim\limits_{t \to \infty} y(t)$. Does the limit depend on $y(0)$?

9. Compute a formula for $\frac{d}{dt} \sin At$ and for $\frac{d}{dt} \cos At$.

10. (Cayley-Hamilton Theorem) Prove if $\varphi(\lambda)$ is the characteristic poly-
    nomial for $A$, $\varphi(A) = 0$. (Hint: Break this down to $P\varphi(J)P^{-1} = 0$
    and use the formulas.)

11. Explain why $\frac{d}{dt} e^{At}$ can be computed termwise.

12. Let $A$ be a $2 \times 2$ matrix with positive eigenvalues. By using functions,
    find the solution to $y'' + Ay = 0$. (Hint: Look at the corresponding
    scalar problem for ideas.)

13. Let $A$ be an $n \times n$ matrix with $\max\limits_{i,j} |a_{ij}| \le m$. Show that $\left| a_{ij}^{(k)} \right| \le$
    $n^k m^k$ for all $k \ge 1$.

14. To solve the differential equation

    $$x^{(n)} + a_{n-1} x^{(n-1)} + \cdots + a_1 x = 0,$$

    set $y_1 = x, y_2 = x^{(1)}, \ldots, y_{n-1} = x^{(n-2)}, y_n = x^{(n-1)}$. Then, using
    the $n+1$ equations, we have

    $$y_1' = y_2$$
    $$y_2' = y_3$$
    $$\cdots$$
    $$y_{n-1}' = y_n$$
    $$y_n' = -a_{n-1} y_n - \cdots - a_1 y_1.$$

    This system can be solved by matrix techniques and $y_1$ gives the
    solution $x$. Do this technique to solve

    $$x'' - 3x' + 2x = 0.$$

15. Two tanks of solution are linked as in Figure 4.13.

    Initially, there are 100 gallons of solution in each tank. The solution
    in tank A contains 50 grams of salt, while there is no salt in tank
    B. Water is pumped into tank A at 20 gallons/min from an outside
    source. Solution is pumped as shown in the diagram.

    Let $y_1(t)$ and $y_2(t)$ denote the grams of salt in tanks $A$ and $B$,
    respectively, at time $t$.

    (a) Model this problem with a system of differential equations.

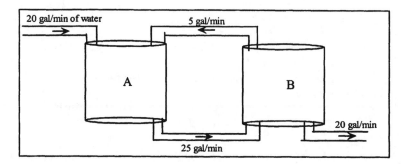

20 gal/min of water

5 gal/min

A

B

20 gal/min

25 gal/min

FIGURE 4.13.

(b) Solve the equations in (a).

(c) Compute $\lim_{t \to \infty} y(t)$.

(d) Explain what the calculation in (c) says about the amount of salt in the tanks as $t$ increases.

16. Let $f$ be a function with a Maclaurin series that converges for all $\tau$.

Let $J = \begin{bmatrix} \lambda & 1 & 0 \\ 0 & \lambda & 1 \\ 0 & 0 & \lambda \end{bmatrix}$. Show, by summing the series, that $f(Jt) =$

$\begin{bmatrix} f(\lambda t) & \frac{tf^{(1)}(\lambda t)}{1!} & \frac{t^2 f^{(2)}(\lambda t)}{2!} \\ 0 & f(\lambda t) & \frac{tf^{(1)}(\lambda t)}{1!} \\ 0 & 0 & f(\lambda t) \end{bmatrix}$.

17. The equation

$$ y' = \begin{bmatrix} 1 & -1 \\ 1 & 1 \end{bmatrix} y, \ y(0) = \begin{bmatrix} 2 \\ 0 \end{bmatrix} $$

has solution $y = e^{(1+i)t} \begin{bmatrix} 1 \\ -i \end{bmatrix} + e^{(1-i)t} \begin{bmatrix} 1 \\ i \end{bmatrix}$.

(a) Show that the imaginary part of $y$ is 0.

(b) Find the solution in terms of real numbers.

18. Let $A$ be an $n \times n$ matrix and $b(t)$ an $n \times 1$ vector of continuous functions.

(a) Show that the solution to

$$ y'(t) = Ay(t) + b(t) $$
$$ y(0) = c $$

is

$$y(t) = e^{At}c + \int_0^t e^{A(t-\tau)}b(\tau)d\tau.$$

(Hint: Mimic the variation of parameter technique of scalar differential equations.)

(b) Solve $y'(t) = \begin{bmatrix} -2 & 1 \\ 1 & -2 \end{bmatrix} y(t) + \begin{bmatrix} 1 \\ 1 \end{bmatrix}$.

19. (Optional) Find the mathematical model for the three-story building diagrammed in Figure 4.14.

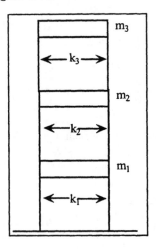

FIGURE 4.14.

20. (MATLAB). Graph the solution to

$$\frac{d}{dt}x(t) = \begin{bmatrix} 0 & .9 \\ -.9 & 0 \end{bmatrix} x(t)$$

$$x(0) = \begin{bmatrix} 1 \\ 1 \end{bmatrix}.$$

(a) By solving and then graphing the solution $x(t)$.
(b) In steps $t = 0, .1, .2, \ldots$, as in Example 4.12.

21. (MATLAB) Solve using the eigenvalue-eigenvector formula.

$$
\begin{aligned}
x_1'(t) &= x_1(t) & +2x_2(t) & \\
x_2'(t) &= 2x_1(t) & + x_2(t) & \\
x_3'(t) &= & x_2(t) & +2x_3(t) \\
x_1(0) &= 10 \\
x_2(0) &= -8 \\
x_3(0) &= 0
\end{aligned}
$$

# 5

# Normed Vector Spaces

In previous chapters we used the standard definition of distance, $d_E$, on Euclidean $n$-space. In this chapter, we extend this work by defining various distances on Euclidean $n$-space and by defining distance on more general vector spaces, as well. Why we use various different ways to measure distance in a vector space will also be explained and shown in various examples.

## 5.1  Vector Norms

In this section we show how to define distance in vector spaces in general. As in Euclidean $n$-space, this is done by first defining the length of a vector.

Recall that the length of a vector $x$ in $R^2$ is

$$\|x\| = \left(x_1^2 + x_2^2\right)^{\frac{1}{2}}.$$

To get the general definition of length of a vector (called a norm in this setting), we use the properties of this length, as given in calculus.

**Definition 5.1** *Let $V$ be a vector space. Suppose there is a way of assigning to each $x$ in $V$, a nonnegative number, written $\|x\|$. We call the assignment function a* norm *(or* vector norm *when we want to distinguish it from other norms that appear later in this book), provided that it satisfies the following properties for all $x$, $y$ in $V$ and scalars $\alpha$.*

  *i. $\|x\| > 0$ if $x \neq 0$ and $\|0\| = 0$*

*ii.* $\|\alpha x\| = |\alpha|\,\|x\|$

*iii.* $\|x+y\| \le \|x\| + \|y\|$   *(This property, called the triangular inequality,
generalizes by induction to* $\|x_1 + \cdots + x_r\| \le \|x_1\| + \cdots + \|x_r\|$ .*)*

*A vector space that has a norm defined on it is called a* normed vector
space.

In a normed vector space $V$ we can define distance $d$ between a pair of
vectors (Points may be a better word when talking about distance.) $x$ and
$y$ as the norm of $x - y$.  (See Figure 5.1.)

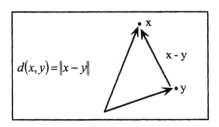

FIGURE 5.1.

In this setting, the distance $d$ is called a *metric*. This metric is translation
invariant, that is, if $a \in V$,

$$d(x+a, y+a) = d(x,y).$$

Thus, $d(x,0) = d(x+a,a)$, i.e., the distance from $x$ to 0 is the same as
the distance from $x + a$ to $a$.

The *classical norms* on Euclidean $n$-space follow. Others are included
in the exercises.

**Theorem 5.1** *Defined for all vectors $x$ in Euclidean $n$-space, the following
are norms.*

*(a)* $\|x\|_1 = \sum\limits_{k=1}^{n} |x_k|$, *called the 1-norm*

*(b)* $\|x\|_2 = \left( \sum\limits_{k=1}^{n} |x_k|^2 \right)^{\frac{1}{2}}$, *called the 2-norm  Note that* $\|x\|_2 = d_E(x,0)$.

*(c)* $\|x\|_\infty = \max\{|x_1|, \ldots, |x_n|\}$, *called the $\infty$-norm*

**Proof.** We prove (a), leaving (b) and (c) as exercises. Since the first
two properties of the definition of a norm are easily verified, we only show

the third property. For it,

$$\|x+y\|_1 = \sum_{k=1}^{n} |x_k + y_k|$$

$$\leq \sum_{k=1}^{n} (|x_k| + |y_k|)$$

$$= \sum_{k=1}^{n} |x_k| + \sum_{k=1}^{n} |y_k|$$

$$= \|x\|_1 + \|y\|_1$$

as required. ∎

**Example 5.1** *Let* $x = (1, -2, 2)^t$. *Then*

$$\|x\|_1 = |1| + |-2| + |2| = 5$$
$$\|x\|_2 = \left(1^2 + (-2)^2 + 2^2\right) = 3$$
$$\|x\|_\infty = \max\{|1|, |-2|, |3|\} = 3.$$

It is interesting to graph the unit "circles" of these norms in $R^2$.

(a) To graph $C_0 = \{x \in R^2 : \|x\|_1 = 1\}$, we graph $\|x\|_1 = 1$, or $|x_1| + |x_2| = 1$. To do this, in the first quadrant we graph $x_1 + x_2 = 1$, in the second quadrant $-x_1 + x_2 = 1$, etc.

The graph of $C_a = \{x \in R^2 : \|x - a\|_1 = 1\}$, where $a = \begin{bmatrix} 2 \\ 2 \end{bmatrix}$, is a translation of the graph $C_0$. Both graphs are shown in Figure 5.2, and they are congruent.

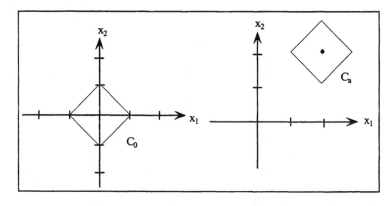

FIGURE 5.2.

(b) The graph of $C_0 = \{x \in R^2 : \|x\|_2 = 1\}$ is shown in Figure 5.3.

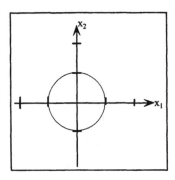

FIGURE 5.3.

(c) The graph of $C_0 = \{x \in R^2 : \|x\|_\infty = 1\}$ is given in Figure 5.4.

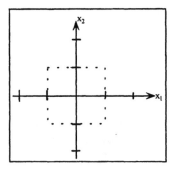

FIGURE 5.4.

Note that only the 2-norm is orientation invariant. That is, if we measure the length of a stick, with one end at the origin, we get the same result regardless of how the stick is placed. The length of the stick, however, will change in the 1-norm, and the $\infty$-norm if, say, we tilt it a bit.

It is also interesting to graph the norms as functions of the entries of the vectors. The graphs of $f(x) = \|x\|$, for the various classical norms, are given in Figure 5.5.

Observe that the only norm showing a smooth surface (so partial derivatives can be taken everywhere) is the 2-norm. We will show the importance of this when we look at least-squares problems. Also, note that the graphs in the previous examples are level curves of these functions.

We might wonder about the necessity of various norms and, thus, various metrics. To provide an answer we can recall that angles can be measured by using degrees or radians. However, in calculus, derivative formulas involving the trigonometric functions are given in radians. If they were done for degrees, those formulas would be more complicated. In the same way, often calculations are more easily done when choosing an appropriate norm.

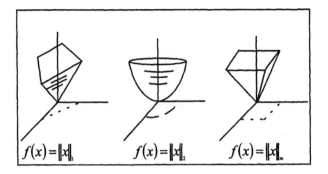

$$f(x) = \|x\|_1, \qquad f(x) = \|x\|_2, \qquad f(x) = \|x\|_\infty.$$

FIGURE 5.5.

And, in some problems, the information obtained by using one norm can be better than that obtained by another.

Still, all norms are equivalent in the following sense. Given any norm $\|\cdot\|$, there are positive scalars $\alpha$ and $\beta$ such that

$$\alpha d_E(x, 0) \le \|x\| \le \beta d_E(x, 0)$$

for all $x$. Thus $\|x\|$ is small if and only if the entries of $x$ are small. We will show this for our classical norms.

**Theorem 5.2** *For all vectors* $x$,

(a) $d_E(x, 0) \le \|x\|_1 \le \sqrt{n} d_E(x, 0)$.

(b) $d_E(x, 0) = \|x\|_2$.

(c) $\frac{1}{\sqrt{n}} d_E(x, 0) \le \|x\|_\infty \le d_E(x, 0)$.

**Proof.** We prove (a), leaving (c) for the reader. For this note that

$$\sum_{k=1}^n |x_k|^2 \le \left( \sum_{k=1}^n |x_k| \right)^2.$$

Thus, taking square roots

$$d_E(x, 0) \le \|x\|_1.$$

And, since by the Cauchy-Schwarz inequality, as given in the exercises,

$$\sum_{k=1}^n 1 \, |x_k| \le \left( \sum_{k=1}^n 1^2 \right)^{\frac{1}{2}} \left( \sum_{k=1}^n |x_k|^2 \right)^{\frac{1}{2}},$$
$$\|x\|_1 \le \sqrt{n} d_E(x, 0)$$

which yields (a). ■

Since $d_E(x_k, x_0) = d_E(x_k - x_0, 0)$, we see that if $x_1, x_2 \ldots$ is a sequence of vectors and $x_0$ a vector, then for any norm $\|\cdot\|$, and corresponding $\alpha$ and $\beta$,

$$\alpha d_E(x_k, x_0) \leq \|x_k - x_0\| \leq \beta d_E(x_k, x_0).$$

Thus, if we establish convergence to $x_0$ in any of our norms, we equivalently have established convergence in the Euclidean distance, and vice versa. Figure 5.6 and 5.7 shows the convergence of $x_1, x_2, \ldots$ to $x_0$ using Euclidean distance and the $\infty$-norm.

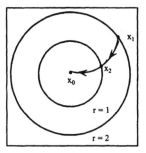

FIGURE 5.6.

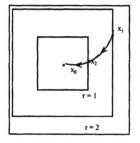

FIGURE 5.7.

## 5.1.1  Optional (Evaluating Models)

Mathematical models are often built to predict or describe some phenomenon. Such models should, when possible, be evaluated. We show how this can be done on a small social science problem.

Social scientists study people: numbers of people in each age group, job category, sex, etc. And they study movements of people in various categories.

Taking some liberties with the real data (adjusting to keep the size of the problem small), we look at sons' occupations versus their fathers' occupations. The occupations are Professional, Supervisor, or Laborer.

A survey was taken in which fathers gave their occupations and that of their first sons. The results are recorded in the table below.

Fathers' Occupations ↓

| Sons' Occupations ↓ | Professional | Supervisory | Labor | Total ** |
|---|---|---|---|---|
| Professional | 15 | 10 | 3 | 28 |
| Supervisory | 9 | 35 | 44 | 88 |
| Labor | 8 | 55 | 221 | 284 |
| Total * | 32 | 100 | 268 | |

*Total number of fathers surveyed in each category
**Total number of sons recorded in each category.

From this table we can compute the transition matrix showing the percentage of sons in each column category.

$$A = \begin{bmatrix} \frac{15}{32} & \frac{10}{100} & \frac{3}{268} \\ \frac{9}{32} & \frac{35}{100} & \frac{44}{268} \\ \frac{8}{32} & \frac{55}{100} & \frac{221}{268} \end{bmatrix}.$$

Note that the first row of $A$ gives the percentage of each categories' sons that end up as professionals. The second and third row of $A$ have corresponding interpretations. Thus, if $f$ is a distribution of fathers in the categories, then

$$s = Af$$

gives the distribution of their sons in the categories. For example in our data

$$f = \begin{bmatrix} 32 \\ 100 \\ 268 \end{bmatrix}, s = \begin{bmatrix} 28 \\ 88 \\ 284 \end{bmatrix}$$

and $Af = s$.

What we now do is use this transition matrix to compute the distribution of the sons' sons in the categories. For this, we calculate

$$\text{sons' sons} = As$$

$$= \begin{bmatrix} 25.1 \\ 85.3 \\ 289.6 \end{bmatrix} \approx \begin{bmatrix} 25 \\ 85 \\ 290 \end{bmatrix}.$$

We can compare our calculations with the actual sons' sons distribution

$$\begin{bmatrix} 30 \\ 90 \\ 280 \end{bmatrix}.$$

And, so we see there is some error. We can measure this error by using norms. The relative (percentage) error is

$$\frac{\left\| \begin{bmatrix} 25 \\ 85 \\ 290 \end{bmatrix} - \begin{bmatrix} 30 \\ 90 \\ 280 \end{bmatrix} \right\|_1}{\left\| \begin{bmatrix} 30 \\ 90 \\ 280 \end{bmatrix} \right\|_1} = \frac{20}{400} = .05$$

or 5%. (We could have used another norm, however, this one was easy to check.) So, our prediction is within 95% of the actual distribution.

It is not unusual to have this kind of % error, and even more, in social science problems.

### 5.1.2   MATLAB (Vector Norms)

The commands to obtain vector norms are natural: norm$(x, 1)$ provides $\|x\|_1$, norm$(x, 2)$ provides $\|x\|_2$, and norm$(x, \text{inf})$ provides $\|x\|_\infty$. For more, type in *help norm*.

## Exercises

1. Let $x = (1, 1, 1)^t$. Compute $\|x\|_1$, $\|x\|_2$, and $\|x\|_\infty$.

2. Find the distance between $(1, 0, 1, 1)^t$ and $(1, 1, 2, 0)^t$ using the
   (a) 1-norm.     (b) 2-norm.
   (c) ∞-norm.

3. Let $x = (3 - 4i, 4 + 3i)^t$. Find the length of $x$ in the
   (a) 1-norm.     (b) 2-norm.
   (c) ∞-norm.

4. Let $x = (1 + 2i, 2 + i)^t$ and $y = (1 + i, 1 - 2i)^t$. Find the distance between $x$ and $y$ in the
   (a) 1-norm.     (b) 2-norm.
   (c) ∞-norm.

5. Draw the unit 'circles' of the 1-norm, 2-norm, and ∞-norm in $R^2$, superimposing one upon the others. Using these pictures, decide the following.

(a) If $\|x\|_1 \leq 1$ is $\|x\|_2 \leq 1$

(b) If $\|x\|_\infty \leq 1$ is $\|x\|_2 \leq 1$

6. Graph the unit 'circles' of the 1-norm and the $\infty$-norm in $R^3$. Is the 'circle' for the 1-norm similar to a rotation of that of the $\infty$-norm, as it is in $R^2$? (Hint: Count vertices.)

7. Prove that if $\|\cdot\|$ is a vector norm,

   (a) $\|-x\| = \|x\|$.    (b) $\|x - y\| = \|y - x\|$.

8. Define $f: R^2 \to R$ by $f(x_1, x_2) = \|x\|_2$ where $x = (x_1, x_2)^t$. Find $\frac{\delta f}{\delta x_1}(x_1, x_2)$ and $\frac{\delta f}{\delta x_2}(x_1, x_2)$.

9. Place a stick in $R^2$ so that one end is at the origin and the other at $(0, 1)^t$. Tilt the stick by $\frac{\pi}{4}$ radian. Find the length of the tilted stick in each of the following.

   (a) 1-norm    (b) 2-norm
   (c) $\infty$-norm

10. (Cauchy-Schwarz inequality) Let $x, y \in R^n$. Prove that $\sum_{k=1}^{n} x_k y_k \leq \|x\|_2 \|y\|_2$ as follows. (This inequality can be recalled from the calculus result $\cos \theta = \frac{x \cdot y}{\|x\|_2 \|y\|_2}$ by noting that $|\cos \theta| \leq 1$.)

   (a) Show $0 \leq \|x + ty\|_2^2 = \|x\|_2^2 + 2t \sum_{k=1}^{n} x_k y_k + t^2 \|y\|_2^2$ where $t$ is scalar.

   (b) Plug $t = - \|x\|_2^2 \Big/ \sum_{k=1}^{n} x_k y_k$.

   (c) Extend the result to complex scalars showing $|x^H y| \leq \|x\|_2 \|y\|_2$.

11. Prove Theorem 5.1

   (a) Part (b). Hint: Use the Cauchy-Schwarz inequality.

   (b) Part (c).

12. Let $x$ and $y$ be in Euclidean $n$-space. Use Theorem 5.2 in the following.

   (a) If $\|x\|_1 < .001$, find bounds on $\|x\|_2$ and $\|x\|_\infty$.

   (b) If $\|x - y\|_\infty < .001$, find bounds on $\|x - y\|_1$ and $\|x - y\|_2$.

13. Define

$$\|x\|_p = (|x_1|^p + \cdots + |x_n|^p)^{\frac{1}{p}}.$$

where $p$ is a positive integer. Prove that $\|x\|_p$, called the $p$-norm, is a vector norm on Euclidean $n$-space. (Just verify norm properties (i) and (ii)).

14. Let $d_1 > 0$ and $d_2 > 0$. Show that $\|x\| = \left(d_1 |x_1|^2 + d_2 |x_2|^2\right)^{\frac{1}{2}}$, called a weighted norm, is a norm on Euclidean $n$-space.

15. Prove part (c) of Theorem 5.2.

16. (Optional) Using MATLAB and eigenvalues and eigenvectors, compute $\lim_{k \to \infty} A^k f$. Explain what this vector tells us about the long-run behavior of the sons' occupations.

17. (MATLAB) The population of a small country is placed in categories 0–9 years old, 10–19 years old, ... The population in the categories in 1970 was

$$(82, 330, 506, 525, 425, 431)^t.$$

The Leslie matrix was found to be

$$\begin{bmatrix} 0 & 0 & .232 & .207 & .036 & 0 \\ .98 & 0 & 0 & 0 & 0 & 0 \\ 0 & .99 & 0 & 0 & 0 & 0 \\ 0 & 0 & .99 & 0 & 0 & 0 \\ 0 & 0 & 0 & .99 & 0 & 0 \\ 0 & 0 & 0 & 0 & .99 & 0 \end{bmatrix}.$$

(a) Compute the population, in the categories in 1980 and 1990.

(b) The actual population in 1990 is given by

$$(187, 81, 330, 506, 524, 424)^t.$$

Find the relative (percentage) error between the estimate computed in (a) and the actual population.

## 5.2   Induced Matrix Norms

In various calculating situations, involving vectors and matrices, we need to pull out $A$ in $\|Ax\|$, similar to pulling out a scalar, $\|ax\| = |a|\,\|x\|$. The matrix norm of this section is designed to have that property.

**Definition 5.2** *Let $\|\cdot\|$ be a vector norm on Euclidean n-space. Define the induced matrix norm for an $n \times n$ matrix $A$ as*

$$\|A\| = \max_{x \neq 0} \frac{\|Ax\|}{\|x\|}. \tag{5.1}$$

*(It can be proved that there is a maximum, as well as a minimum value of $f(x) = \frac{\|Ax\|}{\|x\|}$ when $f$ is evaluated over all $x \neq 0$. Later we will prove this for the classical vector norms.)*

*If $A$ is $m \times n$ and we use the same classical vector norm on $\|x\|$ and $\|Ax\|$, $\|A\|$ is also defined by (5.1). (Note that if $m \neq n$ then $x$ and $Ax$ are in different vector spaces.)*

By the way we defined $\|A\|$, we see that

$$\|Ax\| \leq \|A\|\,\|x\|$$

for all $x$, precisely the property that allows us to pull out $A$ in $\|Ax\|$.

On some problems, the following method to calculate $\|A\|$ is useful.

**Theorem 5.3** $\|A\| = \displaystyle\max_{\|u\|=1} \|Au\|.$

**Proof.** Let $f(x) = \frac{\|Ax\|}{\|x\|}$ for all $x \neq 0$. Then $f(x) = \frac{\|Ax\|}{\|x\|} = \frac{1}{\|x\|}\|Ax\| = \left\|A\frac{x}{\|x\|}\right\|$ and setting $u = \frac{x}{\|x\|}$

$$f(x) = \|Au\|$$
$$= f(u).$$

Thus, we see that every value of $f$ is achieved by some $u$, $\|u\| = 1$.

Furthermore, if $\|u\| = 1$, then setting $x = u$, we have $f(x) = f(u)$. So every value achieved by $u$, $\|u\| = 1$, is also achieved by an $x$, $x \neq 0$. Thus,

$$\max_{x \neq 0} f(x) = \max_{\|u\|=1} f(u) = \|A\|$$

the desired result. ■

An example calculating the induced matrix norm may now be helpful.

**Example 5.2** *Let $A = \begin{bmatrix} 2 & -1 \\ -1 & 2 \end{bmatrix}$. Then, using the vector 2-norm*

$$f(x) = \|Ax\|_2 \tag{5.2}$$
$$= \sqrt{5x_1^2 - 8x_1x_2 + 5x_2^2}.$$

*If $\|x\|_2 = 1$, $x_1^2 + x_2^2 = 1$, so $x_1 = \pm\sqrt{1 - x_2^2}$. Plugging this into (5.2) and using calculus, we can show that $\max f(x) = 3$. So $\|A\|_2 = 3$.*

We now need to show that an induced matrix norm is in fact a norm.

**Theorem 5.4** *Any induced matrix norm is a norm. That is, for any $m \times n$ matrices $A$ and $B$ and for any scalar $\alpha$,*

1. $\|A\| > 0$ *if* $A \neq 0$ *and* $\|0\| = 0$,

2. $\|\alpha A\| = |\alpha| \, \|A\|$ *for any scalar $\alpha$, and*

3. $\|A + B\| \leq \|A\| + \|B\|$.

*In addition, every induced matrix norm has the following properties:*

(a) $\|I\| = 1$.

(b) $\|Ax\| \leq \|A\| \, \|x\|$, *with equality for some $x \neq 0$.*

(c) $\|AB\| \leq \| A\| \, \|B\|$, *assuming the product is defined.*

(d) $\min\limits_{\|x\|=1} \|Ax\| = \frac{1}{\|A^{-1}\|}$, *if $A$ is nonsingular.*

(e) $\|Ax\| \geq \frac{1}{\|A^{-1}\|} \|x\|$, *provided $A$ is nonsingular.*

**Proof.** We first prove that the induced matrix norm is actually a norm. For this, we prove properties (2) and (3), leaving property (1) as an exercise.

Part 1. For (2), using vector norm properties, we have that

$$\|\alpha A\| = \max_{\|u\|=1} \|(\alpha A)\, u\|$$

$$= \max_{\|u\|=1} |\alpha| \, \|Au\|$$

$$= |\alpha| \max_{\|u\|=1} \|Au\|$$

$$= |\alpha| \, \|A\| \, .$$

For (3), we have that

$$\|A + B\| = \max_{\|u\|=1} \|(A + B)\, u\|$$

$$= \max_{\|u\|=1} \|Au + Bu\|$$

$$\leq \max_{\|u\|=1} (\|Au\| + \|Bu\|)$$

$$\leq \max_{\|u\|=1} \|Au\| + \max_{\|u\|=1} \|Bu\|$$

$$= \|A\| + \|B\| \, .$$

Part 2. We now prove three of the remaining properties of the theorem. For (b), by definition

$$\|A\| = \max_{x \neq 0} \frac{\|Ax\|}{\|x\|}.$$

Thus, for any $x \neq 0$,

$$\|A\| \geq \frac{\|Ax\|}{\|x\|}$$

or

$$\|A\| \, \|x\| \geq \|Ax\| \,.$$

And, the latter inequality holds for all $x$. Further, equality holds for some $x \neq 0$ since $f(x) = \frac{\|Ax\|}{\|x\|}$ achieves a maximum at some $x \neq 0$.

For (c), since

$$\|AB\| = \max_{x \neq 0} \frac{\|ABx\|}{\|x\|},$$

we have by (b)

$$\|AB\| \leq \max_{x \neq 0} \frac{\|A\| \, \|Bx\|}{\|x\|},$$

and again by (b)

$$\|AB\| \leq \max_{x \neq 0} \frac{\|A\| \, \|B\| \, \|x\|}{\|x\|}$$
$$= \|A\| \, \|B\| \,.$$

For (d), for any $\|x\| = 1$,

$$1 = \|x\| = \left\|A^{-1} Ax\right\| \leq \left\|A^{-1}\right\| \, \|Ax\|$$

where equality holds for some $x$ by (b). Thus $\frac{1}{\|A^{-1}\|} \leq \|Ax\|$ and since equality holds for some $x$,

$$\frac{1}{\|A^{-1}\|} = \min_{\|x\|=1} \|Ax\|$$

which is the result desired. ∎

Since induced matrix norms are norms, the equivalence of norms result holds. That is, for any induced matrix norm $\|\cdot\|$, there are positive scalars $\alpha$ and $\beta$ such that

$$\alpha d_E(A, 0) \leq \|A\| \leq \beta d_E(A, 0)$$

where $d_E$ is the Euclidean distance on matrices. The consequences of this result are as those for vector norms. For example, if a sequence of matrices $A_1, A_2, \ldots$ is such that

$$\|A_k - A\| \to 0 \text{ as } k \to 0,$$

for some matrix $A$, then

$$\alpha d_E(A_k, A) \to 0 \text{ as } k \to 0.$$

So, $A_k$ tends to $A$ entrywise.

As you might suspect, computing $\|A\|$ by definition can be rather challenging. Remarkably, however, we can find formulas for a few of the induced matrix norms.

**Theorem 5.5** *Let $A$ be an $m \times n$ matrix. Using the classical vector norms, we have the following.*

(a) *For the vector norm $\|\cdot\|_1$, $\|A\|_1 = \max\limits_{j} \sum\limits_{k=1}^{m} |a_{kj}|$, the maximum absolute column sum.*

(b) *For the vector norm $\|\cdot\|_\infty$, $\|A\|_\infty = \max\limits_{i} \sum\limits_{k=1}^{n} |a_{ik}|$, the maximum absolute row sum.*

(c) *For the vector norm $\|\cdot\|_2$, $\|A\|_2 = \max \left[ \lambda \left( A^H A \right) \right]^{\frac{1}{2}}$ where the maximum is taken over the square root of all eigenvalues $\lambda \left( A^H A \right)$ of $A^H A$. (For completeness we included this formula here. It is proved in Chapter 7.)*

**Proof.** We prove (a). There are two parts.

Part 1. We show $\|A\|_1 \le \max\limits_{j} \sum\limits_{k=1}^{m} |a_{kj}|$. For this,

$$\|A\|_1 = \max_{\|u\|_1 = 1} \|Au\|_1 = \max_{\|u\|_1 = 1} \left( \left| \sum_{j=1}^{n} a_{1j} u_j \right| + \cdots + \left| \sum_{j=1}^{n} a_{mj} u_j \right| \right)$$

$$\le \max_{\|u\|_1 = 1} \left( \sum_{j=1}^{n} |a_{1j}| \, |u_j| + \cdots + \sum_{j=1}^{n} |a_{mj}| \, |u_j| \right)$$

$$= \max_{\|u\|_1 = 1} \left( \sum_{j=1}^{n} (|a_{1j}| + \cdots + |a_{mj}|) \, |u_j| \right)$$

$$\le \max_{j} (|a_{1j}| + \cdots + |a_{mj}|) \left( \sum_{j=1}^{n} |u_j| \right)$$

$$= \max_{j} (|a_{1j}| + \cdots + |a_{mj}|) = \max_{j} \sum_{k=1}^{m} |a_{kj}| \ .$$

Part 2. We show there is a $u$, $\|u\|_1 = 1$, where $\|Au\| = \max\limits_{j} \sum\limits_{k=1}^{m} |a_{kj}|$.

For this suppose $\max\limits_{j} \sum\limits_{k=1}^{m} |a_{kj}| = \sum\limits_{k=1}^{m} |a_{kr}|$. Then set $u = e_r$. Using this $u$, since $\|u\|_1 = 1$,

$$\|A\|_1 \geq \|Au\|_1 = |a_{1r}| + \cdots + |a_{mr}| = \sum_{k=1}^{m} |a_{kr}| = \max_{j} \sum_{k=1}^{m} |a_{kj}|.$$

Putting the parts together yields the result. ∎

**Example 5.3** Let $A = \begin{bmatrix} 3 & -1 \\ -2 & 2 \end{bmatrix}$. Then

$$\|A\|_1 = \max\{|3| + |-2|, |-1| + |2|\}$$
$$= \max\{5, 3\} = 5$$
$$\|A\|_\infty = \max\{|3| + |-1|, |-2| + |2|\}$$
$$= \max\{4, 4\} = 4$$
$$\|A\|_2 = \max\left[\lambda\left(A^t A\right)\right]^{\frac{1}{2}}$$
$$= \max\{4.13, .97\} \text{ rounded to the hundreths place}$$
$$= 4.13.$$

We conclude this section by showing what induced matrix norms tell us about a linear transformation,

$$L(x) = Ax.$$

If we look at the image of the unit circle,

1. $\max\limits_{\|x\|=1} \|L(x)\| = \|A\|$ says that the longest vector there has length $\|A\|$.

2. $\min\limits_{\|x\|=1} \|L(x)\| = \frac{1}{\|A^{-1}\|}$ says that the shortest vector there has length $\frac{1}{\|A^{-1}\|}$. (We assume here that $A$ is nonsingular.)

Thus

3. $\dfrac{\max\limits_{\|z\|=1} \|L(z)\|}{\min\limits_{\|y\|=1} \|L(y)\|} = \|A\| \, \|A^{-1}\|$. This number gives us some indication of how much the image of the circle is distorted.

**Example 5.4** Let $L(x) = \begin{bmatrix} 3 & 1 \\ 1 & 3 \end{bmatrix} x$. As indicated in the Figure 5.8 and shown in Chapter 10, the image of the unit circle is an ellipse.

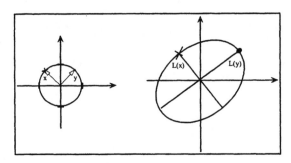

FIGURE 5.8.

*The major axis is in line with the vector* $\begin{bmatrix} 1 \\ 1 \end{bmatrix}$ *and the minor axis in line with* $\begin{bmatrix} -1 \\ 1 \end{bmatrix}$ .

*We now find the maximum and minimum lengths among the image vectors. Since* $L\left(\begin{bmatrix} \frac{\sqrt{2}}{2} \\ \frac{\sqrt{2}}{2} \end{bmatrix}\right) = 4 \begin{bmatrix} \frac{\sqrt{2}}{2} \\ \frac{\sqrt{2}}{2} \end{bmatrix}$ *, the maximum length is 4, which agrees with*

$$\|A\|_2 = 4.$$

*And, since* $L\left(\begin{bmatrix} -\frac{\sqrt{2}}{2} \\ \frac{\sqrt{2}}{2} \end{bmatrix}\right) = 2 \begin{bmatrix} -\frac{\sqrt{2}}{2} \\ \frac{\sqrt{2}}{2} \end{bmatrix}$ *, the minimum length is 2. This is the same as*

$$\frac{1}{\|A^{-1}\|_2} = 2.$$

We can convert these remarks into results about approximations. For this, let $w$ be given and $z$ an approximation of $w$. Then

$$\|L(w) - L(z)\| \leq \|A\| \|w - z\| \tag{5.3}$$

which say that the error between image vectors, $\|L(w) - L(z)\|$, is no more than $\|A\|$ times the error between $w$ and $z$, $\|w - z\|$.

Should we want to bound the error independent of scaling, we would use relative error. The error between $w$ and $z$, relative to $w$, is defined as

$$RE = \frac{\|w - z\|}{\|w\|}.$$

Independence of scaling can be seen in the calculation

$$\frac{\|100w - 100z\|}{\|100w\|} = \frac{\|w - z\|}{\|w\|}.$$

Note that if

$$\frac{\|w - z\|}{\|w\|} \leq 10^{-s},$$

where $s$ is a positive integer, then

$$\|w - z\| \leq 10^{-s} \|w\|.$$

Now, $10^{-s} \|w\|$ moves the decimal in $\|w\|$ to the left $s$ places. Thus, the first digit of $\|w - z\|$ begins (in the worst case) at the $(s + 1) - st$ digit of $\|w\|$. For example, $10^{-3}(12345) = 12.345$, whose first digit begins in the 4-th digit of 12345. So $z$ approximates $w$ to within the $s$-th digit of $\|w\|$, the size of $w$. We will say here that $z$ is an $s$ digit approximation of $w$.

Using Theorem 5.4,

$$\frac{\|L(w) - L(z)\|}{L\|(w)\|} \leq \frac{\|A\| \|w - z\|}{\frac{1}{\|A\|^{-1}} \|w\|} \tag{5.4}$$

$$= c(A) \frac{\|w - z\|}{\|w\|}$$

where

$$c(A) = \|A\| \|A^{-1}\|,$$

$c(A)$ called the *condition number* of $A$. Thus the relative error between image vectors is no greater than $c(A)$ times that of the relative error between the vectors themselves. So $c(A)$ is somewhat like the derivative in calculus. (The derivative indicates how much change in function values we might expect from a change in values.)

As given in the exercises, $c(A) \geq 1$ for all $A$. In addition, although $c(A)$ can be computed using any induced matrix norm, the sizes of the corresponding $c(A)$'s are about the same. For example, it can be shown that

$$\frac{1}{n} \leq \frac{c_\infty(A)}{c_2(A)} \leq n$$

$$\frac{1}{n} \leq \frac{c_1(A)}{c_2(A)} \leq n,$$

where $c_1(A)$, $c_2(A)$, and $c_\infty(A)$ are the condition numbers of $A$ with respect to the induced 1, induced 2, and induced $\infty$ matrix norms, respectively.

## 5.2.1   Optional (Error in Solving $Ax = b$)

For us, a $t$-digit computer is a computer which rounds or truncates all numbers to the first $t$ digits of the number. For example, using rounding, a 3-digit computer gives

$$64872 = 64900$$

and

$$329 + 2.67 = 331.67 = 332.$$

Thus, if a problem is solved on a $t$-digit computer, the numerical $t$-digit solution is in error. For example, suppose we solve, using a 2-digit computer,

$$\begin{bmatrix} 2 & -1 \\ -2 & 6 \end{bmatrix} \begin{bmatrix} x_1 \\ x_2 \end{bmatrix} = \begin{bmatrix} 410 \\ 250 \end{bmatrix}.$$

Applying $R_1 + R_2$, we get

$$\left[ \begin{array}{cc|c} 2 & -1 & 410 \\ 0 & 5 & 660 \end{array} \right]$$

Thus, $x_2 = \frac{660}{5} = 132 = 130$ and by back substituting

$$2x_1 - (130) = 410$$
$$2x_1 = 540$$
$$x_1 = 270.$$

The computed solution $(270, 130)$ can be compared to the actual solution $(271, 132)$, and we can see that we computed the first 2 digits of the entries of the actual solution accurately. Unfortunately, such accuracy is not always the case.

Let $A$ be an $n \times n$ nonsingular matrix and $b$ an $n \times 1$ vector. Consider

$$Ax = b. \tag{5.5}$$

It can be reasoned that if (5.5) is solved by, say, Gaussian elimination with partial pivoting, on a t-digit computer, then the obtained answer $\hat{x}$, satisfies

$$(A + E)\hat{x} = b \tag{5.6}$$

for some $n \times n$ matrix $E$ where

$$\frac{\|E\|_\infty}{\|A\|_\infty} \approx 10^{-t}$$

($\approx$ denotes approximately). Actually, $E$ depends on $n$ and so for small problems, the approximation can be better, while on larger problems, it can be worse. Furthermore, if $x$ is the solution to (5.5),

$$\frac{\|x - \hat{x}\|_\infty}{\|x\|_\infty} \approx c_\infty (A) \frac{\|E\|_\infty}{\|A\|_\infty}.$$

So the condition number shows how much affect the change $E$ in (5.6) has on the solution in (5.5). For example, if $t = 7$ and $c(A) = 10^2$ then

$$\frac{\|x - \hat{x}\|_\infty}{\|x\|_\infty} \approx 10^5 \ .$$

So $\hat{x}$ need not be a 7 digit approximation of $x$. We may have lost 2 digits because of the condition of the problem.

Loosely, if $c(A)$ is small (error magnification is not significant), then $A$ is called *well conditioned*. If, on the other hand, $c(A)$ is large (error magnification is beyond what is desired), then $A$ is called *ill-conditioned*. In between, $A$ is called moderately conditioned. (What is significant and what is tolerable depends on the problem at hand.)

We now look at an example which puts some of the discussion together.

**Example 5.5** *Let* $A = \begin{bmatrix} \frac{10}{3} & \frac{5}{3} & \frac{10}{7} \\ \frac{10}{3} & \frac{30}{7} & \frac{50}{9} \\ \frac{20}{9} & 5 & \frac{50}{7} \end{bmatrix}$ *and* $b = \begin{bmatrix} \frac{5}{3} \\ \frac{9}{7} \\ \frac{2}{3} \end{bmatrix}$. *We solve* $Ax = b$ *using MATLAB.*

*Using format long to get about 15 digits (MATLAB calculates in about 15 digits.), we get*

$$\hat{x} = \begin{bmatrix} 1.26923076923075 \\ -3.19999999999993 \\ 1.93846153846149 \end{bmatrix},$$

*while*

$$x = \begin{bmatrix} \frac{33}{26} \\ \frac{-16}{5} \\ \frac{126}{65} \end{bmatrix} \approx \begin{bmatrix} 1.26923076923079 \\ -3.20000000000000 \\ 1.93846153846154 \end{bmatrix}.$$

*So we can see there is some error.*

*Now*

$$\text{cond}(A,inf) = 1.206666666666643 \times 10^3.$$

*All but the leading digit or so of these digits are unimportant (so even if this computation has lots of its last digits in error, it still gives us what we want). We note that*

$$c(A) \approx 10^3.$$

*So, we might expect to lose a few digits. Now,*

$$\frac{\|x - \hat{x}\|_\infty}{\|x\|_\infty} = 2.1926 \times 10^{-14}$$

$$\approx 10^{-14}.$$

*So we lost about 1 digit. (Our answer was not exactly obtained as a 15-digit computation.)*

Since MATLAB calculations are done in about 15 digits, and normally only the first 5 digits are displayed, unless $c(A)$ is very large, we should have the answer we want.

From the above, we see that computing $c(A)$, and getting, say, $10^s$, tells us about our answer $\hat{x}$. If $s$ is large, it is a red flag that the computed answer may not be a sufficiently close digit approximation of $x$. (In such a case, using iterative improvement, as discussed in Chapter 8 can provide more digits of accuracy.)

### 5.2.2  MATLAB (Matrix Norms and Condition Numbers)

The commands for computing matrix norms are like those for vector norms, namely: $norm(A, 1)$ for $\|A\|_1$, $norm(A, 2)$ for $\|A\|_2$, $norm(A, \text{inf})$ for $\|A\|_\infty$, $norm(A, \text{'fro'})$ for $\|A\|_F = d_E(A, 0)$. For more type in *help norm*.

In addition, we can compute the various condition numbers: *cond* $(A, 1)$ for $c_1(A)$, *cond* $(A, 2)$ for $c_2(A)$, *cond* $(A, \text{inf})$ for $c_\infty(A)$. For more type in *help cond*.

## Exercises

1. Let $\{m_1, \ldots, m_r\}$ be a set of real numbers.

   (a) If $c$ is a positive number, prove that $\max\{cm_1, \ldots, cm_r\} = c\max\{m_1, \ldots, m_r\}$.

   (b) If $\{n_1, \ldots, n_r\}$ is also a set of numbers, prove that

   $$\max\{m_1 + n_1, \ldots, m_r + n_r\}$$
   $$\leq \max\{m_1, \ldots, m_r\} + \max\{n_1, \ldots, n_r\}.$$

2. Prove Theorem 5.4,
   (a) Property 1.    (b) Property a.
   (c) Property e.

3. Prove Theorem 5.5, part b.

4. Let $A = \begin{bmatrix} 4 & 1 \\ 1 & 4 \end{bmatrix}$. Prove that $\|A\|_2 = 5$ by maximizing $f(x) = \|Ax\|_2$ over all $x$, $\|x\|_2 = 1$.

5. Using the formulas in Theorem 5.5, compute both $\|A\|_1$ and $\|A\|_\infty$ for

(a) $A = \begin{bmatrix} 1 & -1 & -3 \\ 2 & 0 & -5 \\ 1 & 4 & -2 \end{bmatrix}.$    (b) $A = \begin{bmatrix} i & 1+2i \\ 3-4i & 6 \end{bmatrix}.$

(c) $A = \begin{bmatrix} 1 & -1 & 3 & 2 \\ 4 & -1 & 0 & -3 \end{bmatrix}.$

6. Using the formula for Theorem 5.5, compute $\|A\|_2$ for

(a) $A = \begin{bmatrix} 1 & 2 \\ 1 & 2 \end{bmatrix}.$    (b) $A = \begin{bmatrix} 1-i & 1+i \\ 1+i & 1+i \end{bmatrix}.$

(c) $A = \begin{bmatrix} 1 & 1 & 1 \\ 1 & 1 & 1 \end{bmatrix}.$

7. Let $A = \begin{bmatrix} 1 & 0 \\ 2 & 1 \end{bmatrix}.$

(a) Graph $C_0$ in the 1-norm.
(b) Graph the image of $C_0$ under $L(x) = Ax$. (Note that $L$ maps edges to edges.)
(c) Compute $\|A\|_1$ from your sketch.
(d) Compute $\|A\|_1$ by the formula in Theorem 5.5.

8. Repeat Exercise 7 for $A = \begin{bmatrix} 3 & 1 \\ 1 & 3 \end{bmatrix}$ and the $\infty$-norm.

9. Let $A = \begin{bmatrix} 1 & 0 \\ 2 & 1 \end{bmatrix}.$

(a) Find the grid view of $L(x) = Ax$.
(b) Using the 1-norm, find the distance between $\begin{bmatrix} 1 \\ 1 \end{bmatrix}$ and $\begin{bmatrix} 2 \\ 2 \end{bmatrix}$, and the distance between $L\left(\begin{bmatrix} 1 \\ 1 \end{bmatrix}\right)$ and $L\left(\begin{bmatrix} 2 \\ 2 \end{bmatrix}\right)$.
(c) Compute $\|A\|_1$ and verify (5.3).
(d) Compute $c_1(A)$ and verify (5.4).

10. Let $A$ be an $n \times n$ matrix and $\|\cdot\|$ an induced matrix norm. Prove that
$$\|A^k\| \le \|A\|^k \text{ for all } k.$$

11. Let $A$ be an $n \times n$ matrix. Prove that
$$\frac{1}{n}\|A\|_F \le \|A\|_1 \le \sqrt{n}\,\|A\|_F$$
Also prove that
$$\frac{1}{n}\|A\|_F \le \|A\|_\infty \le \sqrt{n}\,\|A\|_F.$$

12. Prove using any induced matrix norm, that for nonsingular matrices,
    (a) $c(A) \geq 1$.           (b) $c(I) = 1$.
    (c) $c(A) = c(A^{-1})$.    (d) $c(AB) \leq c(A)c(B)$.

13. Let $A$ be a nonsingular matrix and suppose that $Ax = b$.

    (a) If $Ay = c$, show that

    $$\frac{\|x - y\|}{\|x\|} \leq c(A)\frac{\|b - c\|}{\|b\|}.$$

    (b) If $\hat{x}$ is the computed solution and $r = b - A\hat{x}$, show that

    $$\frac{\|\hat{x} - y\|}{\|x\|} \leq c(A)\frac{\|r\|}{\|b\|}.$$

    Explain what this means.

14. Find $x$ and $y$ in $R^2$ such that $\|x - y\|_2 > 100$ but $\frac{\|x-y\|_2}{\|y\|_2} < .01$.
    (Hint: It may help to work from drawings.)

15. Let $x = \begin{bmatrix} x_1 \\ x_2 \end{bmatrix}$, $y = \begin{bmatrix} y_1 \\ y_2 \end{bmatrix}$. If $\|x - y\|_\infty < 10^{-t}$, then $|x_i - y_i| <$
    $10^{-t}$ for $i = 1, 2$. If $\frac{\|x-y\|_\infty}{\|x\|_\infty} < 10^{-t}$, is it true that $\frac{|x_i-y_i|}{|x_i|} < 10^{-t}$ for
    all $i$?

16. Concerning relative error,

    (a) Suppose $\frac{|123.4-x|}{123.4} \leq 10^{-3}$ for some scalar $x$. Prove that $x$ differs
        from 123.4 in the 4-th digit of 123.4 (counting digits from the
        left).

    (b) Sometimes we say 'loosely' that 123.4 and $x$ agree in the first 3
        digits. To see why this can be wrong, note that

        $$\frac{|1.000 - .999|}{1} \leq 10^{-3}.$$

        Does .999 'differ' from 1.000 in the 4th digit? (Remember there
        are 2 representations of 1 in decimal notation, namely, 1 and
        .999...)

    (c) If

        $$\frac{\left\| \begin{bmatrix} 2 \\ 3 \end{bmatrix} - x \right\|}{\left\| \begin{bmatrix} 2 \\ 3 \end{bmatrix} \right\|} \leq 10^{-3},$$

        show that $\begin{bmatrix} 2 \\ 3 \end{bmatrix}$ and $x$ differ in the 4th digit of $\left\| \begin{bmatrix} 2 \\ 3 \end{bmatrix} \right\|$.

17. Let $A$ be a $2 \times 2$ matrix and $L(x) = Ax$. How should $w$ and $z$ be placed in $R^2$ so that

   (a) $\|L(w) - L(z)\|_2 = \|A\|_2 \|w - z\|_2$.

   (b) $\frac{\|L(w) - L(z)\|_2}{\|L(w)\|_2} = c_2(A) \frac{\|w - z\|_2}{\|w\|_2}$.

18. (Optional) Sometimes, in getting data for a mathematical model, we can only get 3 or 4 digits (with reasonable accuracy). Suppose the model is

$$Ax = b$$

where

$$A = \begin{bmatrix} 1 & 1 & 1 \\ 1 & 2 & 4 \\ 1 & 3 & 9 \end{bmatrix} \text{ and } b = \begin{bmatrix} 123 \\ 456 \\ 869 \end{bmatrix}.$$

In the following problems, use MATLAB.

   (a) Solve $Ax = b$.

   (b) Set

$$\hat{A} = \begin{bmatrix} 1.004 & 0.999 & 1.003 \\ 0.998 & 2.004 & 3.995 \\ 0.995 & 3.004 & 8.995 \end{bmatrix}.$$

   (Perhaps $\hat{A}$ is the actual data and $A$ the rounded data.) Solve $\hat{A}x = b$ and compare to the answer in (a).

   (c) If our mathematical model is a 3-digit approximation, should we accept the 15-digit answer given by MATLAB?

19. (MATLAB) Let $z = (-1, 1)^t$ and $y = (1, 1)^t$. Then $\|z - y\|_2 = 2$. Let $L(x) = \begin{bmatrix} 1 & 1 \\ 0 & \epsilon \end{bmatrix} x$ where $\epsilon = .4$. The grid view of $L$ is below. Note in Figure 5.9 that $\frac{\|L(z) - L(y)\|_2}{\|L(z)\|_2}$ increases as $\epsilon \to 0$. Since

$$\frac{\|L(z) - L(y)\|_2}{\|L(z)\|_2} \leq c_2(A) \frac{\|z - y\|_2}{\|z\|_2},$$

it follows that $c_2(A)$ is increasing as well. (That $c_2(A)$ is large and that $A$ is close to being singular are linked, as we will show in Chapter 7.)

Using the idea above, and MATLAB, find a $2 \times 2$ positive matrix with $c_2(A) > 1000$.

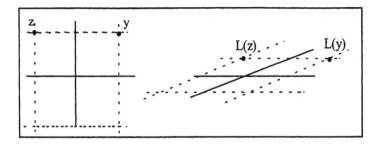

FIGURE 5.9.

## 5.3   Some Special Norms

In this section, we give two special matrix norms that are also useful. For the first of these, recall that the formula for the matrix norm induced from the vector 2-norm is a bit complicated to compute (by hand). However, when we need it (pulling $A$ out of the 2-norm), we can usually use the *Frobenius norm,*

$$\|A\|_F = \left( \sum_{i=1}^{m} \sum_{j=1}^{n} |a_{ij}|^2 \right)^{\frac{1}{2}}.$$

Note that $\|A\|_F = d_E(A, 0)$.

**Example 5.6** *Let* $A = \begin{bmatrix} -1 & 3 \\ 2 & -4 \end{bmatrix}$. *Then*

$$\|A\|_F = \left( |-1|^2 + |3|^2 + |2|^2 + |-4|^2 \right)^{\frac{1}{2}}$$
$$= (30)^{\frac{1}{2}} = 5.48 \text{ rounded to 3 digits.}$$

The proof that this is a norm is exactly the same as that of the vector 2-norm.

We link the Frobenius norm to the induced matrix norm $\|\cdot\|_2$ as follows.

**Theorem 5.6** *Let* $A$ *be an* $m \times n$ *matrix. Then*

*(a)* $\|Ax\|_2 \le \|A\|_F \|x\|_2.$

*(b)* $\|A\|_2 \le \|A\|_F.$

*Further, if* $B$ *is an* $n \times r$ *matrix, then*

*(c)* $\|AB\|_F \le \|A\|_F \|B\|_F.$

**Proof.** We prove (a) leaving (b) and (c) as exercises. By direct calculation,

$$\|Ax\|_2 = \left\| \left[ \sum_{k=1}^{n} a_{ik}x_k \right] \right\|_2 = \left( \sum_{i=1}^{m} \left| \sum_{k=1}^{n} a_{ik}x_k \right|^2 \right)^{\frac{1}{2}}$$

and by the Cauchy-Schwarz inequality, applied to $\sum_{k=1}^{n} a_{ik}x_k$,

$$\leq \left( \sum_{i=1}^{m} \left( \sum_{k=1}^{n} |a_{ik}|^2 \right) \left( \sum_{k=1}^{n} |x_k|^2 \right) \right)^{\frac{1}{2}} = \left( \sum_{i=1}^{m} \sum_{k=1}^{n} |a_{ik}|^2 \|x\|_2^2 \right)^{\frac{1}{2}}$$

$$\leq \left( \sum_{i=1}^{m} \sum_{k=1}^{n} |a_{ik}|^2 \right)^{\frac{1}{2}} \|x\|_2 \leq \|A\|_F \|x\|_2$$

which verifies (a). ∎

Concluding, the Frobenius norm is not an induced matrix norm for any vector norm. For square matrices, all such norms have the property that $\|I\| = 1$; however, $\|I\|_F = n^{\frac{1}{2}}$. Still, it is easy to calculate and useful.

The second norm we give is another induced matrix norm and also turns out to very useful. To define it, let $R$ be an $n \times n$ nonsingular matrix. For any vector norm $\|\cdot\|$ on Euclidean $n$-space, define the vector $R$-norm as

$$\|x\|_R = \|Rx\|.$$

The proof that $\|\cdot\|_R$ is a vector norm is left as an exercise.

**Example 5.7** Let $R = \begin{bmatrix} 1 & 1 \\ 0 & 2 \end{bmatrix}$. *Using the vector 1-norm,*

$$\|x\|_R = \|Rx\|_1 = \left\| \begin{bmatrix} x_1 + x_2 \\ 2x_2 \end{bmatrix} \right\|_1 = |x_1 + x_2| + |2x_2|.$$

*So,*

$$\left\| \begin{bmatrix} 1 \\ 2 \end{bmatrix} \right\|_R = |1 + 2| + |2 \cdot 2| = 7.$$

Now the matrix norm induced by vector norm $\|\cdot\|_R$ is, by definition,

$$\|A\|_R = \max_{x \neq 0} \frac{\|Ax\|_R}{\|x\|_R}.$$

This matrix norm can be computed from the matrix norm induced by $\|\cdot\|$.

**Theorem 5.7** *If $\|\cdot\|$ is a vector norm and $\|\cdot\|_R$ the corresponding vector R-norm, then the induced matrix norm satisfies*

$$\|A\|_R = \|RAR^{-1}\|$$

*for any $n \times n$ matrix $A$.*

**Proof.** By definition,

$$\|A\|_R = \max_{x \neq 0} \frac{\|Ax\|_R}{\|x\|_R}$$

$$= \max_{x \neq 0} \frac{\|RAx\|}{\|Rx\|}.$$

Setting $y = Rx$ and noting that $\|RAx\| = \|RAR^{-1}y\|$, we have

$$\|A\|_R = \max_{y \neq 0} \frac{\|RAR^{-1}y\|}{\|y\|}$$

$$= \|RAR^{-1}\|,$$

the desired result. ∎

An example follows.

**Example 5.8** *Given the $\infty$-vector norm and $R = \begin{bmatrix} 1 & 0 \\ 2 & 1 \end{bmatrix}$, then*

$$\|A\|_R = \|RAR^{-1}\|_\infty.$$

*Thus, if $A = \begin{bmatrix} 1 & -1 \\ 0 & 2 \end{bmatrix}$, then*

$$\|A\|_R = \left\| \begin{bmatrix} 1 & 0 \\ 2 & 1 \end{bmatrix} \begin{bmatrix} 1 & -1 \\ 0 & 2 \end{bmatrix} \begin{bmatrix} 1 & 0 \\ -2 & 1 \end{bmatrix} \right\|_\infty$$

$$= \left\| \begin{bmatrix} 3 & -1 \\ 2 & 0 \end{bmatrix} \right\|_\infty = 4.$$

To conclude this section, we show how norms can be used to bound eigenvalues.

For any $n \times n$ matrix $A$, we define the spectral radius of $A$, denoted $\rho(A)$, by

$$\rho(A) = \max |\lambda|$$

where the maximum is over all eigenvalues $\lambda$ of $A$.

As shown below, any induced matrix norm bounds the *spectral radius*.

**Theorem 5.8** *Let A be an $n \times n$ matrix and $\|\cdot\|$ any vector norm. Then*

$$\rho(A) \le \|A\|$$

*where $\|A\|$ is the induced matrix norm of A.*

**Proof.** Let $\lambda$ be an eigenvalue of $A$. Then

$$Ax = \lambda x$$

for some eigenvector $x$. Thus,

$$\|\lambda x\| = \|Ax\|$$

and by using the properties,

$$|\lambda| \, \|x\| \le \|A\| \, \|x\| \, .$$

And since $x \ne 0$,

$$|\lambda| \le \|A\| \, .$$

Thus, since $\lambda$ was arbitrary,

$$\rho(A) \le \|A\| \, .$$

The result follows. ∎

We demonstate the theorem with an example.

**Example 5.9** *Let $A = \begin{bmatrix} 1 & 1 \\ -1 & 1 \end{bmatrix}$. Then the eigenvalues of A are $1 + i$ and $1 - i$. Further, $\|A\|_1 = \|A\|_\infty = 2$, while $\|A\|_2 = 1.4142$. So $\|A\|_2$ provides the best estimate of $\rho(A)$ here. (See Figure 5.10)*

Actually, as can be seen in Figure 5.9, for a given $n \times n$ matrix, and any positive scalar $\epsilon$, there is 'some' norm equal to, or close to, the spectral radius.

**Theorem 5.9** *Let A be an $n \times n$ matrix and $\epsilon$ a positive scalar. Then there is an induced matrix norm $\|\cdot\|$ such that*

$$\rho(A) \le \|A\| \le \rho(A) + \varepsilon.$$

**Proof.** The first inequality has already been shown, so we need only argue the second. By Theorem 3.7, there is a nonsingular matrix $P$ such that

$$P^{-1}AP = J_\varepsilon$$

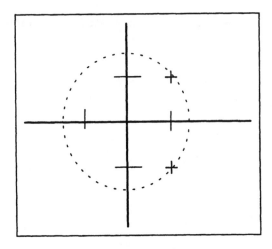

FIGURE 5.10.

where $J_\varepsilon$ is the Jordan form of $A$ with the 1's on the superdiagonal replaced by $\varepsilon$. Now,

$$\begin{aligned}
\|A\|_{P^{-1}} &= \|P^{-1}AP\|_\infty \\
&= \|J_\varepsilon\|_\infty \\
&\leq \rho(A) + \varepsilon,
\end{aligned}$$

the required bound. ∎

A special case of the theorem follows.

**Corollary 5.1** *Let $A$ be an $n \times n$ matrix where all eigenvalues $\lambda$, $|\lambda| = \rho(A)$, are on $1 \times 1$ Jordan blocks of the Jordan form of $A$. Then there exists an induced matrix norm $\|\cdot\|$ such that*

$$\|A\| = \rho(A).$$

**Proof.** Note that in the proof of Theorem 3.7, $\epsilon$'s in $J_\epsilon$ only occur in rows containing eigenvalues $\lambda$ where $|\lambda| < \rho(A)$. Thus $\epsilon$ can be chosen sufficiently small such that $\|J_\epsilon\|_\infty = \rho(A)$. Hence $\|A\|_{P^{-1}} = \rho(A)$. ∎

And, as a consequence, we have a norm proof of the following.

**Corollary 5.2** *If $A$ is an $n \times n$ matrix and $\rho(A) < 1$, then $\lim_{k \to \infty} A^k = 0$.*

**Proof.** Choose $\varepsilon > 0$ such that $\rho(A) + \varepsilon < 1$. Then, using the theorem, let $\|\cdot\|$ be an induced matrix norm such that $\|A\| \leq \rho(A) + \varepsilon$.
Since

$$\|A^k\| \leq \|A\|^k,$$

for all $k$, $\lim\limits_{k \to \infty} \|A^k\| = 0$. Thus, by the equivalence of norms, we have that $\lim\limits_{k \to \infty} A^k = 0$. ∎

One of the important uses for norms is that they provide rates of convergence for sequences of vectors. For example, consider the sequence generated by

$$x(k+1) = Ax(k) + b \tag{5.7}$$

where $A$ is an $n \times n$ matrix and $b$ an $n \times 1$ vector. We know that if $\|\cdot\|$ is an induced matrix norm and $\|A\| < 1$, then $\rho(A) < 1$, so by Neumann's formula $x(0), x(1), \ldots$ converges to, say, $x$. Thus, calculating the limit of the sides of (5.7), we get

$$x = Ax + b. \tag{5.8}$$

Subtracting (5.8) from (5.7), we get

$$x(k+1) - x = A(x(k) - x).$$

Thus,

$$\|x(k+1) - x\| \leq \|A\| \, \|x(k) - x\|.$$

(So, if, for example $\|A\| = .9$, then $\|x(k+1) - x\|$ is no more than .9 of $\|x(k) - x\|$, the previous calculation.) And, so the convergence rate of $x(0), x(1), \ldots$ to $x$ is, using the norm, $\|A\|$ per iterate or, overall, using this norm, the convergence rate is $\|A\|^k$ after $k$ iterations. Note that by Theorem 5.9 and Corollary 5.1, we can get this rate to be either $\rho(A)$ or slightly more.

## 5.3.1   Optional (Splitting Techniques)

In this optional, we see how to convert the problem of solving a system of linear equations into a difference equation.

To see this, let $A$ be an $n \times n$ matrix and $b$ an $n \times 1$ vector. To solve

$$Ax = b \tag{5.9}$$

we can use the direct method of Gaussian elimination with partial pivoting. Of course, due to round off, the numerical answer, say, $\hat{x}$, may be incorrect in the last few digits, an perhaps more, depending on $c(A)$.

Another way to solve (5.9) is by converting it into a difference equation. To do this, let $D = \text{diag}(a_{11}, \ldots, a_{nn})$ and $B = A - D$. (Thus $A$ is split into $D$ and $B$.) Then (5.9) becomes

$$(B + D)x = b$$

or by rearranging

$$Dx = -Bx + b.$$

If $D$ is nonsingular,

$$x = -D^{-1}Bx + D^{-1}b.$$

The difference equation arises by setting $x(0)$ as some constant vector and inductively defining $x(1), x(2), \dots$ by

$$x(k+1) = -D^{-1}Bx(k) + D^{-1}b. \qquad (5.10)$$

Now, if $A$ is diagonally dominant $(|a_{ii}| > |a_{i1}| + \cdots + |a_{i,i-1}| + |a_{i,i+1}| + \cdots + |a_{in}|$ for all $i$, a situation which often occurs when numerically solving differential and partial differential equations), then $\|D^{-1}B\|_\infty < 1$. Thus, the sequence converges to, say, $x$. Calculating the limit of the sides of (5.10), we get

$$x = -D^{-1}Bx + D^{-1}b.$$

Rearranging, we have

$$Ax = b.$$

So $x$ is the solution to (5.9).

Note that

$$\|x(k+1) - x\| \le \|A\| \|x(k) - x\|.$$

Thus, if a numerically computed $\hat{x}(k)$ is in error, then the next $\hat{x}(k+1) = A\hat{x}(k) + b$ satisfies

$$\|\hat{x}(k+1) - x\| \le \|A\| \|\hat{x}(k) - x\|$$

and so the difference $\|\hat{x}(k+1) - x\|$ is no more than $\|A\| \|\hat{x}(k) - x\|$. Thus, even if small errors are made in computing the iterates, $x$ continues to convergence.

Another feature about iterative methods is that we can continue to compute $x(k)$'s until we have the desired number of digits of accuracy. To help understand this remark, we provide an example.

**Example 5.10** *Let* $A = \begin{bmatrix} 3 & 2 \\ 1 & 4 \end{bmatrix}$ *and* $b = \begin{bmatrix} 1 \\ 2 \end{bmatrix}$. *We solve* $Ax = b$ *by splitting.*

*Here,* $D = \begin{bmatrix} 3 & 0 \\ 0 & 4 \end{bmatrix}$ *and* $B = \begin{bmatrix} 0 & 2 \\ 1 & 0 \end{bmatrix}$. *So* $D^{-1}B = \begin{bmatrix} 0 & \frac{2}{3} \\ \frac{1}{4} & 0 \end{bmatrix}$ *and* $D^{-1}b = \begin{bmatrix} \frac{1}{3} \\ \frac{2}{4} \end{bmatrix}$. *Since,* $\rho(D^{-1}B) = .4082$

$$x(k+1) = -D^{-1}Bx(k) + D^{-1}b$$

*converges at the rate of* $(.4082)^k$ *to the solution to* $Ax = b$.

*Starting with* $x(0) = \begin{bmatrix} 1 \\ 1 \end{bmatrix}$, *we have*

| $x(1)$ | $x(2)$ | $x(3)$ |
|---|---|---|
| $\begin{bmatrix} -0.3333 \\ 0.2500 \end{bmatrix}$ | $\begin{bmatrix} 0.1667 \\ 0.5833 \end{bmatrix}$ | $\begin{bmatrix} 0.0556 \\ 0.4583 \end{bmatrix}$ |

| $x(4)$ | $x(5)$ | $x(6)$ |
|---|---|---|
| $\begin{bmatrix} 0.0278 \\ 0.5139 \end{bmatrix}$ | $\begin{bmatrix} -0.0093 \\ 0.4931 \end{bmatrix}$ | $\begin{bmatrix} 0.0046 \\ 0.5023 \end{bmatrix}$ |

| $x(7)$ | $x(8)$ | $x(9)$ |
|---|---|---|
| $\begin{bmatrix} -0.0015 \\ 0.4988 \end{bmatrix}$ | $\begin{bmatrix} 0.0008 \\ 0.5004 \end{bmatrix}$ | $\begin{bmatrix} -0.0003 \\ 0.5001 \end{bmatrix}$ |

| $x(10)$ | $x(11)$ |
|---|---|
| $\begin{bmatrix} 0.0001 \\ 0.5001 \end{bmatrix}$ | $\begin{bmatrix} 0.0000 \\ 0.5000 \end{bmatrix}$ |

*And, after the 11-th iterate, all entries have the same digits and so* $x = \begin{bmatrix} 0.0000 \\ 0.5000 \end{bmatrix}$ *as can be checked.*

*On larger problems, it may be that the last digit doesn't converge, e.g., when the amount of convergence and the amount of error balance. In that case, we take as the approximate solution the answer to those digits that do agree.*

There are many well-known splitting methods. For example, the point Jacobi and the Gauss-Seidel methods are two of the better known splitting methods.

## 5.3.2  MATLAB (Code for Iterative Solutions)

The MATLAB codes for the calculations in this section follow.

### 1. Code for Example 5.10

```
b = [1/3; 1/2];
B = [0 - 2/3; -1/4 0];
x = [1; 1];
for k = 1 : 12
    x = B * x + b
end
```

## 2. A More General Code for Iterative Solutions

```
y = [1; 1]
x = [0; 0] ;                              % This is x (0).
c = 1;                                    % Used as a counter.
while (norm(x − y)/norm(x)              % Continues until 14 digit
                    > 10 ∧ (−14))           approximation is reached.
    y = x;                               % This keeps x (k).
    x = B * x + b                        % This calculates x (k + 1).
    c = c + 1                            % Updates the counter.
        if c > 10000
            break
        end                             % If we can't get a 14
                                            digit approximation, we
                                            need to stop somewhere.

end
```

# Exercises

1. Find the Frobenius norm for each matrix.

   (a) $A = \begin{bmatrix} 2 & -1 & 0 \\ -2 & 3 & 4 \\ 0 & 2 & 1 \end{bmatrix}$

   (b) $A = \begin{bmatrix} i & 2-i \\ 1+2i & 2 \end{bmatrix}$

   (c) $A = \begin{bmatrix} 1-i \\ 2+3i \end{bmatrix}$

2. Prove Theorem 5.6.

   (a) Part b. (Hint: Use Part (a) and consider the ratios $\frac{\|Ax\|_2}{\|x\|_2}$.)

   (b) Part c.

3. Let $R$ be a nonsingular matrix. Prove that for any vector norm $\|\cdot\|$, $\|\cdot\|_R$ is a vector norm.

4. Find $\|x\|_R$ for

   (a) $x = \begin{bmatrix} 1 \\ -1 \end{bmatrix}$, $R = \begin{bmatrix} 1 & 1 \\ 0 & 1 \end{bmatrix}$ and the vector 1-norm.

   (b) $x = \begin{bmatrix} 2 \\ -3 \end{bmatrix}$, $R = \begin{bmatrix} 1 & 3 \\ 0 & 1 \end{bmatrix}$, and the vector ∞-norm.

5. Find $\|A\|_R$ for

(a) $A = \begin{bmatrix} 1 & 1 \\ 1 & 1 \end{bmatrix}$, $R = \begin{bmatrix} 1 & 2 \\ 0 & 1 \end{bmatrix}$, and the vector 1-norm.

(b) $A = \begin{bmatrix} 1 & 2 \\ 1 & 2 \end{bmatrix}$, $R = \begin{bmatrix} 1 & -1 \\ 1 & 1 \end{bmatrix}$, and the vector $\infty$-norm.

6. Let $A = \begin{bmatrix} 1 & 3 \\ 2 & 2 \end{bmatrix}$.

(a) Find the eigenvalues of $A$. Plot these in $R^2$.

(b) Using the 1-norm and the $\infty$-norms, find bounds on the eigenvalues. Draw circles of radii $\|A\|_1$ and $\|A\|_\infty$ about the origin, and observe that these circles contain the eigenvalues of $A$. What norm gives the better result?

7. Let $A = \begin{bmatrix} 1 & t \\ 0 & 1 \end{bmatrix}$. Find $\rho(A)$, $\|A\|_1$, and $\lim_{t\to\infty} (\|A\|_1 - \rho(A))$.

(a) Is the bound $\rho(A) \le \|A\|_1$ always a good one?

(b) Let $R = \begin{bmatrix} 1 & 0 \\ 0 & t^2 \end{bmatrix}$. Calculate, using the 1-norm to define $\|\cdot\|_R$, $\lim_{t\to\infty} (\|A\|_R - \rho(A))$.

(c) Can you improve on the bound $\|A\|_R$ of (b) by using $R = \begin{bmatrix} 1 & 0 \\ 0 & t \end{bmatrix}$?

8. Give an example of a $2 \times 2$ matrix $A$ such that $\rho(A) = 1$ and

(a) $\lim_{l\to\infty} A^k$ exists.

(b) $\lim_{l\to\infty} A^k$ doesn't exist.

9. Let $A$ be an $n \times n$ matrix. Prove

(a) $\frac{1}{n}\|A\|_\infty \le \|A\|_F \le n\|A\|_\infty$.

(b) $\frac{1}{n}\|A\|_1 \le \|A\|_F \le n\|A\|_1$.

(c) $\|e^{At}\| \le e^{\|A\|t}$ for any induced matrix norm $\|\cdot\|$.

10. For the following difference equations, find $x$ such that $x(1), x(2), \dots$ converges to $x$. What is the rate of convergence, using $\|\cdot\|_1$ and $\|\cdot\|_\infty$, for the following?

(a) $x(k+1) = \begin{bmatrix} .1 & .2 \\ .2 & .1 \end{bmatrix} x(k) + \begin{bmatrix} 1 \\ 1 \end{bmatrix}$

$$(b) \ x(k+1) = \begin{bmatrix} .6 & .1 & 0 \\ 0 & .5 & 1 \\ 0 & 0 & .4 \end{bmatrix} x(k) + \begin{bmatrix} 1 \\ 2 \\ 1 \end{bmatrix}$$

11. Let $D = \begin{bmatrix} 1 & 0 \\ 0 & t \end{bmatrix}$. Find a $t$ so that $\|A\|_{D^{-1}} = \|D^{-1}AD\|_\infty < 1$.

(a) $A = \begin{bmatrix} .9 & 1 \\ 0 & .9 \end{bmatrix}$   (b) $A \begin{bmatrix} .7 & .3 \\ .1 & .7 \end{bmatrix}$

12. (MATLAB) Consider the school problem diagrammed in Figure 5.11. Provide the mathematical model, $x(k+1) = Ax(k) + b$, for this process.

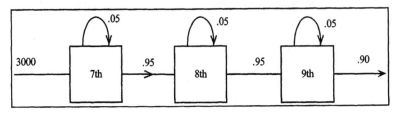

FIGURE 5.11.

(a) Provide the mathematical model, $x(k+1) = Ax(k)+b$, for this process.

(b) Find the steady state vector $x$, $x = \lim_{k \to \infty} x(k)$, and explain what the entries in this vector mean.

(c) Using the convergence rate, and $x(0) = 0$, find the smallest $k$ such that $\|x - x(k)\|_\infty \le (.1)\|x\|_\infty$, so the process is within 10% of steady state.

13. (MATLAB) Consider $x(k+1) = Ax(k)$ where

$$A = \begin{bmatrix} 0 & 0 & .99 \\ 0 & 0 & .99 \\ .99 & 0 & 0 \end{bmatrix}.$$

Find
(a) $\|A\|_1$.   (b) $\|A\|_\infty$.
(c) $\|A\|_2$.

Which of these norms gives the convergence at the best rate?

14. (MATLAB) Solve $\begin{bmatrix} 3 & 1 & -1 \\ -1 & 4 & 2 \\ 0 & 2 & 5 \end{bmatrix} x = \begin{bmatrix} 1 \\ 2 \\ 3 \end{bmatrix}$ by splitting as described in Optional.

15. (MATLAB) Find, by tinkering, a matrix $A$ such that $\|A\|_1 < \|A\|_2$.

## 5.4  Inner Product Norms and Orthogonality

In this section we define a dot-product, called an inner product in this setting, on vector spaces. We show how to use inner products to define norms. As with dot products, inner products can also be used to define orthogonality, and orthogonality can be used as it was in calculus.

To define an inner product, we use the properties of a dot product in Euclidean 2-space. Recall that for vectors $x$ and $y$ in this space,

$$x \cdot y = x_1\bar{y}_1 + x_2\bar{y}_2,$$

or perhaps in different notation

$$(x,y) = x_1\bar{y}_1 + x_2\bar{y}_2.$$

**Definition 5.3** *Let $V$ be a vector space. Suppose that there is a way of assigning to each pair of vectors $x$ and $y$ in $V$ a scalar, written $(x,y)$. This function,$(\cdot,\cdot)$, is an inner product on $V$ provided it satisfies the following properties for all vectors and all scalars.*

*i. $(x,x) > 0$ if $x \neq 0$ and $(0,0) = 0$*

*ii. $(x,y) = \overline{(y,x)}$*

*iii. $(\alpha x, y) = \alpha(x,y)$ and $(x, \alpha y) = \bar{\alpha}(x,y)$*

*iv. $(x, y+z) = (x,y) + (x,z)$ and $(x+y, z) = (x,z) + (y,z)$*

*A vector space $V$ that has an inner product defined on it is called an inner product space.*

Actually the second properties in (iii) and (iv) can be proved from the remaining given properties. We include them since they are companions to the first properties.

Some classical inner products and inner product spaces follow.

**Example 5.11** *On Euclidean n-space, an inner product is*

$$(x,y) = \sum_{i=1}^{n} x_i\bar{y}_i. \text{ (In the real case, } \bar{y}_i = y_i.) \qquad (5.11)$$

**Example 5.12** *On $m \times n$ matrix space, an inner product is*

$$(A,B) = \sum_{i=1}^{m}\sum_{j=1}^{n} a_{ij}\bar{b}_{ij}. \qquad (5.12)$$

**Example 5.13** *On $C[a, b]$, define*

$$(f, g) = \int_a^b f(t) g(t) \, dt. \tag{5.13}$$

*This is an inner product.*

Note that all of these inner products arise, like the dot product, by multiplying corresponding entries (second entries conjugated) and summing those products.

As with dot products, an inner product can be used to determine lengths of vectors, that is,

$$\|x\| = (x, x)^{\frac{1}{2}}.$$

And, mimicking the proof for the vector 2-norm, we can show that $\|\cdot\|$ is a norm.

The following example will show how to use this norm in calculating.

**Example 5.14** *Let $A = \begin{bmatrix} 1 & -1 \\ 0 & 1 \end{bmatrix}$. Then, using the inner product of Example 5.12,*

$$\|A\| = (A, A)^{\frac{1}{2}}$$
$$= (1 \cdot 1 + (-1)(-1) + 0 \cdot 0 + 1 \cdot 1)^{\frac{1}{2}}$$
$$= \sqrt{3}.$$

*And if $B = \begin{bmatrix} 2 & 1 \\ 3 & -2 \end{bmatrix}$,*

$$d(A, B) = \|A - B\| = \left\| \begin{bmatrix} -1 & -2 \\ -3 & 3 \end{bmatrix} \right\|$$
$$= \left( \begin{bmatrix} -1 & -2 \\ -3 & 3 \end{bmatrix}, \begin{bmatrix} -1 & -2 \\ -3 & 3 \end{bmatrix} \right)^{\frac{1}{2}}$$
$$= \left( (-1)^2 + (-2)^2 + (-3)^2 + 3^2 \right)^{\frac{1}{2}}$$
$$= \sqrt{23}.$$

Orthogonality of two vectors $x$ and $y$ is defined in the natural way, namely, if

$$(x, y) = 0$$

then $x$ and $y$ are *orthogonal*. (Note that if $(x, y) = 0$, then $(y, x) = 0$, so the order of $x$ and $y$ isn't important.)

Using orthogonality, we have the Pythagorean theorem for any inner product space.

**Lemma 5.1** *Let $V$ be an inner product space and $x, y \in V$. If $(x, y) = 0$, then $\|x + y\|^2 = \|x\|^2 + \|y\|^2$, as depicted in Figure 5.12.*

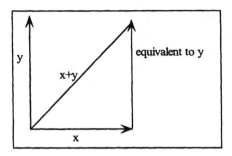

FIGURE 5.12.

**Proof.** Since

$$\|x + y\|^2 = (x + y, x + y)$$
$$= \|x\|^2 + (x, y) + (y, x) + \|y\|^2$$
$$= \|x\|^2 + \|y\|^2,$$

we have the result. ∎

A major use of orthogonality is in calculating coefficients of linear combinations of pair-wise orthogonal nonzero vectors. An example of how follows.

**Lemma 5.2** *Let $q_1, \ldots, q_n$ be pairwise orthogonal nonzero vectors in Euclidean $n$-space. Then $q_1, \ldots, q_n$ are linearly independent.*

**Proof.** Let $(\beta_1, \ldots, \beta_n)$ be a solution to the pendent equation for the vectors $q_1, \ldots, q_n$. Thus,

$$\beta_1 q_1 + \cdots + \beta_n q_n = 0.$$

Now,

$$(q_k, \beta_1 q_1) + \cdots + (q_k, \beta_n q_n) = (q_k, 0).$$

As given in the exercises, $(q_k, 0) = 0$, so we have

$$\beta_k (q_k, q_k) = 0.$$

Since $(q_k, q_k) > 0$, $\beta_k = 0$. And, since $k$ was arbitrary, $\beta_1 = \cdots = \beta_n = 0$. Thus $q_1, \ldots, q_n$ are linearly independent. ∎

Any set of pair-wise orthogonal vectors is called an *orthogonal set*; while if, in addition, each vector has length, or norm, 1, the set is called an *orthonormal set*.

Orthogonality is often used in closest point (approximation) work. To show how, we provide a brief review of calculus.

In calculus, we use dot products to find the component of a vector $x$ on a vector $u$. (See Figure 5.13.) If $c$ is that component, as given in calculus, $c = \frac{x \circ u}{u \circ u}$. Observe that $x - cu$ is orthogonal to $u$ and that $cu$ is the closest-

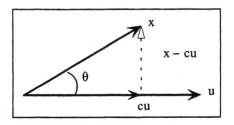

FIGURE 5.13.

point in span $\{u\}$ to $x$.

Using components, if $u_1$ and $u_2$ are orthogonal in $R^3$ and $x \in R^3$, then $x - c_1 u_1 - c_2 u_2$ is orthogonal to $u_1$ and $u_2$; it is orthogonal to span $\{u_1, u_2\}$. Here $c_1 = \frac{x \circ u_1}{u_1 \circ u_1}$ and $c_2 = \frac{x \circ u_2}{u_2 \circ u_2}$ are the components of $x$ on $u_1$ and $u_2$, respectively. Thus, $c_1 u_1 + c_2 u_2$ is the closest point in span $\{u_1, u_2\}$ to $x$. (Perhaps making a small 3-D model, Figure 5.14, will help.)

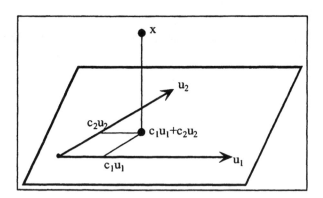

FIGURE 5.14.

We now give this result for any inner product space.

**Theorem 5.10** *Let $V$ be an inner product space and $u_1, \ldots, u_m$ pairwise orthogonal nonzero vectors in $V$. Let $x \in V$ and from it define the corre-*

*sponding Fourier sum, using* $u_1, \ldots, u_m$, *as*

$$x_f = c_1 u_1 + \cdots + c_m u_m$$

*where the component $c_k$ of $x$ on $u_k$ is $c_k = \frac{(x, u_k)}{(u_k, u_k)}$. Then $x - x_f$ is orthogonal to each of $u_1, \ldots, u_m$. (See Figure 5.15.)*

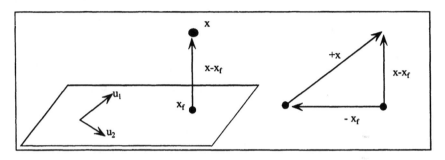

FIGURE 5.15.

**Proof.** To prove that $x - x_f$ is orthogonal to $u_k$, show by direct calculation that $(x - x_f, u_k) = 0$. ∎

We now apply this theorem to find orthogonal vectors $u_1, u_2, \ldots, u_m$, which span the same space as given linearly independent vectors $x_1, \ldots, x_m$. To do this, set

$$u_1 = x_1$$
$$u_2 = x_2 - x_f = x_2 - c_1 u_1$$
$$\text{where} \quad c_1 = \frac{(x_2, u_1)}{(u_1, u_1)}$$
$$u_3 = x_3 - x_f = x_3 - c_1 u_1 - c_2 u_2$$
$$\text{where } c_1 = \frac{(x_3, u_1)}{(u_1, u_1)}, \ c_2 = \frac{(x_3, u_2)}{(u_2, u_2)}.$$

In general

$$u_k = x_k - x_f$$
$$= x_k - c_1 u_1 - c_2 u_2 - \cdots - c_{k-1} u_{k-1}$$

where $c_1 = \frac{(x_k, u_1)}{(u_1, u_1)}, \ldots, c_{k-1} = \frac{(x_k, u_{k-1})}{(u_{k-1}, u_{k-1})}$. (So we can compute all $u_k$'s by formula.)

That span $\{u_1, \ldots, u_m\} = \{x_1, \ldots, x_m\}$ is left as an exercise.

As you may recall from linear algebra, this method is called the *Gram-Schmidt process*. An example demonstrating the Gram-Schmidt process follows.

**Example 5.15** *Let* $x_1 = \begin{bmatrix} 1 \\ 1 \\ 1 \end{bmatrix}$, $x_2 = \begin{bmatrix} 1 \\ 1 \\ 0 \end{bmatrix}$, *and* $x_3 = \begin{bmatrix} 1 \\ 0 \\ 0 \end{bmatrix}$. *Then*

*i.* $u_1 = x_1 = \begin{bmatrix} 1 \\ 1 \\ 1 \end{bmatrix}$.

*ii. For* $u_2$ *we need to calculate the corresponding* $x_f$. *Here*

$$x_f = c_1 u_1 = \frac{(x_2, u_1)}{(u_1, u_1)} u_1 = \frac{2}{3} u_1$$

*and so*

$$u_2 = x_2 - x_f = x_2 - \frac{2}{3} u_1 = \begin{bmatrix} \frac{1}{3} \\ \frac{1}{3} \\ -\frac{2}{3} \end{bmatrix}. \quad \textit{(We can visually check to see}$$

*if* $(u_1, u_2) = 0$.)

*iii. To get* $u_3$, *we need to find the corresponding* $x_f$. *So*

$$x_f = c_1 u_1 + c_2 u_2 = \frac{(x_3, u_1)}{(u_1, u_1)} u_1 + \frac{(x_3, u_2)}{(u_2, u_2)} u_2$$

$$= \frac{1}{3} u_1 + \frac{1}{2} u_2 = \begin{bmatrix} \frac{1}{2} \\ \frac{1}{2} \\ 0 \end{bmatrix}.$$

*Hence* $u_3 = x_3 - x_f = x_3 - \frac{1}{3} u_1 - \frac{1}{2} u_2 = \begin{bmatrix} \frac{1}{2} \\ -\frac{1}{2} \\ 0 \end{bmatrix}$. *Observe that* $u_3$

*is orthogonal to* $u_1$ *and* $u_2$.

If $A = [x \ldots x_n]$, then applying the Gram-Schmidt process to the columns of $A$ leads to an important factorization of $A$, called the $QR$-factorization, namely

$$A = QR \tag{5.14}$$

where the columns of $Q$ form an orthonormal set and $R$ is an upper triangular matrix. We show how this can be done in an example.

**Example 5.16** *Let* $A = [x_1 x_2 x_3]$ *where the* $x_i$*'s are given in the previous example. We use the calculations of the previous example,*

$$u_1 = x_1$$

$$u_2 = x_2 - \frac{2}{3}u_1$$

$$u_3 = x_3 - \frac{1}{3}u_1 - \frac{1}{2}u_2$$

*Solving for the* $x_k$*'s yields*

$$u_1 = x_1$$

$$\frac{2}{3}u_1 + u_2 = x_2$$

$$\frac{1}{3}u_1 + \frac{1}{2}u_2 + u_3 = x_3.$$

*Writing these equations as a matrix equation, using backward multiplication on columns, yields*

$$[u_1 u_2 u_3] \begin{bmatrix} 1 & \frac{2}{3} & \frac{1}{3} \\ 0 & 1 & \frac{1}{2} \\ 0 & 0 & 1 \end{bmatrix} = [x_1 x_2 x_3].$$

*Now, normalizing the* $u_i$*'s, we have on the left side*

$$[u_1 u_2 u_3] \begin{bmatrix} \frac{1}{\|u_1\|} & 0 & 0 \\ 0 & \frac{1}{\|u_2\|} & 0 \\ 0 & 0 & \frac{1}{\|u_3\|} \end{bmatrix} \begin{bmatrix} \|u_1\| & 0 & 0 \\ 0 & \|u_2\| & 0 \\ 0 & 0 & \|u_3\| \end{bmatrix} \begin{bmatrix} 1 & \frac{2}{3} & \frac{1}{3} \\ 0 & 1 & \frac{1}{2} \\ 0 & 0 & 1 \end{bmatrix}$$

*or*

$$\begin{bmatrix} \frac{u_1}{\|u_1\|} & \frac{u_2}{\|u_2\|} & \frac{u_3}{\|u_3\|} \end{bmatrix} \begin{bmatrix} \|u_1\| & \frac{2}{3}\|u_1\| & \frac{1}{3}\|u_1\| \\ 0 & \|u_2\| & \frac{1}{2}\|u_2\| \\ 0 & 0 & \|u_3\| \end{bmatrix} = [x_1 x_2 x_3].$$

*Finally, plugging in the numbers, we get the QR-factorization,*

$$\begin{bmatrix} \frac{1}{\sqrt{3}} & \frac{1}{\sqrt{6}} & \frac{1}{\sqrt{2}} \\ \frac{1}{\sqrt{3}} & \frac{1}{\sqrt{6}} & -\frac{1}{\sqrt{2}} \\ \frac{1}{\sqrt{3}} & \frac{-2}{\sqrt{6}} & 0 \end{bmatrix} \begin{bmatrix} \sqrt{3} & \frac{2}{3}\sqrt{3} & \frac{1}{3}\sqrt{3} \\ 0 & \frac{\sqrt{6}}{3} & \frac{1}{3}\sqrt{6} \\ 0 & 0 & \frac{\sqrt{2}}{2} \end{bmatrix} = \begin{bmatrix} 1 & 1 & 1 \\ 1 & 1 & 0 \\ 1 & 0 & 0 \end{bmatrix}.$$

*(The method of factoring* $A = QR$ *given here is not good in numerical computations involving rounding. In Chapter 8, we give a good method.)*

Using components, we now give the closest point (approximation) theorem.

**Theorem 5.11** *Let $V$ be an inner product space and $u_1, \ldots, u_m$ pairwise orthogonal vectors in $V$. Let $x \in V$. Then the closest vector in the subspace $W = \text{span}\{u_1, \ldots, u_m\}$ to $x$ is precisely $x_f$, the Fourier sum using $u_1, \ldots, u_m$.*

**Proof.** Let $v \in W$. Since $x_f \in W$, so is $x_f - v$. (See Figure 5.16.) Thus, we can write

$$x_f - v = \alpha_1 u_1 + \cdots + \alpha_m u_m$$

for some scalars $\alpha_1, \ldots, \alpha_m$. Since, by the lemma, $x - x_f$ is orthogonal to

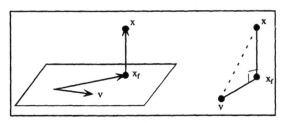

FIGURE 5.16.

$u_1, \ldots, u_m$, by direct calculation we can show it is orthogonal to $x_f - v$. Thus by the Pythagorean Theorem,

$$\|x - v\|^2 = \|x - x_f\|^2 + \|x_f - v\|^2.$$

Rearranging leads to

$$\|x - x_f\|^2 = \|x - v\|^2 - \|x_f - v\|^2 \text{ and so}$$
$$\|x - x_f\|^2 \leq \|x - v\|^2$$

where equality can hold only if $v = x_f$. Since this inequality holds for all $v \in W$, $x_f$ is precisely the closest vector in $W$ to $x$. ∎

Given a subspace $W$ of Euclidean $n$-space, Theorem 5.11 can be used to find a matrix $P$ such that $Px$ is the closest point in $W$ to $x$, for any $x$. This $P$ is called an *orthogonal projection matrix*.

We find $P$ in $R^n$ since this is the matrix we use most often. ($C^n$ is done in exactly the same way.) To find $P$, let $u_1, \ldots, u_r$ be an orthonormal basis (a basis of orthogonal vectors each having length one) for $W$ and $U = [u_1, \ldots, u_r]$. Then, we need $P$ to satisfy $Px = x_f$ Recall that

$$x_f = (x, u_1) u_1 + \cdots + (x, u_r) u_r.$$

Thus by backward multiplication,

$$x_f = [u_1, \ldots, u_r] \begin{bmatrix} (x, u_1) \\ \cdots \\ (x, u_r) \end{bmatrix}.$$

Noting $(x, u_i) = u_i^t x$, we have

$$= [u_1, \ldots, u_r] \begin{bmatrix} u_1^t \\ \cdots \\ u_r^t \end{bmatrix} x$$

$$= U\, U^t x.$$

Thus,

$$P = UU^t$$

is the orthogonal projection matrix desired.

An example can help.

**Example 5.17** *Let $W$ be the plane given by $z = 0$. Since* $\begin{bmatrix} 1 \\ 1 \\ 0 \end{bmatrix}$ *and* $\begin{bmatrix} 1 \\ 0 \\ 0 \end{bmatrix}$

*satisfy this equation, $W = \text{span} \left\{ \begin{bmatrix} 1 \\ 1 \\ 0 \end{bmatrix}, \begin{bmatrix} 1 \\ 0 \\ 0 \end{bmatrix} \right\}$. Note that the basis for*

*$W$ is not an orthonormal basis.*

*Applying Gram-Schmidt to* $\begin{bmatrix} 1 \\ 1 \\ 0 \end{bmatrix}, \begin{bmatrix} 1 \\ 0 \\ 0 \end{bmatrix}$ *, we have the orthogonal basis*

*given below.*

$$v_1 = \begin{bmatrix} 1 \\ 1 \\ 0 \end{bmatrix}$$

$$v_2 = \begin{bmatrix} \frac{1}{2} \\ -\frac{1}{2} \\ 0 \end{bmatrix}.$$

*These vectors can be normalized to obtain an orthonormal basis, namely*

$$u_1 = \begin{bmatrix} \frac{1}{\sqrt{2}} \\ \frac{1}{\sqrt{2}} \\ 0 \end{bmatrix}$$

$$u_2 = \begin{bmatrix} \frac{1}{\sqrt{2}} \\ -\frac{1}{\sqrt{2}} \\ 0 \end{bmatrix}.$$

*Now*

$$U = [u_1 u_2] = \begin{bmatrix} \frac{1}{\sqrt{2}} & \frac{1}{\sqrt{2}} \\ \frac{1}{\sqrt{2}} & -\frac{1}{\sqrt{2}} \\ 0 & 0 \end{bmatrix}$$

*and*

$$P = UU^t = \begin{bmatrix} 1 & 0 & 0 \\ 0 & 1 & 0 \\ 0 & 0 & 0 \end{bmatrix}.$$

*And, since* $P \begin{bmatrix} x_1 \\ x_2 \\ x_3 \end{bmatrix} = \begin{bmatrix} x_1 \\ x_2 \\ 0 \end{bmatrix}$, *it is geometrically clear that* $P$ *projects* $R^3$ *onto the plane* $W$.

### 5.4.1 Optional (Closest Matrix from Symmetric Matrices)

Sometimes in work we have data which is not precise, perhaps obtained by measurement or experiment. If it is known that the actual problem involves a symmetric matrix, the data may only give some close approximation to it. When this occurs, we can replace the approximate matrix by the closest symmetric matrix by using Fourier sums. We show how this is done for $2 \times 2$ matrices.

The $2 \times 2$ symmetric matrices form a subspace of $R^{2 \times 2}$ of dimension 3. An orthogonal basis for subspace is

$$\begin{bmatrix} 1 & 0 \\ 0 & 0 \end{bmatrix}, \begin{bmatrix} 0 & 0 \\ 0 & 1 \end{bmatrix}, \begin{bmatrix} 0 & 1 \\ 1 & 0 \end{bmatrix}.$$

For any $A = \begin{bmatrix} a & b \\ c & d \end{bmatrix}$, we compute its closest symmetric matrix by applying Theorem 5.11. This matrix is

$$A_f = a \begin{bmatrix} 1 & 0 \\ 0 & 0 \end{bmatrix} + d \begin{bmatrix} 0 & 0 \\ 0 & 1 \end{bmatrix} + \frac{b+c}{2} \begin{bmatrix} 0 & 1 \\ 1 & 0 \end{bmatrix}$$
$$= \begin{bmatrix} a & \frac{b+c}{2} \\ \frac{b+c}{2} & d \end{bmatrix}.$$

Of course this result can be extended to $n \times n$ matrices.

### 5.4.2 MATLAB (Orth and the Projection Matrix)

MATLAB can be used to find an orthonormal basis for a vector space spanned by given vectors. And, the orthogonal projection matrix can be computed from it.

We do Example 5.17 in MATLAB. Recall, $W = \text{span} \left\{ \begin{bmatrix} 1 \\ 1 \\ 0 \end{bmatrix}, \begin{bmatrix} 1 \\ 0 \\ 0 \end{bmatrix} \right\}.$

We use the spanning vectors to form a matrix $A = \begin{bmatrix} 1 & 1 \\ 1 & 0 \\ 0 & 0 \end{bmatrix}.$

For the orthonormal basis, use the command *orth* (A) which gives an orthonormal basis for the span of the columns of A.

orth(A)

$$\text{ans} = \begin{bmatrix} 0.8507 & 0.5257 \\ 0.5257 & -0.8507 \\ 0 & 0 \end{bmatrix}$$

The first two columns of this matrix form an orthonormal basis. Note, however, it is not the basis obtained by Gram-Schmidt. This kind of computation is usually done with a $QR$ decomposition or a singular value decomposition. (See Chapters 7 and 8.)

Now, for the orthogonal projection matrix we use,

$Q = orth\,(A)$;

$P = Q * Q'$            ($Q'$ is the transpose of $Q$.)

$$\text{ans} = \begin{bmatrix} 1 & 0 & 0 \\ 0 & 1 & 0 \\ 0 & 0 & 0 \end{bmatrix}.$$

This is, as expected, what we obtained in Example 5.17.

For more, type in *help orth*.

# Exercises

1. Find the inner product of each of the following:

   (a) $x = (1, -1, 1)^t$, $y = (2, 0, -1)^t$.

   (b) $x = (i, 1, 1 - i)^t$, $y = (2 - 3i, 2i, 1 + i)^t$.

   (c) $A = \begin{bmatrix} 1 & 0 & 2 \\ 0 & -1 & 3 \end{bmatrix}$, $B = \begin{bmatrix} 2 & -1 & -3 \\ 1 & 4 & -2 \end{bmatrix}$.

   (d) $f(t) = t$, $g(t) = 1$, where $a = -1$ and $b = 1$.

2. Find the distance between the vectors in Exercise 1, parts (a), (b), (c), and (d).

3. Decide which pair of vectors are orthogonal.

   (a) $x = (1, -1)^t$, $y = (1, 1)^t$

   (b) $x = (1, i, 1 - i)^t$, $y = (i, -1 + i, 1)^t$

   (c) $A = \begin{bmatrix} 1 & -1 \\ 1 & 1 \end{bmatrix}$, $B = \begin{bmatrix} -1 & 1 \\ 1 & 1 \end{bmatrix}$

   (d) $f(t) = \cos t$, $g(t) = \sin t$, where $a = -\pi$ and $b = \pi$

4. Prove that the expressions in (a) Example 5.11, (b) Example 5.12, and (c) Example 5.13 are inner products.

5. In the definition of inner product, prove that the second properties of (iii) and (iv) can be proved from the remaining properties.

6. For any inner product space $V$, prove that $(0, x) = 0$ for any $x \in V$.

7. In an inner product space, the *angle* $\theta$, $0 \le \theta < \pi$, between two nonzero vectors $x$ and $y$ satisfies the equation

$$\cos \theta = \frac{(x, y)}{\|x\| \, \|y\|}.$$

(Recall here that $\|x\| = (x, x)^{\frac{1}{2}}$, and $\|y\| = (y, y)^{\frac{1}{2}}$.) Find the angles between the following vectors.

(a) $x = (1, 0)^t$, $y = (1, 1)^t$    Check your answer geometrically.

(b) $x = (1, 1, 1, 1)^t$, $y = (1, 0, 0, 1)^t$

(c) $x = \begin{bmatrix} 1 & 0 \\ 0 & 1 \end{bmatrix}$, $y = \begin{bmatrix} 1 & 0 \\ 0 & 0 \end{bmatrix}$

8. Normalize the following vectors.

(a) $(1, 1, 1)^t$    (b) $(1 + i, 2 - 3i)^t$

(c) $A = \begin{bmatrix} 1 & -1 & 2 \\ -3 & 0 & 4 \end{bmatrix}$    (d) $f(t) = t$ where $a = -1, b = 1$

9. Let $u_1, \ldots, u_m$ be an orthonormal set. Prove that

$$\|\alpha_1 u_1 + \cdots + \alpha_m u_m\|_2^2 = |\alpha_1|^2 + \cdots + |\alpha_m|^2.$$

10. Apply the Gram-Schmidt process to

(a) $x_1 = (1, 1, 1)^t$, $x_2 = (1, 1, 0)^t$, $x_3 = (1, -1, 2)^t$.

(b) $A = \begin{bmatrix} 1 & 1 \\ 1 & 0 \end{bmatrix}$, $B = \begin{bmatrix} 1 & 1 \\ 0 & 1 \end{bmatrix}$, $C = \begin{bmatrix} 0 & 1 \\ 1 & 1 \end{bmatrix}$.

(c) $f(t) = 1$, $g(t) = t$, $h(t) = t^2$, where $a = 0$, $b = 1$.

11. In the Gram-Schmidt process, prove that

(a) span $\{x_1, x_2, x_3\}$ = span $\{u_1, u_2, u_3\}$.

(b) span $\{x_1, \ldots, x_m\}$ = span $\{u_1, \ldots, u_m\}$.

12. Find the orthogonal projection matrix that orthogonally projects

(a) $R^2$ onto the line parametrically described by $x_1 = t$, $x_2 = t$ where $-\infty < t < \infty$. (Check your work geometrically.)

(b) $R^3$ onto the plane given by $x_1 + x_2 + x_3 = 0$.

13. Apply the Gram-Schmidt process to $\begin{bmatrix} 1 \\ -1 \\ 1 \end{bmatrix}$, $\begin{bmatrix} 1 \\ 0 \\ 1 \end{bmatrix}$, $\begin{bmatrix} 0 \\ 1 \\ 0 \end{bmatrix}$. Explain geometrically why $u_3 = 0$.

14. In $span \left\{ \begin{bmatrix} 1 \\ 1 \\ 0 \end{bmatrix}, \begin{bmatrix} 1 \\ 1 \\ 1 \end{bmatrix} \right\}$, find the closest vector to $\begin{bmatrix} 1 \\ 2 \\ 3 \end{bmatrix}$.

15. Find the polynomial of degree 1 (or less) closest to $1+t+t^2$, on $[0, 1]$.

16. Consider the line parametrically described by $x_1 = t$, $x_2 = 2t$ where $-\infty < t < \infty$.

   (a) Find the orthogonal projection matrix from $R^2$ onto this line.

   (b) Use the orthogonal projection matrix to compute the closest point on this line to $\begin{bmatrix} 1 \\ 1 \end{bmatrix}$. (Check your answer geometrically.)

17. Let $P$ be an orthogonal projection matrix. Show that $P^2 = P$.

18. Find the formula for the orthogonal projection matrix from $C^n$ to $W$ where $W$ has $u_1, \ldots, u_r$ as an orthonormal basis.

19. Does the set of orthogonal projection matrices, defined on $R^2$, form a subspace of $R^{2\times2}$?

20. (Optional). Using Fourier sums, find the closest upper triangular matrix to $\begin{bmatrix} a & b \\ c & d \end{bmatrix}$.

21. (MATLAB). Use MATLAB to solve the following problem. Let $W \subseteq R^3$ be the vectors in the solution to

$$x_1 + 2x_2 - x_3 = 0.$$

   (a) Find an orthonormal basis for $W$. (Find null$(A)$, $A = [1\ 2\ -1]$.)

   (b) Find the corresponding orthogonal projection matrix $P$.

   (c) Use $P$ to find the closest point in $W$ to $[1, 1, 1]^t$.

   (d) How close is $[1, 1, 1]^t$ to $W$?

22. (MATLAB). Let $L(x) = Ax$ where $A = \begin{bmatrix} 1 & 1 & 1 \\ -1 & 1 & 0 \\ 1 & 1 & 1 \end{bmatrix}$.

   (a) Find an orthogonal basis for range $L$.

   (b) Find the orthogonal projection matrix of $R^3$ onto range $L$.

# 6
# Unitary Similarity

A unitary matrix $U$ is a special matrix, which as a linear transformation $L(x) = Ux$, preserves figures in the space. The grid view of these transformations shows no shearing nor scaling. They appear as rotations or reflections of the space. And, since these matrices do not distort figures, they are excellent for obtaining simple coordinate views of curves, surfaces, and other geometrical objects. In addition, since these matrices do not magnify error, they are also important in developing numerical algorithms which provide good answers.

## 6.1  Unitary Matrices

This section concerns the Euclidean $n$-space with inner product

$$(x,y) = x_1\bar{y}_1 + \cdots + x_n\bar{y}_n.$$

(Recall for real numbers, $\bar{y}_i = y_i$.) It is often helpful to write this inner product as a matrix product,

$$(x,y) = y^H x. \tag{6.1}$$

In fact, much of what we do in this section can be directly observed by using (6.1).

We study $n \times n$ matrices $U$ such that $L(x) = Ux$ preserves figures (including lengths and orthogonality). Thinking in terms of the grid view, since the columns of $U$ form the axes for the new grid in the range of $L$, those columns should form an orthonormal set (as $e_1$ and $e_2$ did in $R^2$). Otherwise the geometry is distorted. See Figure 6.1.

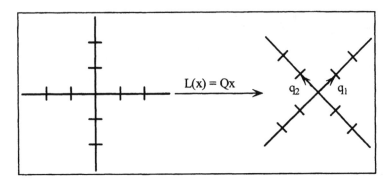

FIGURE 6.1.

**Definition 6.1** *An $n \times n$ matrix $U$ is* unitary *if the columns of $U$ are pair-wise orthogonal and length one. If the entries in the matrix are real numbers, we call the matrix* orthogonal, *and, as is customary, use $Q$ instead of $U$.*

Two examples demonstrate what $L(x) = Qx$, or simply $Q$, does to $R^2$.

**Example 6.1** *(Rotation) Let $Q = \begin{bmatrix} \cos\theta & -\sin\theta \\ \sin\theta & \cos\theta \end{bmatrix}$. By definition, this matrix is orthogonal. The grid view of $L(x) = Qx$ is given in Figure 6.2. This transformation rotates the plane $\theta$ degrees. More generally, the $n \times n$*

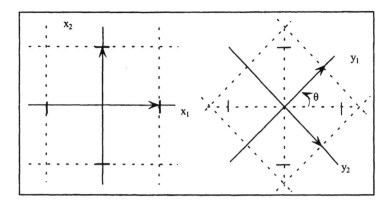

FIGURE 6.2.

*matrix*

$$
Q = \begin{bmatrix}
1 & 0 & \cdots & & & & \cdots 0 \\
0 & 1 & \cdots & & & & \cdots 0 \\
& & \cdots & & & & \\
0 & 0 & \cdots & \cos\theta & -\sin\theta & \cdots 0 & \quad r \\
& & \cdots & & & & \\
0 & 0 & \cdots & \sin\theta & \cos\theta & \cdots 0 & \quad s \\
& & \cdots & & & & \\
0 & 0 & \cdots & & & & \cdots 1
\end{bmatrix} \quad r \quad ,
$$

*called a Givens matrix or plane rotation, rotates the $x_r x_s$-plane in $R^n$, leaving all other coordinates alone. (In $R^3$, rotating in the $x_1 x_2$-plane keeps the $x_3$-axis fixed and rotates $R^3$ about it. In higher dimensions, this may be a bit hard to imagine.)*

**Example 6.2** *(Reflection)  Let* $H = \begin{bmatrix} \cos\theta & \sin\theta \\ \sin\theta & -\cos\theta \end{bmatrix}$, *the only orthogonal matrix, other than that in Example 6.1, with first column* $\begin{bmatrix} \cos\theta \\ \sin\theta \end{bmatrix}$. *The grid view of $L(x) = Qx$ is given in Figure 6.3. It is clear that $L$ is not*

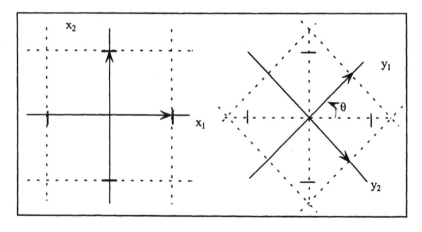

FIGURE 6.3.

*a rotation. However, this transformation can be seen as a reflection (flip) of the plane about the line $\ell$ bisecting the $x_1$-axis and the $y_1$-axis. (Perhaps overlaying a transparency, drawing the axes on it and then reflecting as described will help.)*

*Alternately, $L$ can be described as inverting the plane through $\ell$ so that each $x$ ends up at its mirror image $x'$ as shown in Figure 6.4. We use this latter view to find the matrix for this transformation on $R^n$.*

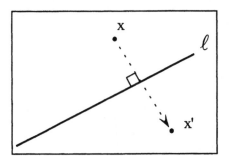

FIGURE 6.4.

For this, let $u \in R^n$ be such that $\|u\|_2 = 1$. (In analogy, $u$ will be the direction of the inversion.) Define

$$W = \{x : x \in R^n \text{ and } (u, x) = 0\}.$$

(W generalizes $\ell$. $R^n$ will be inverted through $W$.) As given in the exercises, $W$ is a subspace.

We show how to reflect $R^n$ through $W$ parallel to $u$. To do this, let $P$ be the orthogonal projection matrix from $R^n$ onto the line, span $\{u\}$. Thus,

$$P = u\, u^t.$$

(Note that $P$ is $n \times n$.) Given $x \in R^n$, $Px$ is the projection of $x$ onto the line determined by $u$. Thus, the inversion of $x$ through $W$ parallel to $u$ is the vector $x'$ where $x' = x - 2Px$ as shown in in Figure 6.5. From this, we

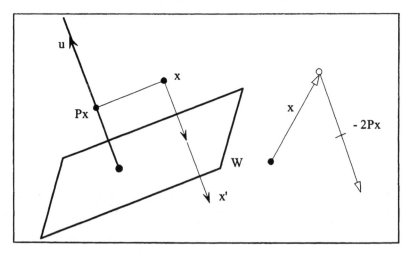

FIGURE 6.5.

*have that*

$$L(x) = Hx,$$

*where*

$$H = I - 2u\,u^t \qquad (6.2)$$

*reflects $R^n$ through $W$ parallel to $u$.*

*For example, if we want to reflect $R^2$ parallel to $u = \begin{bmatrix} -\frac{1}{\sqrt{2}} \\ \frac{1}{\sqrt{2}} \end{bmatrix}$, through*

$W = \{x : (x, u) = 0\} = \text{span}\left\{ \begin{bmatrix} 1 \\ 1 \end{bmatrix} \right\}$, *we would use*

$$H = I - 2 \begin{bmatrix} -\frac{1}{\sqrt{2}} \\ \frac{1}{\sqrt{2}} \end{bmatrix} \overset{\left[ -\frac{1}{\sqrt{2}}\ \frac{1}{\sqrt{2}} \right]}{}$$

$$= I - 2 \begin{bmatrix} \frac{1}{2} & -\frac{1}{2} \\ -\frac{1}{2} & \frac{1}{2} \end{bmatrix}$$

$$= \begin{bmatrix} 0 & 1 \\ 1 & 0 \end{bmatrix}.$$

*The matrix $H$, as defined in (6.2), is called a* Householder matrix .

To see how much of matrix space orthogonal matrices comprise, we can graph a piece of this space from $R^{2\times 2}$, which contains the plane rotations. For this, we look at the 3-dimensional subspace of matrices of the form $\begin{bmatrix} a & b \\ c & a \end{bmatrix}$ and graph the matrices $\begin{bmatrix} \cos\theta & -\sin\theta \\ \sin\theta & \cos\theta \end{bmatrix}$, $0 \le \theta < 2\pi$, in this space. The graph appears in Figure 6.6.

Actually the orthogonal matrices are on the sphere of radius $\sqrt{2}$ about the origin. But interestingly, they do not constitute all of this space since $(1, 1, 0, 0)^t$ (which corresponds to $\begin{bmatrix} 1 & 1 \\ 0 & 0 \end{bmatrix}$) is also on this sphere. However, what we need to see is that the orthogonal matrices cover only a small part of matrix space.

Definition 6.1 can be formulated as a matrix equation.

**Lemma 6.1** *Let $U$ be an $n \times n$ matrix. Then $U$ is unitary if and only if $U$ satisfies the unitary equation,*

$$U^H U = I.$$

**Proof.** We argue both implications of the biconditional.

Part a. For the direct implication, suppose $U$ is unitary. Then

$$U^H U = \begin{bmatrix} u_1^H u_1 \cdots u_1^H u_n \\ \cdots \\ u_n^H u_1 \cdots u_n^H u_n \end{bmatrix}$$

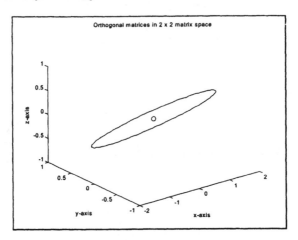

FIGURE 6.6.

where the columns of $U$ are $u_1, \ldots, u_n$. Using (6.1), $U^H U = I$.

Part b. For the converse, suppose $U$ satisfies the unitary equation $U^H U = I$. In terms of entries, the equation is

$$
\begin{bmatrix} u_1^H u_1 \cdots u_1^H u_n \\ \cdots \\ u_n^H u_1 \cdots u_n^H u_n \end{bmatrix} = \begin{bmatrix} 1 & 0 & \cdots & 0 \\ & & \cdots & \\ 0 & 0 & \cdots & 1 \end{bmatrix} \tag{6.3}
$$

where $u_1, \ldots, u_n$ are the columns of $U$. Thus, by (6.1), the columns of $U$ are pair-wise orthogonal and of length one, and so $U$ is unitary. ∎

An immediate consequence of this theorem is that if $U$ is unitary, then $U^{-1} = U^H$.

This lemma is usually used to develop results about unitary matrices rather than to decide if a particular matrix is unitary. We show this in the results below.

**Theorem 6.1** *Let $T$ be a triangular matrix that is unitary. Then $T$ is a diagonal matrix with $|t_{kk}| = 1$ for all $k$.*

**Proof.** Since $T$ is unitary (and so $T^{-1} = T^H$), $T$ satisfies the equation,

$$
TT^H = T^H T. \tag{6.4}
$$

We now argue the $3 \times 3$ case, since that argument is easily extended to the general case. We suppose $T$ is lower triangular.

Writing out (6.4) entrywise, we have

$$\begin{bmatrix} t_{11} & 0 & 0 \\ t_{21} & t_{22} & 0 \\ t_{31} & t_{32} & t_{33} \end{bmatrix} \begin{bmatrix} \bar{t}_{11} & \bar{t}_{21} & \bar{t}_{31} \\ 0 & \bar{t}_{22} & \bar{t}_{32} \\ 0 & 0 & \bar{t}_{33} \end{bmatrix}$$

$$= \begin{bmatrix} \bar{t}_{11} & \bar{t}_{21} & \bar{t}_{31} \\ 0 & \bar{t}_{22} & \bar{t}_{32} \\ 0 & 0 & \bar{t}_{33} \end{bmatrix} \begin{bmatrix} t_{11} & 0 & 0 \\ t_{21} & t_{22} & 0 \\ t_{31} & t_{32} & t_{33} \end{bmatrix}.$$

Comparing the 1,1-entries in the products, we get

$$t_{11}\bar{t}_{11} = \bar{t}_{11}t_{11} + \bar{t}_{21}t_{21} + \bar{t}_{31}t_{31} \text{ or}$$
$$|t_{11}|^2 = |t_{11}|^2 + |t_{21}|^2 + |t_{31}|^2.$$

Thus, $t_{21} = t_{31} = 0$. Comparing the 2,2-entries, then the 3,3-entries establishes that $T$ is a diagonal matrix. Finally, since $T$ is unitary, the columns of $T$ must be length 1 and so each $|t_{kk}| = 1$. ∎

For the arithmetic of unitary matrices, we have the following.

**Theorem 6.2** *Two properties of unitary matrices follow.*

(a) *If $U_1$ and $U_2$ are unitary, so is $U_1 U_2$.*

(b) *If $U$ is unitary, so is $U^H$.*

**Proof.** We argue Part (a), leaving Part (b) as an exercise.
Let $U_1, U_2$ be unitary. Then, checking the unitary equation,

$$(U_1 U_2)^H (U_1 U_2) = U_2^H U_1^H U_1 U_2 = U_2^H U_2 = I.$$

Thus, by Lemma 6.1, $U_1 U_2$ is unitary. ∎

Considering the grid view of $L(x) = Ux$, we would expect the following geometric properties.

**Theorem 6.3** *Let $U$ be an $n \times n$ unitary matrix. Then for all $x$ and $y$, in Euclidean n-space,*

(a) $\|Ux\|_2 = \|x\|_2$. *(Length is unchanged.)*

(b) $(Ux, Uy) = (x, y)$. *(If $U$ is an orthogonal matrix then using (a) and (b), we leave it as an exercise to show that $U$ preserves angles.)*

*In addition,*

(c) $|\det U| = 1$. *(It is known that for a polygonal shape $X$, the volume of $AX$ is $|\det A|$ times the volume of $X$. Thus, under $L(x) = Ux$, $L$ preserves volume.)*

**Proof.** We argue two parts.
Part a. Using Lemma 6.1,

$$\|Ux\|_2^2 = (Ux)^H (Ux) = x^H U^H U z = x^H x = \|x\|_2^2.$$

Thus, $\|Ux\|_2 = \|x\|_2$.

Part b. This is done as in Part (a). ∎

Concerning calculation, the norm and condition number of a unitary matrix are as expected from the grid view. The unit circle is not distorted.

**Theorem 6.4** *Let $U$ be a unitary matrix. Then $\|U\|_2 = 1$ and $c_2(U) = 1$.*

**Proof.** Both parts are calculations.

$$\|U\|_2 = \max_{\|x\|_2=1} \|Ux\|_2$$

$$= \max_{\|x\|_2=1} \|x\|_2 = 1$$

$$c_2(U) = \|U\|_2 \|U^{-1}\|_2 = \|U\|_2 \|U^H\|_2 = \|U^H\|_2 = 1$$

using the first calculation and that $U^H$ is unitary. ∎

Recalling Section 2 of Chapter 5, and the Optional there, we see that $U$ neither magnifies error nor relative error. So, unitary matrices are ideal for use in numerical computations (where rounding error occurs). An additional norm result follows.

**Theorem 6.5** *Let $A$ be an $m \times n$ matrix. Let $U_1$ and $U_2$ be $m \times m$ and $n \times n$ unitary matrices, respectively. Then*

(a) $\|U_1 A U_2\|_2 = \|A\|_2$.

(b) $\|U_1 A U_2\|_F = \|A\|_F$.

**Proof.** We argue both parts.
Part a. By definition,

$$\|U_1 A U_2\|_2 = \max_{\|x\|_2=1} \|U_1 A U_2 x\|_2$$

$$= \max_{\|x\|_2=1} \left[ (U_1 A U_2 x)^H (U_1 A U_2 x) \right]^{\frac{1}{2}}$$

$$= \max_{\|x\|_2=1} \left[ x^H U_2^H A^H A U_2 x \right]^{\frac{1}{2}}$$

and setting $y = U_2 x$ and noting that $\|y\|_2 = \|U_2 x\|_2 = \|x\|_2$

$$= \max_{\|y\|_2 = 1} \left[ y^H A^H A y \right]^{\frac{1}{2}}$$

$$= \max_{\|y\|_2 = 1} \|Ay\|_2$$

$$= \|A\|_2 .$$

Part b. Note that for any $m \times n$ matrix $B = [b_1 \ldots b_n]$, where $b_i$ is the $i$-th column of $B$,

$$\|U_1 B\|_F^2 = \|[U_1 b_1 \ldots U_1 b_n]\|_F^2 = \|U_1 b_1\|_2^2 + \cdots + \|U_1 b_m\|_2^2$$
$$= \|b_1\|_2^2 + \cdots + \|b_m\|_2^2 = \|[b_1 \ldots b_m]\|_F^2 = \|B\|_F^2 .$$

Similarly, looking at rows, $\|B U_2\|_F = \|B\|_F$. So putting together,

$$\|U_1 A U_2\|_F = \|A U_2\|_F = \|A\|_F ,$$

the desired result. ■

### 6.1.1   Optional (Symmetry)

Orthogonal matrices can be used to describe symmetry in designs. (See Figure 6.7.)

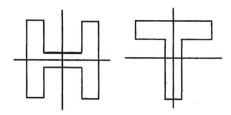

FIGURE 6.7.

For example, the letter $H$ has rotational symmetry (rotate 180°) and reflectional symmetry (about the x-axis and about the y-axis). The letter T has reflectional symmetry (about the $y$ -axis). Such symmetries are sometimes discussed on Sesame Street (a syndicated television series for children).

Symmetries can be classified using orthogonal matrices. To see this, consider a square in the plane, as shown in Figure 6.8. Note that the 4 rotations $\begin{bmatrix} \cos\theta & -\sin\theta \\ \sin\theta & \cos\theta \end{bmatrix}$ for $\theta = 0, \frac{\pi}{2}, \pi, \frac{3\pi}{2}$ leave this figure fixed. And

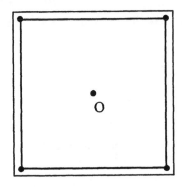

FIGURE 6.8.

the 4 reflections

$$\begin{bmatrix} 1 & 0 \\ 0 & -1 \end{bmatrix}, \begin{bmatrix} -1 & 0 \\ 0 & 1 \end{bmatrix}, \begin{bmatrix} 0 & 1 \\ 1 & 0 \end{bmatrix}, \begin{bmatrix} 0 & -1 \\ -1 & 0 \end{bmatrix}$$

also leave it fixed. We call this set of rotations and reflections $D_4$ (the dihedral group of the square). Note that the star in Figure 6.9 has precisely the same symmetry, namely $D_4$. (It can be shown that if a figure has $k$

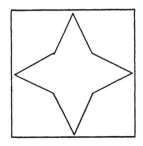

FIGURE 6.9.

reflections of symmetry, then it has $k$ rotations of symmetry, counting the identity, and we say the symmetry is $D_k$.)

Not all figures have reflectional symmetry. For example, the letter $Z$, in Figure 6.10, has only rotational symmetry $\begin{bmatrix} 1 & 0 \\ 0 & 1 \end{bmatrix}, \begin{bmatrix} \cos \pi & -\sin \pi \\ \sin \pi & \cos \pi \end{bmatrix}$. Here, we say the symmetry is $C_2$, indicating there are only 2 rotations from $Z$ onto $Z$. $C_k$ is defined similarly.

In nature, symmetry is all over. As an example, a daisy has lots of rotational and reflectional symmetry (in applications, mathematics need not fit precisely, so there may be some variation from the precise mathematical description of symmetry.)

In computer graphics, noticing symmetry can save time. For example, if the Mandelbrot set is graphed, only that part above the $x$-axis is required,

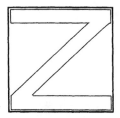

FIGURE 6.10.

since the set has reflectional symmetry about the $x$-axis. (Changing the sign of the second entries of the computed vectors gets the bottom half.) Thus, the amount of time to graph this set is cut in half. (It could be many minutes, depending on the computer.)

## 6.1.2  MATLAB (Code for Picture of Orthogonal Matrices in 2 × 2 Matrix Space)

We graph $\begin{bmatrix} \cos\theta & -\sin\theta \\ \sin\theta & \cos\theta \end{bmatrix}$, which is in the 2-dimensional subspace of

matrices having the form $\begin{bmatrix} a & -b \\ b & a \end{bmatrix}$ in the 3-dimensional subspace of ma-

trices having the form $\begin{bmatrix} a & c \\ d & a \end{bmatrix}$. We identify this subspace with $\begin{bmatrix} \sqrt{2}a \\ c \\ d \end{bmatrix}$

and graph $\begin{bmatrix} \sqrt{2}\cos\theta \\ -\sin\theta \\ \sin\theta \end{bmatrix}$ for $0 \leq \theta \leq 2\pi$.

**Code for Picture of Orthogonal Matrices**

```
theta = linspace(0, 2 * pi, 100);
plot3(sqrt(2)*cos(theta),          % Plot3 draws
          -sin(theta), sin (theta))      curves in R³.
```

# Exercises

1. Compute $(x, y)$, using $y^H x$, for the following.

(a) $x = \begin{bmatrix} -1 \\ 2 \\ 1 \end{bmatrix}$, $y = \begin{bmatrix} 1 \\ -1 \\ 2 \end{bmatrix}$

(b) $x = \begin{bmatrix} i \\ 1-i \\ 2i \end{bmatrix}$, $y = \begin{bmatrix} 2-i \\ i \\ 1+2i \end{bmatrix}$

2. Give the grid view of each.

(a) $Q = \begin{bmatrix} \frac{1}{\sqrt{2}} & -\frac{1}{\sqrt{2}} \\ \frac{1}{\sqrt{2}} & \frac{1}{\sqrt{2}} \end{bmatrix}$

(b) $Q = I - 2uu^t$ where $u = \begin{bmatrix} \frac{1}{\sqrt{2}} \\ \frac{1}{\sqrt{2}} \end{bmatrix}$

3. Give a $2 \times 2$ orthogonal matrix that does the following.

   (a) Rotates the plane $30°$

   (b) Reflects the plane about the axis $y = 2x$

4. Find the $3 \times 3$ Householder matrix that reflects $R^3$ parallel to $u = \begin{bmatrix} 0 \\ 0 \\ 1 \end{bmatrix}$. What is $W$?

5. Reflect the clock given in Figure 6.11 about the $x_1$-axis. Is the orientation (1-2-3-1 clockwise) still the same?

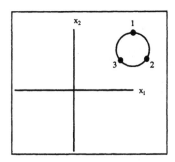

FIGURE 6.11.

6. Prove that if $U$ is unitary, then

   (a) $U$ is nonsingular and    (b) $U^{-1} = U^H$.

7. Prove Theorem 6.2, part (b).

8. Prove Theorem 6.3, part (c).

9. Let $H$ be a Householder matrix. Prove each of the following.

   (a) $H$ is orthogonal

   (b) $H = H^t$

10. Let $A$ be an $n \times n$ matrix and $U_1, U_2$ be $n \times n$ unitary matrices. Prove $c(U_1 A U_2) = c(A)$ for the induced 2-norm and the Frobenius norm.

11. Let $Q$ be an $n \times n$ orthogonal matrix, and $x, \hat{x}, \in R^n$. Prove that if $y = Qx, \hat{y} = Q\hat{x}$, then the angle between $y$ and $\hat{y}$ is the same as that between $x$ and $\hat{x}$.

12. Find $\theta$ such that
$$\begin{bmatrix} \cos\theta & -\sin\theta \\ \sin\theta & \cos\theta \end{bmatrix} \begin{bmatrix} 3 \\ 4 \end{bmatrix} = \begin{bmatrix} 5 \\ 0 \end{bmatrix}.$$

13. Find $u$ ($\|u\|_2 = 1$) so that
$$(I - 2uu^t) \begin{bmatrix} 3 \\ 4 \end{bmatrix} = \begin{bmatrix} 5 \\ 0 \end{bmatrix}.$$

14. Let $U$ be a unitary matrix. Is it true that $\|U\|_2 = \|U\|_F$? (Give an example if it is false.)

15. Let $Q$ be the rotation matrix of Example 6.1. Let $x = \begin{bmatrix} r\cos\phi \\ r\sin\phi \end{bmatrix}$, expressed in polar coordinates. Show that $Qx = \begin{bmatrix} r\cos(\theta + \phi) \\ r\sin(\theta + \phi) \end{bmatrix}$.
(Use trigonometry identities.) What does this say that $Q$ does to $R^2$?

16. Find the matrix $F$ for each flag on the left of Figure 6.12 and Figure 6.13. Then, find an orthogonal matrix $Q$, if possible, which rotates or reflects the plane, bringing the flag on the left to that on the right. Show your answer is correct by showing plot $(QF)$.

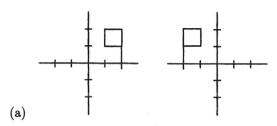

(a)

FIGURE 6.12.

17. Prove that if $Q$ is orthogonal, then $L(x) = Qx$ maps a sphere $S$ of radius $r$ about $z$ ($S = \{x : \|x - z\|_2 = r\}$) into a sphere of radius $r$ about $L(z)$.

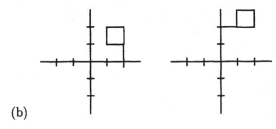

(b)

FIGURE 6.13.

18. Prove that the matrices in Example 6.1 and 6.2 are the only $2 \times 2$ orthogonal matrices.

19. (Optional)  Classify the symmetries as $C_k$ or $D_k$ for the figures in Figure 6.14.

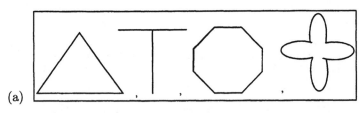

(a)

FIGURE 6.14.

(b) What symmetry do you see in a kaleidoscope?

20. (MATLAB) Two parts.

(a) Let $P$ be the matrix for the propeller shown in Figure 6.15. Find the matrix $Q$ that rotates the propeller $\frac{\pi}{4}$ radian. Plot $QP$ to see the new configuration.

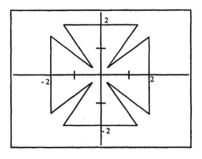

FIGURE 6.15.

Find the matrix $C$ for the cube, of side 2, shown in Figure 6.16. (The center of the cube is at the origin, and the faces are parallel to the planes determined by the axes.) Rotate the cube $\frac{\pi}{4}$ radian in the $x_1x_2$-plane and then $\frac{\pi}{4}$ radian in the $x_2x_3$-plane. Find the matrix $Q$ that provides this motion and plot $QC$.

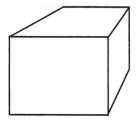

FIGURE 6.16.

## 6.2   Schur Decompositions

In Chapter 3, we factored $A = PJP^{-1}$, where $J$ is a Jordan form, and in the following chapter we saw some of its uses. In this section we look at another version of this factorization, the case where we require $P$ to be unitary.

As shown in the picture in Section 1, unitary matrices comprise a small part of matrix space so we expect that in such a factorization, we will not achieve a form as simple as $J$. To see what we might be able to do, we can use the $QR$ factorization (Gram-Schmidt process) to write $P = UR$ where $U$ is unitary and $R$ upper triangular. Then, substituting,

$$A = PJP^{-1} = URJR^{-1}U^H.$$

Since $R$, $J$, and $R^{-1}$ are upper triangular, so is the product $RJR^{-1}$. Hence, we can write

$$A = UTU^H$$

where $U$ is unitary and $T = RJR^{-1}$ is upper triangular. Thus, $A$ is not only similar to an upper triangular matrix $T$, it is unitarily similar to $T$. And as a consequence of similarity, the eigenvalues of $A$ are on the main diagonal of $T$.

In general, if $A$ and $B$ are $n \times n$ matrices such that $A = UBU^H$ for some unitary matrix $U$, we say that $A$ and $B$ are *unitarily similar*. (And if $A$, $B$, and $U$ have real entries, we say $A$ and $B$ are *orthogonally similar*.)

We intend to show a direct way (without resorting to the Jordan form) of proving that every square matrix is unitarily similar to an upper triangular matrix. To develop this unitary matrix version of Jordan's theorem, we need the following lemma, which gives a simple way to extend an orthogonal set to an orthogonal basis.

**Lemma 6.2** *Let $u_1, \ldots, u_r$ be pair-wise orthogonal nonzero vectors in Euclidean $n$-space. Then there are vectors $u_{r+1}, \ldots, u_n$ such that $u_1, \ldots, u_r$, $u_{r+1}, \ldots, u_n$ are pair-wise nonzero orthogonal vectors, and thus these vectors form a basis for Euclidean $n$-space.*

**Proof.** We intend to solve for $u_{r+1}, \ldots, u_n$ one at a time.
Consider

$$\begin{bmatrix} u_1^H \\ \cdots \\ u_r^H \end{bmatrix} x = 0. \tag{6.5}$$

Since $u_1, \ldots, u_r$ are pair-wise orthogonal nonzero vectors, and thus are linearly independent, $[u_1 \ldots u_r]$ has rank $r$ and hence so does

$$[u_1 \ldots u_r]^H = \begin{bmatrix} u_1^H \\ \cdots \\ u_r^H \end{bmatrix}.$$

If $r < n$, any echelon form of (6.5) has a free variable, and so there is a nonzero solution, say $u_{r+1}$, to (6.5).

Note that since $u_k^H u_{r+1} = 0$ ($u_k^H u_{r+1} = (u_{r+1}, u_k)$), $u_k$ is orthogonal to $u_{r+1}$ for all $k \le r$. Thus $u_1, \ldots, u_r, u_{r+1}$ are pair-wise orthogonal nonzero vectors.

Now we continue the procedure to find $u_{r+2}$ and then $u_{r+3}$ until $u_n$ is found. Thus we obtain $n$ pair-wise orthogonal nonzero vectors. By Lemma 5.2, these vectors give a basis for Euclidean $n$-space. ∎

An example of the technique follows.

**Example 6.3** *Let $u_1 = (1, 1, 1)^t$. To extend $u_1$ to an orthogonal basis, we solve*

$$u_1^t x = 0.$$

*The augmented matrix is*

$$\begin{bmatrix} 1 & 1 & 1 \mid 0 \end{bmatrix}.$$

*There are two free variables, and a solution is $u_2 = \begin{bmatrix} 1 \\ -1 \\ 0 \end{bmatrix}$. Now solve*

$$\begin{bmatrix} u_1^t \\ u_2^t \end{bmatrix} x = 0.$$

*The augmented matrix is*

$$\left[\begin{array}{ccc|c} 1 & 1 & 1 & 0 \\ 1 & -1 & 0 & 0 \end{array}\right].$$

*Applying* $-R_1 + R_2$, *we have*

$$\left[\begin{array}{ccc|c} 1 & 1 & 1 & 0 \\ 0 & -2 & -1 & 0 \end{array}\right].$$

*Thus with* $\alpha$ *an arbitrary scalar,*

$$x_3 = \alpha$$
$$x_2 = -\frac{1}{2}\alpha$$
$$x_1 = -\frac{1}{2}\alpha.$$

*For* $\alpha = 2$ *(We can choose any nonzero* $\alpha$*.), we have*

$$u_3 = \left[\begin{array}{c} -1 \\ -1 \\ 2 \end{array}\right].$$

*Thus* $u_1, u_2, u_3$ *forms an orthogonal basis.*
*If we normalize these vectors, we have*

$$\frac{u_1}{\|u_1\|_2} = \left[\begin{array}{c} \frac{1}{\sqrt{3}} \\ \frac{1}{\sqrt{3}} \\ \frac{1}{\sqrt{3}} \end{array}\right]$$

$$\frac{u_2}{\|u_2\|} = \left[\begin{array}{c} \frac{1}{\sqrt{2}} \\ -\frac{1}{\sqrt{2}} \\ 0 \end{array}\right]$$

$$\frac{u_3}{\|u_3\|} = \left[\begin{array}{c} -\frac{1}{\sqrt{6}} \\ -\frac{1}{\sqrt{6}} \\ \frac{2}{\sqrt{6}} \end{array}\right],$$

*an orthonormal basis.*

Now, to get the idea of how to find a unitary matrix $U$ such that $A = UTU^H$ where $T$ is upper triangular, we look at a small case.

Let $A$ be a $3 \times 3$ matrix. Let $\lambda$ be an eigenvalue for $A$ with corresponding eigenvector $u_1$, of unit length. Extend $u_1$ to $u_1, u_2, u_3$ an orthonormal basis. Then, by partitioned and backward multiplication

$$A\,[u_1u_2u_3] = [\lambda_1 u_1 \; Au_2 \; Au_3]$$

$$= [u_1u_2u_3] \begin{bmatrix} \lambda_1 & \alpha_1 & \alpha_2 \\ 0 & \beta_1 & \beta_2 \\ 0 & \gamma_1 & \gamma_2 \end{bmatrix}$$

where the $\alpha_i$'s, $\beta_i$'s, and $\gamma_i$'s satisfy $Au_2 = \alpha_1 u_1 + \beta_1 u_2 + \gamma_1 u_3$ and $Au_3 = \alpha_2 u_1 + \beta_2 u_2 + \gamma_2 u_3$. Thus, setting $U_1 = [u_1u_2u_3]$, $b = [\alpha_1\alpha_2]$, and $B = \begin{bmatrix} \beta_1 & \beta_2 \\ \gamma_1 & \gamma_2 \end{bmatrix}$, we have

$$U_1^H A U_1 = \begin{bmatrix} \lambda_1 & b \\ 0 & B \end{bmatrix},$$

a start toward the triangular matrix.

Now, we continue with $B$. Let $\lambda_2$ be an eigenvalue of $B$ with $v_1$ a corresponding eigenvector of unit length. Extend $v_1$ to an orthonormal basis, say $v_1$, $v_2$. Then

$$B\,[v_1v_2] = [\lambda_2 v_1 \; Bv_2]$$

$$= [v_1v_2] \begin{bmatrix} \lambda_2 & \alpha \\ 0 & \beta \end{bmatrix}$$

where $\alpha$ and $\beta$ are determined by $Bv_2 = \alpha v_1 + \beta v_2$. Setting $V = [v_1v_2]$, we have

$$V^H B V = \begin{bmatrix} \lambda_2 & \alpha \\ 0 & \beta \end{bmatrix}.$$

Now, to get this back to $3 \times 3$ matrices, set $U_2 = \begin{bmatrix} 1 & 0 \\ 0 & V \end{bmatrix}$. Then

$$U_2^H U_1^H A U_1 U_2 = U_2^H \begin{bmatrix} \lambda_1 & b \\ 0 & B \end{bmatrix} U_2$$

$$= \left[ \begin{array}{c|c} \lambda_1 & bV \\ \hline 0 & V^H B V \end{array} \right]$$

$$= \left[ \begin{array}{c|cc} \lambda_1 & bV & \\ \hline 0 & \lambda_2 & \alpha \\ 0 & 0 & \beta \end{array} \right].$$

Setting $U = U_1U_2$, a unitary matrix, we have

$$U^H B U = \left[ \begin{array}{c|cc} \lambda_1 & bV & \\ \hline 0 & \lambda_2 & \alpha \\ 0 & 0 & \lambda_3 \end{array} \right]$$

where we set $\beta = \lambda_3$. Using similarity, $\lambda_1$, $\lambda_2$, and $\lambda_3$ are the eigenvalues of $A$.

More formally, we have Schur's Theorem.

**Theorem 6.6** *Let $A$ be an $n \times n$ matrix. Then there is an $n \times n$ unitary matrix $U$ such that*

$$A = UTU^H,$$

*where $T$ is an upper triangular matrix.*

**Proof.** The proof is done in steps showing how.

Step 1. (Finding $U_1$) Let $\lambda$ be an eigenvalue of $A$ with $x$ a corresponding eigenvector. Set $u_1 = \frac{x}{\|x\|_2}$ and extend $u_1$ to an orthonormal basis, $u_1, u_2, \ldots, u_n$ by Lemma 6.2. Set $U_1 = [u_1 \ldots u_n]$. Then

$$AU_1 = U_1 \begin{bmatrix} \lambda & & \\ 0 & c_2 \ldots c_n \\ \ldots & \\ 0 & \end{bmatrix}$$

where the entries in column $c_k$, are found by solving

$$Au_k = U_1 c_k.$$

So, deflating,

$$U_1^H A U_1 = \begin{bmatrix} \lambda & b \\ 0 & B \end{bmatrix}$$

where $B$ is an $(n-1) \times (n-1)$ matrix. Thus, we have the first row staggered.

Step 2. (Finding $U_k$) Suppose

$$U_{k-1}^H \cdots U_1^H A U_1 \cdots U_{k-1} = \begin{bmatrix} C_1 & C_2 \\ 0 & C \end{bmatrix}$$

where $C_1$ is a $(k-1) \times (k-1)$ upper triangular matrix. Now, repeating step 1 on $C$, we obtain a unitary matrix $W$, such that

$$W^H C W = \begin{bmatrix} \beta & \hat{b} \\ 0 & \hat{B} \end{bmatrix}$$

where $\beta$ is a scalar. Set

$$U_k = \begin{bmatrix} I & 0 \\ 0 & W \end{bmatrix}$$

an $n \times n$ unitary matrix. Then, deflating (to get a smaller matrix in the lower right corner),

$$U_k^H \left( U_{k-1}^H \cdots U_1^H A U_1 \cdots U_{k-1} \right) U_k = \begin{bmatrix} C_1 & C_2 W \\ 0 & W^H C W \end{bmatrix}$$

which has $k$ staggered rows.

Step 3. (Finding $U$) Set

$$U = U_1 \cdots U_{n-1}.$$

Then, $U$ is unitary and

$$U^H A U = T$$

an upper triangular matrix. ■

A numerical example follows.

**Example 6.4** Let $A = \begin{bmatrix} 0 & -1 & 2 \\ 1 & 0 & 0 \\ 0 & 0 & 1 \end{bmatrix}$. We find $U$ and $T$ in steps.

Step 1. (Finding $U_1$) The eigenvalues of $A$ are $1, i, -i$. Let $\lambda = 1$. A corresponding eigenvector is $x = \begin{bmatrix} 1 \\ 1 \\ 1 \end{bmatrix}$, so $u_1 = \begin{bmatrix} \frac{1}{\sqrt{3}} \\ \frac{1}{\sqrt{3}} \\ \frac{1}{\sqrt{3}} \end{bmatrix}$. Extending to an orthogonal basis, we have

$$u_2 = \begin{bmatrix} \frac{1}{\sqrt{2}} \\ -\frac{1}{\sqrt{2}} \\ 0 \end{bmatrix}, u_3 = \begin{bmatrix} \frac{1}{\sqrt{6}} \\ \frac{1}{\sqrt{6}} \\ -\frac{2}{\sqrt{6}} \end{bmatrix}$$

and so

$$U_1 = \begin{bmatrix} \frac{1}{\sqrt{3}} & \frac{1}{\sqrt{2}} & \frac{1}{\sqrt{6}} \\ \frac{1}{\sqrt{3}} & -\frac{1}{\sqrt{2}} & \frac{1}{\sqrt{6}} \\ \frac{1}{\sqrt{3}} & 0 & -\frac{2}{\sqrt{6}} \end{bmatrix}.$$

Deflating we have

$$U_1^H A U_1 = \begin{bmatrix} 1 & \frac{\sqrt{6}}{3} & -\sqrt{2} \\ 0 & 0 & -\sqrt{3} \\ 0 & \frac{\sqrt{3}}{3} & 0 \end{bmatrix}$$

*Step 2.  (Finding $U_2$)  Now*

$$C = \begin{bmatrix} 0 & -\sqrt{3} \\ \frac{\sqrt{3}}{3} & 0 \end{bmatrix}$$

*Note that $\lambda = i$ is an eigenvalue of $C$ and has a corresponding eigenvector*
$x = \begin{bmatrix} -\sqrt{3} \\ i \end{bmatrix}$. *Thus $v_1 = \begin{bmatrix} -\frac{\sqrt{3}}{2} \\ \frac{i}{2} \end{bmatrix}$ . Take $v_2 = \begin{bmatrix} -\frac{i}{2} \\ \frac{\sqrt{3}}{2} \end{bmatrix}$ and set $W =$*

$\begin{bmatrix} -\frac{\sqrt{3}}{2} & -\frac{i}{2} \\ \frac{i}{2} & \frac{\sqrt{3}}{2} \end{bmatrix}$. *Thus $U_2 = \begin{bmatrix} 1 & 0 \\ 0 & W \end{bmatrix} = \begin{bmatrix} 1 & 0 & 0 \\ 0 & -\frac{\sqrt{3}}{2} & -\frac{i}{2} \\ 0 & \frac{i}{2} & \frac{\sqrt{3}}{2} \end{bmatrix}$. Then*

$$U_2^H U_1^H A U_1 U_2 = \begin{bmatrix} 2 & \frac{-\sqrt{2}}{2} - \frac{\sqrt{2}}{2}i & \frac{-\sqrt{6}}{2} - \frac{\sqrt{6}}{6}i \\ 0 & i & \frac{2\sqrt{3}}{2} \\ 0 & 0 & -i \end{bmatrix} = T.$$

*Step 3.  (Finding $U$)  Set*

$$U = U_1 U_2 = \begin{bmatrix} \frac{1}{\sqrt{3}} & \frac{-\sqrt{6}}{4} + \frac{\sqrt{6}}{12}i & \frac{\sqrt{2}}{4} - \frac{\sqrt{2}}{4}i \\ \frac{1}{\sqrt{3}} & \frac{\sqrt{6}}{8} + \frac{\sqrt{6}}{12}i & \frac{\sqrt{2}}{4} + \frac{\sqrt{2}}{4}i \\ \frac{1}{\sqrt{3}} & \frac{-\sqrt{6}}{6}i & \frac{-\sqrt{2}}{2} \end{bmatrix}.$$

The proof of Schur's theorem also shows that if $A$ has real entries and real eigenvalues, then $A$ is orthogonally similar to a triangular matrix. Even without the real eigenvalues hypothesis, a real version of Schur's Theorem can be given.

**Corollary 6.1** *Let $A$ be an $n \times n$ matrix with real entries. Then there is an orthogonal matrix $Q$, such that*

$$A = Q \begin{bmatrix} T_{11} & T_{12} & \cdots & T_{1r} \\ 0 & T_{22} & \cdots & T_{2r} \\ & & \cdots & \\ 0 & 0 & \cdots & T_{rr} \end{bmatrix} Q^t$$

*where each $T_{kk}$ is $1 \times 1$ or $2 \times 2$.*

*(a) If $T_{kk}$ is $1 \times 1$, $T_{kk} = [\lambda]$ where $\lambda$ is a real eigenvalue of $A$.*

*(b) If $T_{kk}$ is $2 \times 2$, then its eigenvalues are nonreal, complex conjugate eigenvalues $(\lambda$ and $\bar{\lambda})$ of $A$.*

**Proof.** Exercise. ∎

As we have seen, calculations involving a diagonal matrix are much easier than those involving a triangular matrix. Thus we now show when a matrix is unitarily diagonalizable (unitarily similar to a diagonal matrix).

Let $A$ be an $n \times n$ matrix. If

$$A^H A = AA^H$$

then $A$ is called a *normal* matrix. (Examples of normal matrices include Hermitian matrices and symmetric matrices.) The simple condition given above determines precisely those matrices that are unitarily diagonalizable.

**Corollary 6.2** *Let $A$ be an $n \times n$ matrix and $A = U^H T U$ a Schur decomposition. Then $A$ is normal if and only if $T$ is diagonal.*

**Proof.** The proof is as in Theorem 6.1, and so it is left as an exercise. ∎

If $A$ is normal, then we know, by the previous corollary, that $A$ is similar to a diagonal matrix $D$. Thus,

$$A = PDP^{-1}$$

where $D$ must be formed from the eigenvalues of $A$ and $P$ the corresponding eigenvectors, a calculation simpler than the Schur form technique. Actually, this $P$ can be adjusted to form a unitary matrix $U$ where

$$A = UDU^H.$$

To do this, suppose $p_r, p_{r+1}, \dots, p_s$ are those columns in $P$ that are eigenvectors for the eigenvalue $\lambda_i$. Apply the Gram-Schmidt process to these eigenvectors to obtain $u_r, u_{r+1}, \dots, u_s$, an orthonormal set of eigenvectors for $\lambda_i$. Replace $p_r, p_{r+1}, \dots, p_s$ in $P$ with $u_r, u_{r+1}, \dots, u_s$. Doing this for all eigenvalues of $A$ yields a matrix $U$. To prove that this $U$ is unitary, we need only show that if $u_i$ and $u_j$ belong to different eigenvalues then $(u_i, u_j) = 0$.

**Lemma 6.3** *Let $A$ be an $n \times n$ normal matrix. Let $\lambda$ and $\beta$ be eigenvalues of $A$ with corresponding eigenvectors $x$ and $y$, respectively. If $\lambda \neq \beta$, then $(x, y) = 0$.*

**Proof.** Since $A$ is normal, we can factor $A = U^H DU$ as assured by Corollary 6.2. For simplicity, we will now assume $n = 3$ and $D = \begin{bmatrix} \lambda & 0 & 0 \\ 0 & \lambda & 0 \\ 0 & 0 & \beta \end{bmatrix}$.

(The general argument is an extension of this case.)

Now, if $x$ and $y$ are eigenvectors belonging to $\lambda$ and $\beta$ respectively,

$$Ax = \lambda x, \quad Ay = \beta y.$$

Thus

$$U^H DU x = \lambda x, \quad U^H DU y = \beta y$$

or, rearranging

$$DU x = \lambda U x, \quad DU y = \beta U y.$$

Set

$$w = U x, \quad z = U y.$$

So we have

$$D w = \lambda w, \quad D z = \beta z.$$

This says that the eigenspace (See $D$ above.) for $\lambda$ is span $\{e_1, e_2\}$ and for $\beta$ is span $\{e_3\}$. Thus, we have that

$$(w, z) = z^H w = 0.$$

Now,

$$(x, y) = \left(U^H w, U^H z\right) = (w, z) = 0$$

which is the desired result. ∎

Below, we give an example of unitarily diagonalizing a symmetric matrix, using the eigenvalue-eigenvector approach.

**Example 6.5** *We unitarily diagonalize* $A = \begin{bmatrix} 1 & 1 & 1 \\ 1 & 1 & 1 \\ 1 & 1 & 1 \end{bmatrix}$. *Here* $\lambda_1 = 3$,

$\lambda_2 = 0$, $\lambda_3 = 0$. *Corresponding eigenvectors are* $p_1 = \begin{bmatrix} 1 \\ 1 \\ 1 \end{bmatrix}$, $p_2 = $

$\begin{bmatrix} 1 \\ -1 \\ 0 \end{bmatrix}$, *and* $p_3 = \begin{bmatrix} 1 \\ 0 \\ -1 \end{bmatrix}$, *respectively. So* $P = \begin{bmatrix} 1 & 1 & 1 \\ 1 & -1 & 0 \\ 1 & 0 & -1 \end{bmatrix}$ *di-*

*agonalizes* $A$. *Adjusting* $P$ *to an orthogonal matrix, we apply the Gram-Schmidt process to the eigenvectors for* $\lambda_1$, *and then to those for* $\lambda_2$. *This*

*gives* $q_1 = \begin{bmatrix} \frac{1}{\sqrt{3}} \\ \frac{1}{\sqrt{3}} \\ \frac{1}{\sqrt{3}} \end{bmatrix}$, $q_2 = \begin{bmatrix} \frac{1}{\sqrt{2}} \\ -\frac{1}{\sqrt{2}} \\ 0 \end{bmatrix}$, *and* $q_3 = \begin{bmatrix} \frac{1}{\sqrt{6}} \\ \frac{1}{\sqrt{6}} \\ -\frac{2}{\sqrt{6}} \end{bmatrix}$, *respectively. So*

$Q = \begin{bmatrix} \frac{1}{\sqrt{3}} & \frac{1}{\sqrt{2}} & \frac{1}{\sqrt{6}} \\ \frac{1}{\sqrt{3}} & -\frac{1}{\sqrt{2}} & \frac{1}{\sqrt{6}} \\ \frac{1}{\sqrt{3}} & 0 & -\frac{2}{\sqrt{6}} \end{bmatrix}$. *And* $A = QDQ^t$ *where* $D = \begin{bmatrix} 3 & 0 & 0 \\ 0 & 0 & 0 \\ 0 & 0 & 0 \end{bmatrix}$.

We now show that, as seen in the example, symmetric matrices are always orthogonally similar to a diagonal matrix. (Complex numbers are not necessary for the factorization.) To do this, we need to show that symmetric matrices have real eigenvalues so that the previous work can be done using only real numbers. Actually, we can show a bit stronger result.

**Lemma 6.4** *Let $A$ be an $n \times n$ Hermitian matrix. Then each eigenvalue of $A$ is real.*

**Proof.** Let $\lambda$ be an eigenvalue, and $x$ a corresponding eigenvector, of $A$. Then

$$Ax = \lambda x.$$

Multiplying through by $x^H$ yields

$$x^H A x = \lambda x^H x. \tag{6.6}$$

Taking the conjugate transpose of both sides, we have

$$x^H A^H x = \overline{\lambda} x^H x. \tag{6.7}$$

Since $A^H = A$, (6.7) can be written as

$$x^H A x = \overline{\lambda} x^H x. \tag{6.8}$$

Now equating the right sides of (6.6) and (6.8) yields

$$\lambda x^H x = \overline{\lambda} x^H x$$

Recall that $x^H x = \|x\|_2^2 > 0$. Thus,

$$\lambda = \overline{\lambda}.$$

This implies $\lambda$ is a real scalar. ∎

So, we have the desired result below.

**Corollary 6.3** *Let $A$ be a symmetric matrix. Then $A$ is orthogonally similar to a diagonal matrix.*

**Proof.** Apply Corollary 6.2 and Lemma 6.4. ∎

## 6.2.1   Optional (Motion in Principal Axes)

If we take a spring-mass system, as shown in Figure 6.17, where $m_1 = m_2 = $

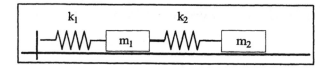

FIGURE 6.17.

1, as given in Chapter 4, the positions $x_1(t)$ and $x_2(t)$ of the two particles satisfy

$$x''(t) + \begin{bmatrix} k_1 + k_2 & -k_2 \\ -k_2 & k_2 \end{bmatrix} x(t) = 0. \tag{6.9}$$

Notice that the matrix $K = \begin{bmatrix} k_1 + k_2 & -k_2 \\ -k_2 & k_2 \end{bmatrix}$ is symmetric due to the springs exerting the same force to the left as to the right.

To solve (6.9), we orthogonally diagonalize $K$, say $K = QDQ^t$ where $D = \text{diag}(\lambda_1, \lambda_2)$ and $Q = [q_1 q_2]$. Plugging $QDQ^t$ into (6.9) and rearranging, we have

$$y''(t) + Dy(t) = 0 \tag{6.10}$$

where $y(t) = Q^t x(t)$ or

$$Qy(t) = x(t). \tag{6.11}$$

Equation (6.11) can be interpreted as a change of coordinates from those determined from the basis $Y = \{q_1, q_2\}$ to the original vectors. Traditionally, the vectors $q_1$ and $q_2$ are called principal axes for the $Y$-coordinate system.

Equation (6.10) describes the motion of the particles with respect to the $Y$-coordinates. In this coordinate system we have, from (6.10),

$$y_1''(t) + \lambda_1 y_1 = 0$$
$$y_2''(t) + \lambda_2 y_2 = 0.$$

Solving these equations, we get

$$y_1(t) = \alpha_1 \cos\left(\sqrt{\lambda_1}t\right) + \beta_1 \sin\left(\sqrt{\lambda_1}t\right)$$
$$y_2(t) = \alpha_2 \cos\left(\sqrt{\lambda_2}t\right) + \beta_2 \sin\left(\sqrt{\lambda_2}t\right).$$

Using that

$$\alpha \cos\left(\sqrt{\lambda}t\right) + \beta \sin\left(\sqrt{\lambda}t\right) = r \cos\left(\sqrt{\lambda}t + \delta\right)$$

where $r = \sqrt{\alpha^2 + \beta^2}$ and $\delta$ satisfies

$$\cos \delta = \frac{\alpha}{r}, \quad \sin \delta = \frac{-\beta}{r}, \tag{6.12}$$

we have

$$y_1(t) = r_1 \cos\left(\sqrt{\lambda_1}t + \delta_1\right)$$
$$y_2(t) = r_2 \cos\left(\sqrt{\lambda_1}t + \delta_2\right)$$

where $r_1, \delta_1$ and $r_2, \delta_2$ are given by (6.12). And we see that with respect to the $y_1$-axis (determined from $q_1$) the motion is like $\cos$ (amplitude $= r_1$, period $= 2\pi/\sqrt{\lambda_1}$) as it is with the $y_2$-axis. Understanding the motion with respect to these axes gives us some view of $y(t)$, and thus $x(t)$.

To firm up the discussion, we look at an example.

**Example 6.6** *Let* $K = \begin{bmatrix} 3 & -2 \\ -2 & 2 \end{bmatrix}$. *Then*

$$x'' + Kx = 0$$

*can be described, in terms of the principal axes as*

$$y'' + Dy = 0$$

*where* $\lambda_1 = 4.5616$ *and* $\lambda_2 = 0.4384$.
*For simplicity, suppose* $y(0) = \begin{bmatrix} 1 \\ 1 \end{bmatrix}$ *and* $y'(0) = 0$. *Then*

$$y_1(t) = \cos\left(\sqrt{\lambda_1}t\right)$$
$$= \cos(2.1358t) \qquad (period\ = 2.9419)$$
$$y_2(t) = \cos\left(\sqrt{\lambda_2}t\right)$$
$$= \cos(0.6621t) \qquad (period\ = 9.4895).$$

*Thus, if we graph* $(y_1(t), y_2(t))^t$, *which is the same as graphing* $(x_1(t), x_2(t))^t$, *we can see its shadow on the* $y_1$-*axis as* $y_1(t)$ *and on the* $y_2$-*axis as* $y_2(t)$.
*In looking at Figure 6.18, we see that* $y_1(t)$ *achieves 1 about three times from its initial position while* $y_2(t)$ *achieves it about once (agreeing with their periods). So, we have some view of what is going on.*

To see how intricate the graph is, we look at it for $t = 0$ to $t = 300$. (See Figure 6.19.)

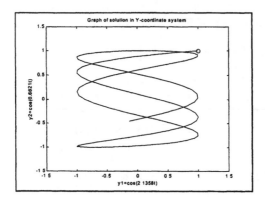

FIGURE 6.18.

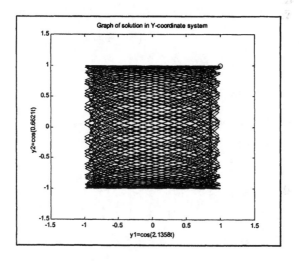

FIGURE 6.19.

## 6.2.2  MATLAB (Schur)

For a given $n \times n$ matrix $A$ the command for the Schur form $T$ is given by schur $(A)$. To obtain $U$ and $T$, use $[U, T] = schur\,(A)$. If $A$ is normal, $T$ will be diagonal.

Since MATLAB doesn't provide Jordan forms, if $A$ is defective (or nearly so), the Shur factorization can often be used in its place. Numerically, this factorization can be found rather accurately. We provide an exercise solving systems of differential equations in this manner.

To see more, type in *help schur*.

**Code for Graphics of Example 6.5**

```
t = linspace(0, 4 * pi, 100);
plot (cos (2.1358 * t),
          cos (0.6621 * t))
hold
plot (1, 1, 'O')
axis([−1.5 1.5 − 1.5 1.5])       % To get the graph off the
                                         edges of the picture.
```

The second picture changes line 1 to
t = linspace(0, 300, 600);

# Exercises

1. Find a unitary matrix $U$ such that $U^H A U$ is upper triangular.

   (a) $A = \begin{bmatrix} 1 & 2 \\ 1 & 2 \end{bmatrix}$     (b) $A = \begin{bmatrix} 0 & 1 \\ -1 & 0 \end{bmatrix}$

   (c) $A = \begin{bmatrix} -2 & 4 & 2 \\ 1 & -2 & 1 \\ -3 & 6 & -3 \end{bmatrix}$

2. Which matrices are normal?

   (a) $\begin{bmatrix} 1 & 2 \\ 2 & 1 \end{bmatrix}$     (b) $\begin{bmatrix} 1 & 2 \\ 1 & 2 \end{bmatrix}$

   (c) $\begin{bmatrix} 1 & i \\ -i & 2 \end{bmatrix}$     (d) $\begin{bmatrix} 1 & 1 & 1 \\ 1 & 1 & 1 \\ 1 & 1 & 1 \end{bmatrix}$

3. Prove Corollary 6.1 for $4 \times 4$ matrices. Assume $A$ has no real eigenvalues.

4. Prove

   (a) Corollary 6.2.

   (b) Lemma 6.3.

5. Unitarily diagonalize $A$ using eigenvalues and eigenvectors.

   (a) $A = \begin{bmatrix} 2 & 1 & 1 \\ 1 & 2 & 1 \\ 1 & 1 & 2 \end{bmatrix}$     (b) $A = \begin{bmatrix} 0 & i \\ -i & 0 \end{bmatrix}$

6. Prove that a unitary matrix is normal.

7. Let $A = \begin{bmatrix} 2 & 0 \\ 1 & 3 \end{bmatrix}$. Find two different Schur decompositions of $A$.

8. Find a matrix that is diagonalizable but not unitarily diagonalizable.

9. Prove that if $A$ is normal, so are the following.

   (a) $A^H$

   (b) $U^H A U$ where $U$ is unitary

10. Is the product of two symmetric matrices symmetric?

11. Let $T$ be a triangular matrix which is not diagonal. Show that $T$ is not normal.

12. Give a direct proof, using the proof of Schur's theorem as a guide, that $A$ is unitarily similar to a lower triangular matrix.

13. Prove $A$ is unitarily similar to a lower triangular matrix by using Schur's Theorem on $A^H$.

14. For the spring-mass system of Chapter 4, Section 1, let the masses be $m_1 = 1$, $m_2 = 1$ and the spring constants $k_1 = 1$, $k_2 = 1$. So the equation is

$$x'' + Kx = 0. \tag{6.13}$$

where $K = \begin{bmatrix} 2 & -1 \\ -1 & 1 \end{bmatrix}$ and $x = \begin{bmatrix} x_1(t) \\ x_2(t) \end{bmatrix}$.

   (a) Find the solution to (6.13).

   (b) Adjust the solution so that $x(0) = \begin{bmatrix} 1 \\ 1 \end{bmatrix}$, $x'(0) = \begin{bmatrix} -1 \\ 1 \end{bmatrix}$.

   (c) Draw the position of the masses at time $t = 1, 2$, and 3.

15. (Optional) The graph of the orthogonal matrices $\begin{bmatrix} \cos\theta & \sin\theta \\ \sin\theta & -\cos\theta \end{bmatrix}$
    (the reflections) is a circle as was the graph of the rotations. Do these circles intersect in $R^{2\times2}$?

16. (MATLAB) Solve

$$x' = \begin{bmatrix} 2 & 0 \\ 1 & 2 \end{bmatrix} x \tag{6.14}$$

$$x(0) = \begin{bmatrix} 1 \\ 1 \end{bmatrix}.$$

   (a) Try the eigenvalue-eigenvector approach.

(b) Use the Shur form and solve

$$y' = Ty$$

where $y = Q^t x$. Thus

$$y(0) = Q^t x(0).$$

(Recall, $z'(t) = \lambda z(t) + f(t)$ has solution $z(t) = z(0) e^{\lambda t} + e^{\lambda t} \int_0^t e^{-\lambda \tau} f(\tau) \, d\tau$.) Convert this solution back to that of (6.14).

# 7

# Singular Value Decomposition

In this chapter we show a decomposition of a matrix $A$ (called a singular value decomposition),

$$A = U\Sigma V^H$$

where $U$ and $V$ are unitary and $\Sigma$ a diagonal matrix. The way this decomposition is used (See Figure 7.1.) is often like that for the previous decompositions. However, the kinds of problems on which a singular value

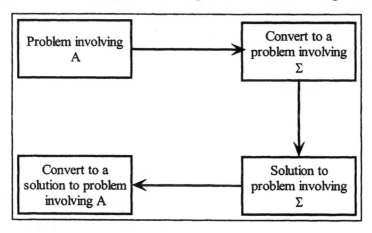

FIGURE 7.1.

decomposition is used are different from those solved by previous decompositions. The problems solved here usually involve maximizing or minimizing lengths or distances (which includes approximations), or involve shapes or figures in geometry.

## 7.1   Singular Value Decomposition Theorem

In this section, we show how to obtain a singular value decomposition of a matrix $A$. To get the idea of how this is done, we first look at the special case where $A$ is nonsingular.

Since $A^H A$ is Hermitian, using Schur's theorem,

$$V^H A^H A V = D$$

for some unitary matrix $V = [v_1 \ldots v_n]$, where $v_i$ is the $i$-th column of $V$, and diagonal matrix $D$. Recall that $D = \text{diag}(\lambda_1, ..., \lambda_n)$ where the $\lambda_i$'s are the eigenvalues of $A^H A$. Rearranging we have

$$(AV)^H (AV) = D. \tag{7.1}$$

The key idea for the decomposition comes from making the appropriate observations in (7.1). To do this, recall that for vectors $x$ and $y$,

$$y^H x = (x, y).$$

Thus, (7.1) tells us that the columns of $AV$ are orthogonal, and so $AV$ is almost orthogonal. And, it says that the square of the length of the $i$-th column $Av_i$, is $\lambda_i$ and so its length is $\sqrt{\lambda_i}$. Thus setting $\sigma_i = \sqrt{\lambda_i}$ for all $i$ and scaling the columns of $AV$, we have that

$$U = \begin{bmatrix} \dfrac{Av_1}{\sigma_1} & \cdots & \dfrac{Av_n}{\sigma_n} \end{bmatrix}$$

is unitary. And, we have

$$AV = [Av_1 \ldots Av_n]$$
$$= \begin{bmatrix} \dfrac{Av_1}{\sigma_1} & \cdots & \dfrac{Av_n}{\sigma_n} \end{bmatrix} \begin{bmatrix} \sigma_1 & 0 & \cdots & 0 \\ & & \cdots & \\ 0 & 0 & \cdots & \sigma_n \end{bmatrix}$$
$$= U\Sigma.$$

where $\Sigma = \text{diag}(\sigma_1, \ldots, \sigma_n)$. Thus, we have the decomposition,

$$A = U\Sigma V^H.$$

To develop this work more generally, we proceed as follows. Let $\Sigma = [\sigma_{ij}]$ be an $m \times n$ matrix. If $\sigma_{ij} = 0$ for $i \neq j$, we call $\Sigma$ a *rectangular diagonal* matrix and write

$$\Sigma = diag\,(\sigma_1, \ldots, \sigma_s)$$

where $s = \min\{m, n\}$ and $\sigma_i = \sigma_{ii}$ for all $i$.

**Example 7.1** *As examples of rectangular diagonal matrices, we have the following.*

(a) $\Sigma = \begin{bmatrix} 3 & 0 & 0 \\ 0 & 4 & 0 \end{bmatrix}$    (b) $\Sigma = \begin{bmatrix} -1 & 0 \\ 0 & 2 \\ 0 & 0 \end{bmatrix}$

(c) $\Sigma = \begin{bmatrix} -6 & 0 \\ 0 & 2 \end{bmatrix}$

The major theorem in this section describes a singular value decomposition (SVD) of an arbitrary matrix.

**Theorem 7.1** *Let $A$ be an $m \times n$ matrix and $s = \min\{m, n\}$. If $A$ has rank $r$, then*

(a) *There is an $m \times m$ unitary matrix $U$, an $n \times n$ unitary matrix $V$, and an $m \times n$ diagonal matrix $\Sigma = diag\,(\sigma_1, \ldots, \sigma_s)$, such that*

$$A = U\Sigma V^H$$

*where*

$$\sigma_1 \geq \cdots \geq \sigma_r > 0 = \sigma_{r+1} = \cdots = \sigma_s,$$

*The scalars $\sigma_1, \ldots, \sigma_s$ are called* singular values *and are the square roots of the nonzero eigenvalues of $A^H A$, ordered by size.*

(b) *The decomposition can be expanded as*

$$A = \sigma_1 u_1 v_1^H + \cdots + \sigma_r u_r v_r^H$$

*where, expressed in terms of their columns, $U = [u_1 \ldots u_m]$ and $V = [v_1 \ldots v_n]$.*

**Proof.** We prove both parts.

Part a. We give the proof in steps, showing how $U$, $V$, and $\Sigma$ are found.

Step 1. (Finding $V$) Note that $A^H A$ is an $n \times n$ Hermitian matrix. Let $V$ be an $n \times n$ unitary matrix that diagonalizes the $n \times n$ matrix $A^H A$, i.e.,

$$V^H \left(A^H A\right) V = D \tag{7.2}$$

where $D = \text{diag}(\lambda_1,\ldots,\lambda_n)$ and $\lambda_1 \geq \lambda_2 \geq \cdots \geq \lambda_n$. (Recall Hermitian matrices have real eigenvalues.) Observe that (7.2) can be written as

$$(AV)^H (AV) = D$$

or setting $W = AV$,

$$W^H W = D.$$

**Step 2.** (Finding $\Sigma$) Using that $w_i$ is the $i$-th column of $W$, we have $\lambda_i = w_i^H w_i = (w_i, w_i) \geq 0$. Thus, $\lambda_1 \geq \cdots \geq \lambda_r > 0 = \lambda_{r+1} = \cdots = \lambda_n$ for some integer $r$. Now, let $\Sigma$ be the $m \times n$ rectangular diagonal matrix, $\Sigma = \text{diag}(\sigma_1,\ldots,\sigma_s)$ where

$$\sigma_i = \begin{cases} \sqrt{\lambda_i} & \text{if } \lambda_i > 0 \\ 0 & \text{otherwise.} \end{cases}$$

Note that these $\sigma_i$ are completely determined by the eigenvalues of $A^H A$ and thus from $A$.

**Step 3.** (Finding $U$) By (7.2), the columns of $AV$ are pairwise orthogonal. (Some of the last columns could be 0.) And the $i$-th column of $AV$ has length $\sigma_i$. Normalizing the nonzero columns, we have

$$u_i = \frac{1}{\sigma_i} A v_i$$

for all $i, i \leq r$. Extend $u_1,\ldots,u_r$ to an orthonormal basis, say,

$$u_1,\ldots,u_r,\ldots,u_r$$

and set

$$U = [u_1 \ldots u_m].$$

Then, using that $Av_i = \sigma_i u_i$ for $i \leq r$ and that $\sigma_{r+1} = \cdots = \sigma_s = 0$, and backward multiplication,

$$AV = U\Sigma$$

and so

$$A = U\Sigma V^H.$$

Finally, $r = \text{rank}\,\Sigma = \text{rank}\,A$.

**Part b.** Write

$$A = U\Sigma V^H = U((\sigma_1 E_1) + \cdots + (\sigma_s E_s)) V^H$$

where $E_i$ is the $m \times n$ matrix having a 1 in the $ii$-th position and 0's elsewhere. So

$$A = U\left(\sigma_1 E_1\right) V^H + \cdots + U\left(\sigma_s E_s\right) V^H$$
$$= \sigma_1 u_1 v_1^H + \cdots + \sigma_s u_s v_s^H$$
$$= \sigma_1 u_1 v_1^H + \cdots + \sigma_r u_r v_r^H.$$

This is the desired result. ∎

An example may help.

**Example 7.2** Let $A = \begin{bmatrix} 1 & 1 & 1 \\ 1 & 1 & 1 \end{bmatrix}$. We find an SVD of A and its expansion. We do this in steps.

Step 1. (Finding V) Orthogonally diagonalize $A^t A$. Since $A^t A =$
$\begin{bmatrix} 1 & 1 \\ 1 & 1 \\ 1 & 1 \end{bmatrix} \begin{bmatrix} 1 & 1 & 1 \\ 1 & 1 & 1 \end{bmatrix} = \begin{bmatrix} 2 & 2 & 2 \\ 2 & 2 & 2 \\ 2 & 2 & 2 \end{bmatrix}$, we find $V = \begin{bmatrix} \frac{1}{\sqrt{3}} & \frac{1}{\sqrt{2}} & \frac{1}{\sqrt{6}} \\ \frac{1}{\sqrt{3}} & -\frac{1}{\sqrt{2}} & \frac{1}{\sqrt{6}} \\ \frac{1}{\sqrt{3}} & 0 & -\frac{2}{\sqrt{6}} \end{bmatrix}$

and $D = \begin{bmatrix} 6 & 0 & 0 \\ 0 & 0 & 0 \\ 0 & 0 & 0 \end{bmatrix}$.

Step 2. (Finding $\Sigma$) Since $\lambda_1 = 6$, $\lambda_2 = \lambda_3 = 0$ and A is $2 \times 3$, we have $\sigma_1 = \sqrt{6}$, $\sigma_2 = 0$, and $\Sigma = \begin{bmatrix} \sqrt{6} & 0 & 0 \\ 0 & 0 & 0 \end{bmatrix}$.

Step 3. (Finding U) To do this, compute

$$AV = \begin{bmatrix} \frac{3}{\sqrt{3}} & 0 & 0 \\ \frac{3}{\sqrt{3}} & 0 & 0 \end{bmatrix}.$$

Set $u_1 = \frac{1}{\sigma_1} A v_1 = \begin{bmatrix} \frac{1}{\sqrt{2}} \\ \frac{1}{\sqrt{2}} \end{bmatrix}$. Extend to an orthonormal basis by setting $u_2 = \begin{bmatrix} -\frac{1}{\sqrt{2}} \\ \frac{1}{\sqrt{2}} \end{bmatrix}$. Then $U = \begin{bmatrix} \frac{1}{\sqrt{2}} & -\frac{1}{\sqrt{2}} \\ \frac{1}{\sqrt{2}} & \frac{1}{\sqrt{2}} \end{bmatrix}$. Since

$$AV = U\Sigma, \quad A = U\Sigma V^t =$$

$$\begin{bmatrix} \frac{1}{\sqrt{2}} & -\frac{1}{\sqrt{2}} \\ \frac{1}{\sqrt{2}} & \frac{1}{\sqrt{2}} \end{bmatrix} \begin{bmatrix} \sqrt{6} & 0 & 0 \\ 0 & 0 & 0 \end{bmatrix} \begin{bmatrix} \frac{1}{\sqrt{3}} & \frac{1}{\sqrt{2}} & \frac{1}{\sqrt{6}} \\ \frac{1}{\sqrt{3}} & -\frac{1}{\sqrt{2}} & \frac{1}{\sqrt{6}} \\ \frac{1}{\sqrt{3}} & 0 & -\frac{2}{\sqrt{6}} \end{bmatrix}^t.$$

*Step 4.* For the expansion, we have

$$A = U\Sigma V^t = \sqrt{6} \begin{bmatrix} \frac{1}{\sqrt{2}} \\ \frac{1}{\sqrt{2}} \end{bmatrix} \begin{bmatrix} \frac{1}{\sqrt{3}} & \frac{1}{\sqrt{3}} & \frac{1}{\sqrt{3}} \end{bmatrix}$$

$$+ 0 \begin{bmatrix} -\frac{1}{\sqrt{2}} \\ \frac{1}{\sqrt{2}} \end{bmatrix} \begin{bmatrix} \frac{1}{\sqrt{2}} & -\frac{1}{\sqrt{2}} & 0 \end{bmatrix}$$

$$= \sqrt{6} \begin{bmatrix} \frac{1}{\sqrt{2}} \\ \frac{1}{\sqrt{2}} \end{bmatrix} \begin{bmatrix} \frac{1}{\sqrt{3}} & \frac{1}{\sqrt{3}} & \frac{1}{\sqrt{3}} \end{bmatrix}.$$

In Chapter 3 we gave an expression for $\|A\|_2$ mentioning that we would prove this later. This proof can now be given.

**Corollary 7.1** *Let $A$ be an $m \times n$ matrix. Then*
(a) $\|A\|_2 = \sigma_1$,
(b) $c_2(A) = \|A\|_2 \|A^{-1}\|_2 = \frac{\sigma_1}{\sigma_n}$, *when $A$ is $n \times n$ and nonsingular.*

**Proof.** We prove both parts.
Part a. We prove this part for $n \times n$ matrices leaving the general case as an exercise.
Let $A = U\Sigma V^H$ be a singular value decomposition of $A$. Recall that $\|A\|_2 = \|U\Sigma V^H\|_2 = \|\Sigma\|_2$ since $U$ and $V^H$ are unitary.
Now,

$$\|\Sigma\|_2 = \max_{\|x\|_2=1} \|\Sigma x\|_2$$

$$= \max_{\|x\|_2=1} \left( \sum_{k=1}^{n} |\sigma_i x_i|^2 \right)^{\frac{1}{2}} \leq \max_{\|x\|_2=1} \left( \sum_{k=1}^{n} |\sigma_1|^2 |x_i|^2 \right)^{\frac{1}{2}}$$

$$= |\sigma_1| \max_{\|x\|_2=1} \left( \sum_{k=1}^{n} |x_i|^2 \right)^{\frac{1}{2}} = |\sigma_1|.$$

So, $\|\Sigma\|_2 \leq \sigma_1$. Further, since $\|e_1\|_2 = 1$, $\|\Sigma\|_2 \geq \|\Sigma e_1\|_2 = \sigma_1$. Thus $\|\Sigma\|_2 = \sigma_1$.
Part b. As in Part (a), we can show that $\|A^{-1}\|_2 = \frac{1}{\sigma_n}$. Thus, putting together, $c_2(A) = \|A\|_2 \|A^{-1}\|_2 = \frac{\sigma_1}{\sigma_n}$. ∎

Perhaps the best known use of singular value decompositions is that they can be used to least-squares solve problems. To see how, let $A$ be an $m \times n$ matrix and b an $m \times 1$ vector, and consider the system of linear equations

$$Ax = b. \tag{7.3}$$

In many problems, it is known that (7.3) has no solution. In any case, we can look for a 'least square' solution, that is, a vector $x$ such that

$$\|Ax - b\|_2 \tag{7.4}$$

is the smallest possible. (Thus, the left and right side of (7.3) are as close as they can be.)

To find a least-squares solution, since multiplying by unitary matrices doesn't change lengths, $\|Ax - b\|_2 = \|\Sigma V^H x - U^H b\|_2$ for all $x$, so (7.4) has the same least-squares solution as

$$\Sigma V^H x = U^H b.$$

Thus, simplifying $V^H x = y$ and $U^H b = c$, we least-squares solve

$$\Sigma y = c$$

or

$$\sigma_1 y_1 = c_1$$
$$\cdots$$
$$\sigma_r y_r = c_r$$
$$0 = c_{r+1}$$
$$\cdots$$
$$0 = c_m.$$

To get the left and right sides as close as possible, we need

$$y_1 = c_1/\sigma_1$$
$$\cdots$$
$$y_r = c_r/\sigma_r$$

and we assign

$$y_{r+1} = 0$$
$$\cdots$$
$$y_n = 0.$$

(Actually, $y_{r+1}, \ldots, y_n$ could be assigned any values. They are free variables.)

Thus,

$$V^H x = \begin{bmatrix} c_1/\sigma_1 \\ \cdots \\ c_r/\sigma_r \\ 0 \\ \cdots \\ 0 \end{bmatrix}$$

and

$$x = V \begin{bmatrix} c_1/\sigma_1 \\ \cdots \\ c_r/\sigma_r \\ 0 \\ \cdots \\ 0 \end{bmatrix}.$$

Note, if $r < n$, there are infinitely many least-squares solutions. The one we computed is the least-squares solution of smallest norm. Others can also be computed.

**Example 7.3** *Consider*

$$Ax = b$$

*where* $A = \begin{bmatrix} 1 & -1 \\ -1 & 1 \end{bmatrix}$ *and* $b = \begin{bmatrix} 2 \\ 1 \end{bmatrix}$. *Note that* $Ax = x_1 \begin{bmatrix} 1 \\ -1 \end{bmatrix} +$ $x_2 \begin{bmatrix} -1 \\ 1 \end{bmatrix}$ *over all* $x$ *yields* span $\left\{ \begin{bmatrix} 1 \\ -1 \end{bmatrix}, \begin{bmatrix} -1 \\ 1 \end{bmatrix} \right\}$, *and observe in Figure 7.2 that* $b$ *is not in this span. So there is no solution to the equation*

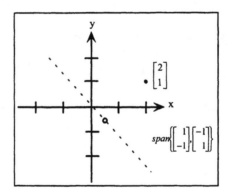

FIGURE 7.2.

$Ax = b$.

*To least-squares solve this equation, we factor* $A = U\Sigma V^H$, *yielding* $U = \begin{bmatrix} \frac{1}{\sqrt{2}} & \frac{1}{\sqrt{2}} \\ -\frac{1}{\sqrt{2}} & \frac{1}{\sqrt{2}} \end{bmatrix}$, $\Sigma = \begin{bmatrix} 2 & 0 \\ 0 & 0 \end{bmatrix}$, *and* $V^H = \begin{bmatrix} \frac{1}{\sqrt{2}} & -\frac{1}{\sqrt{2}} \\ \frac{1}{\sqrt{2}} & \frac{1}{\sqrt{2}} \end{bmatrix}$. *We multiply*

$$Ax = b$$

*through by* $U^H$ *to get*

$$\Sigma V^H x = U^H b.$$

*Simplifying, set $y = V^H x$ and solve*

$$\Sigma y = U^H b$$

*or*

$$\begin{bmatrix} 2 & 0 \\ 0 & 0 \end{bmatrix} y = \begin{bmatrix} \frac{1}{\sqrt{2}} \\ \frac{3}{\sqrt{2}} \end{bmatrix},$$

*which yields*

$$2y_1 = \frac{1}{\sqrt{2}}$$

$$0y_2 = \frac{3}{\sqrt{2}}.$$

*We set $y_2 = 0$ and by solving, $y_1 = \frac{1}{2\sqrt{2}}$. Hence, $y = \begin{bmatrix} \frac{1}{2\sqrt{2}} \\ 0 \end{bmatrix}$. Thus,*

$$V^H x = \begin{bmatrix} \frac{1}{\sqrt{2}} \\ 0 \end{bmatrix} \text{ and so } x = V \begin{bmatrix} \frac{1}{\sqrt{2}} \\ 0 \end{bmatrix} = \begin{bmatrix} \frac{1}{4} \\ -\frac{1}{4} \end{bmatrix}.$$

*Checking visually, since*

$$Ax = \begin{bmatrix} \frac{1}{2} \\ -\frac{1}{2} \end{bmatrix},$$

*we see that $Ax$, marked with an o in the graph, is the closest vector.*

Although the SVD approach is a very accurate method (in terms of computer computations) for finding least-squares solutions, other methods, such as the $QR$-decompositions, are often used instead. (They are faster.)

Concluding this section, we show that least-squares solutions can be obtained by solving the classical normal equations.

**Theorem 7.2** *Let $A$ be an $m \times n$ matrix and $b$ an $m \times 1$ vector. The least-square solutions to*

$$Ax = b$$

*can be found by solving the normal equations*

$$A^H Ax = A^H b.$$

**Proof.** Let $U \Sigma V^H$ be a singular value decomposition of $A$. Then, the least-squares solutions to

$$Ax = b$$

are the least-squares solutions to

$$\Sigma V^H x = U^H b. \tag{7.5}$$

Multiplying through by $\Sigma^H$ yields

$$\Sigma^H \Sigma V^H x = \Sigma^H U^H b. \tag{7.6}$$

We now need to make an observation. For it, let $\Sigma = \begin{bmatrix} 3 & 0 & 0 \\ 0 & 2 & 0 \\ 0 & 0 & 0 \end{bmatrix}$ and

$U^H v = \begin{bmatrix} 4 \\ 7 \\ 5 \end{bmatrix}$. Then (7.5) is

$$\begin{bmatrix} 3 & 0 & 0 \\ 0 & 2 & 0 \\ 0 & 0 & 0 \end{bmatrix} V^H x = \begin{bmatrix} 4 \\ 7 \\ 5 \end{bmatrix}$$

and (7.6) is

$$\begin{bmatrix} 9 & 0 & 0 \\ 0 & 4 & 0 \\ 0 & 0 & 0 \end{bmatrix} V^H x = \begin{bmatrix} 12 \\ 14 \\ 0 \end{bmatrix}.$$

Now getting the sides as close as possible, we observe that the 'solutions' to (7.6) are precisely the 'least-squares solutions' to (7.5).

Now, multiplying (7.6) through by $V$, which won't change the solutions, yields,

$$V \Sigma^H \Sigma V^H x = V \Sigma^H U^H b \tag{7.7}$$

and inserting $U^H U$ yields

$$V \Sigma^H U^H U \Sigma V^H x = V \Sigma^H U^H b$$

or

$$A^H A x = A^H b.$$

Thus, the solutions to these normal equations are the least-squares solutions to $Ax = b$. ∎

**Example 7.4** *Using the data of the previous example, Example 7.3, we have*

$$A = \begin{bmatrix} 1 & -1 \\ -1 & 1 \end{bmatrix} \ and \ b = \begin{bmatrix} 2 \\ 1 \end{bmatrix}.$$

*To least-squares solve*

$$Ax = b,$$

*we look at the normal equations*

$$A^t A x = A^t b,$$

*or*

$$\begin{bmatrix} 2 & -2 \\ -2 & 2 \end{bmatrix} x = \begin{bmatrix} 1 \\ -1 \end{bmatrix}$$

*which has solutions. Using Gaussian elimination, we get*

$$x = \begin{bmatrix} \frac{1}{2} + \alpha \\ \alpha \end{bmatrix} = \begin{bmatrix} \frac{1}{2} \\ 0 \end{bmatrix} + \alpha \begin{bmatrix} 1 \\ 1 \end{bmatrix},$$

*where $\alpha$ is an arbitrary scalar, as the least-squares solutions to $Ax = b$.*

*For $\alpha = -\frac{1}{4}$, $x = \begin{bmatrix} \frac{1}{4} \\ -\frac{1}{4} \end{bmatrix}$ our least-squares solution, with smallest norm, found in the previous example.*

### 7.1.1 Optional (Physical Problems involving Least-Squares Solutions)

We look at three problems showing how least-squares solutions arise.

1. Data Problem: Suppose it is observed that the number of chirps $c$ that a cricket makes is a function $f$ of the temperature $t$ of his environment, i.e., $f(t) = c$. To determine $f$, we collect data, say,

| temperature | $t_1$ | $t_2$ | $\cdots$ | $t_m$ |
|---|---|---|---|---|
| chirps | $c_1$ | $c_2$ | $\cdots$ | $c_m$ |

and plot this data as in Figure 7.3. The data (considered approximately correct for $f(t) = c$) appears to be slight deviations of points on a line and thus we assume $f(t) = \alpha + \beta t$ and try to determine $\alpha$ and $\beta$.

To determine $\alpha$ and $\beta$ so the line 'best' fits the data, we write out the equations determined from the table by plugging in the data. So we have

$$\alpha + \beta t = c$$
$$\alpha + \beta t_1 = c_1$$
$$\alpha + \beta t_2 = c_2$$
$$\cdots$$
$$\alpha + \beta t_m = c_m$$

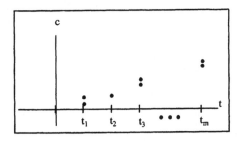

FIGURE 7.3.

or in matrix form

$$\begin{bmatrix} 1 & t_1 \\ 1 & t_2 \\ & \cdots \\ 1 & t_m \end{bmatrix} \begin{bmatrix} \alpha \\ \beta \end{bmatrix} = \begin{bmatrix} c_1 \\ c_2 \\ \cdots \\ c_m \end{bmatrix}.$$

(Here, some of the $t_i$'s could be the same.) Now, we want $\alpha$ and $\beta$ so that the difference between the two sides is as small as possible. Thus, we want $(\alpha + \beta t_1 - c_1)^2 + \cdots + (\alpha + \beta t_m - c_m)^2$ as small as possible. (Recall that the minimum of a square root can be found by finding the minimum of the radicand.) This assures us that we are getting the minimum of the sum of the squared distances between the line points $(t_i, \alpha + \beta t_i)$ and the data points $(t_i, c_i)$. (See Figure 7.4.) From this we see that we want a least-squares solution to the

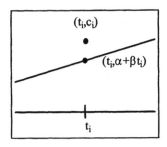

FIGURE 7.4.

equation.

2. Least-squares fitting curves: Here we extend the work given in 1 by looking at data. For this recall that in Chapter 3, Section 1, we saw how to find a polynomial that fits through data. For example, if the data is

| $x$ | 1 | 2 | 3 | 4 | 5 |
|---|---|---|---|---|---|
| $y$ | 0.9 | 4.2 | 8.7 | 16.2 | 24.5 |

we can find a polynomial $p$, of degree at most 4, which passes through the data. The data and this polynomial are shown in Figure 7.5, the graph of the polynomial being shown with the dashed curve.

In looking at this polynomial, especially at the ends, we might wonder if the polynomial describes the relationship of this data. Perhaps, by just looking at the data, we might feel that the data, which probably has some error, is more like a quadratic. Thus, we least-squares fit a polynomial of degree 2 to this data. Since $q(x) = ax^2 + bx + c$ is the general form for a quadratic, we can least-squares solve for the coefficients by solving

$$
\begin{array}{rrrr}
ax^2 & +bx & +c = & y \\
a & +b & +c = & 0.9 \\
4a & +2b & +c = & 4.2 \\
9a & +3b & +c = & 8.7 \\
16a & +4b & +c = & 16.2 \\
25a & +5b & +c = & 24.5
\end{array}
$$

MATLAB gives this function by using the command polyfit$(x, y, 2)$, and using this we find

$$q(x) = 0.9286x^2 + 0.33486x - 0.3600.$$

The graph of this polynomial is also in Figure 7.5, and we probably would agree that this fits the data better than does $p$. Of course, in making such a decision, knowing where the data is from, and a sense or feel for that problem is helpful.

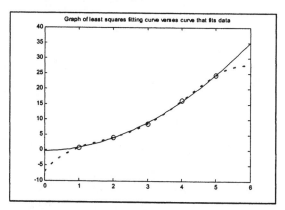

FIGURE 7.5.

3. Space problem: Note that

$$x + y + z = 1$$

is a subspace of dimension 2 in $R^3$. And, note that $(1,1,1)^t$ is not on this subspace. We want to find the closest point on the subspace to $(1,1,1)^t$. Observe in Figure 7.6 that $(1,-1,0)^t$ and $(1,0,-1)^t$ span

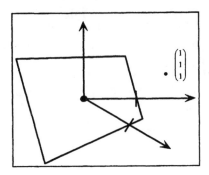

FIGURE 7.6.

the subspace. Thus, we want to find those scalars $x_1$ and $x_2$, which cause the left side of

$$x_1 \begin{bmatrix} 1 \\ -1 \\ 0 \end{bmatrix} + x_2 \begin{bmatrix} 1 \\ 0 \\ -1 \end{bmatrix} = \begin{bmatrix} 1 \\ 1 \\ 1 \end{bmatrix}$$

to be as close to the right side as possible. So, we need to find the least-squares solution to

$$\begin{bmatrix} 1 & 1 \\ -1 & 0 \\ 0 & -1 \end{bmatrix} \begin{bmatrix} x_1 \\ x_2 \end{bmatrix} = \begin{bmatrix} 1 \\ 1 \\ 1 \end{bmatrix}.$$

Of course, this problem could also be solved by Fourier sums or by using the orthogonal projection matrix.

## 7.1.2   MATLAB (Least-Squares Solutions to $Ax = b$)

For the computations used in this section, we can use the following.

1. Least-squares solution to $Ax = b$ where $A$ is $m \times n$ and $m \neq n$: to produce a least-squares solution, we can use $A \backslash b$. We would like to describe our problems here so that the columns of $A$ are linearly independent.

2. Singular Value Decomposition: To compute the $SVD$ of a matrix $A$, use the command $[U, S, V] = svd(A)$. Recall $A = U\Sigma V^t$. The $S$ given is $\Sigma$. If all we need is $\Sigma$, the command is $svd(A)$, which gives the singular values for $A$.

Type in *help mldivide*, *help svd* for further information.  It may also be interesting to check documentation on how SVD is computed.  See Bibliography.

### Code for Picture in Optional

$$x = \begin{bmatrix} 1 & 2 & 3 & 4 & 5 \end{bmatrix};$$
$$y = \begin{bmatrix} .9 & 4.2 & 8.7 & 16.2 & 24.5 \end{bmatrix};$$
$p = $polyfit$(x, y, 4);$
$q = $polyfit$(x, y, 2);$
$xi = $linspace$(0, 6, 50);$
$z = $polyval$(p, xi);$
$w = $polyval$(q, xi);$
plot$(x, y, \text{'O'}, xi, z, \text{':'}, xi, w)$

## Exercises

1. Solve $\begin{bmatrix} 1 & 1 \\ 1 & 1 \end{bmatrix} x = \begin{bmatrix} 1 \\ 2 \end{bmatrix}$ by using the

   (a) SVD approach.
   (b) Normal equations approach.

2. Least-squares fit a line through

   | $x$ | 1 | 2 | 2 |
   |-----|---|---|---|
   | $y$ | 1 | 1 | 2 |

3. Write out the system of linear equations whose least-squares solutions gives the quadratic $(y = ax^2 + bx + c)$ which least-squares fits

   | $x$ | 0 | 1 | 2 | 4 |
   |-----|---|---|---|---|
   | $y$ | 0 | 1 | 3 | 6 |

4. Find the point on the line $y = x$ closest to the point $\begin{bmatrix} 2 \\ 1 \end{bmatrix}$ using

   (a) The least-squares approach.
   (b) The orthogonal projection matrix.
   (c) The Fourier sum approach.

5. Find an SVD for the following.

   (a) $A = \begin{bmatrix} 1 & 1 & 1 \\ 1 & 1 & 1 \\ 1 & 1 & 1 \end{bmatrix}$     (b) $A = \begin{bmatrix} 1 & 0 \\ 0 & 1 \\ 0 & 1 \end{bmatrix}$

   (c) $A = \begin{bmatrix} 1 & 1 & 1 \\ 1 & 1 & 1 \end{bmatrix}$

6. Derive the SVD theorem starting with

$$U(AA^H)U^H = \Sigma$$

and proceed to find $V$.

7. Let $A = \begin{bmatrix} a_{11} & a_{12} \\ a_{21} & a_{22} \end{bmatrix}$ and $b = \begin{bmatrix} b_1 \\ b_2 \end{bmatrix}$. Let $x = \begin{bmatrix} x_1 \\ x_2 \end{bmatrix}$ and $f(x_1, x_2) = \| Ax - b \|_2$.

   (a) To find the critical points of $f$, we would solve

   $$\frac{\delta f}{\delta x_1}(x_1, x_2) = 0$$
   $$\frac{\delta f}{\delta x_2}(x_1, x_2) = 0.$$

   Show that the solutions $x$ to these equations satisfy

   $$A^t Ax = A^t b.$$

   (b) Can we do the same thing with $\|\cdot\|_1$? Explain.

8. Let $A$ be an $m \times n$ matrix. Suppose $U_1 \Sigma V_1^H$ and $U_2 \Sigma V_2^H$ are singular value decompositions of $A$. Is $U_1 = U_2$, $V_1 = V_2$?

9. Let $A$ be an $n \times n$ symmetric matrix with eigenvalues $\lambda_1, \ldots, \lambda_n$. Prove that the singular values of $A$ are $|\lambda_1|, \ldots, |\lambda_n|$.

10. Let $A$ be a Hermitian matrix. Suppose $A = UDU^H$ is a unitary diagonalization of $A$. Is $UDU^H$ an $SVD$ of $A$?

11. If $U\Sigma V^H$ is a singular value decomposition of $A$, prove that $V\Sigma^H U^H$ is one for $A^H$. (Thus, we can convert an $m \times n$ problem into $n \times m$ problem.)

12. Explain when we can get an SVD using only real numbers.

13. Prove Corollary 7.1,

   (a) Part (a) for $m \times n$ matrices.
   (b) Part (b).

14. Rewrite $A = U\Sigma V^H$ into $\hat{U}D\hat{V}^H$ by truncating columns of $U$ and rows of $V^H$, where $D = diag(\sigma_1, \ldots, \sigma_r)$ is $r \times r$. Here $A = \begin{bmatrix} 1 & 1 & 0 \\ 1 & 0 & 1 \\ 2 & 1 & 1 \end{bmatrix}$.

15. Prove that the least-squares solution computed, from (7.3) on, is the least-squares solution of smallest norm.

16. Let $A = U\Sigma V^H$ be an SVD of $A$.

    (a) Prove the columns of $V$ (called right singular vectors) are eigenvectors for $A^H A$.

    (b) Prove the columns of $U$ (called the left singular vectors) are eigenvectors of $AA^H$.

    (c) Show by an example, that if left and right singular vectors of $A$ are found, $U$ and $V^H$ formed from these singular vectors, then $U\Sigma V^H$ need not be $A$.

17. (MATLAB) Two parts:

    (a) Compute the least-squares solutions to

    i. $\begin{bmatrix} 1 & 2 \\ 1 & 3 \\ 1 & 4 \end{bmatrix} x = \begin{bmatrix} 1 \\ 3 \\ 1 \end{bmatrix}$.

    ii. $\begin{bmatrix} 1 & 1 \\ 1 & 1 \end{bmatrix} x = \begin{bmatrix} .9 \\ 1.1 \end{bmatrix}$.

    (b) Find the SVD of

    i. $\begin{bmatrix} 1 & -1 \\ 2 & 1 \\ -2 & 3 \end{bmatrix}$.    ii. $\begin{bmatrix} .1 & -.3 & 2 \\ 4 & 3.1 & 0 \end{bmatrix}$.

18. (MATLAB) Consider the data given below.

| $x$ | -1 | 0 | 1 | 2 | 3 |
|---|---|---|---|---|---|
| $y$ | .9 | .1 | 1.1 | 3.9 | 9.1 |

    (a) Find the polynomial that passes through this data.
    (b) Find the quadratic polynomial that least-squares fits this data.
    (c) Plot the graph of both polynomials and the data.

19. (MATLAB) Suppose we want to estimate the value of a spring constant, say, for one of our spring-mass problems. By stretching the spring and recording the force to do so, we collect the following data.

| $f$ | .4 | .8 | 1.3 | 1.7 |
|---|---|---|---|---|
| $d$ | .2 | .4 | .6 | .8 |

Using Hooke's Law: Force = spring constant times displacement, use least-squares and the data above to estimate the value of the spring constant. (Recall measurements can be in error.)

## 7.2  Applications of the SVD Theorem

The SVD theorem is a remarkably strong tool in matrix theory. In this section, we look at a few additional uses of this theorem. In these applications we assume that the matrix $A$ has SVD as described in Theorem 7.1.

**1. Distance to the closest** rank k $m \times n$ **matrix**

Given $m$ and $n$, define

$$\text{Rank } k = \{B : B \text{ is an } m \times n \text{ matrix having rank } k\}.$$

(Note that Rank $k$ is not a subspace.) Define the distance and relative distance from an $m \times n$ matrix $A$ to Rank $k$ as

$$d(A, \text{Rank } k) = \min \|A - B\|_2$$
$$d_{rel}(A, \text{Rank } k) = \min \frac{\|A - B\|_2}{\|A\|_2}$$

where the minimums are over all $B \in$ Rank $k$. (We will show minimums exist.) The result we want follows.

**Theorem 7.3** *Let $A$ be an $m \times n$ matrix of* rank $r$. *Then, if $k < r$,*

*(a)* $d(A, \text{Rank } k) = \sigma_{k+1}$.

*(b)* $d_{rel}(A, \text{Rank } k) = \frac{\sigma_{k+1}}{\sigma_1}$.

**Proof.** We prove both parts.
Part a. Note that if $B \in$ Rank $k$, since multiplying by unitary matrices doesn't change distance,

$$\|A - B\|_2 = \|U\Sigma V^H - B\|_2$$
$$= \|\Sigma - U^H BV\|_2$$
$$= \|\Sigma - C\|_2$$

where $C = U^H BV$ and $C \in$ Rank $k$. Thus

$$d(A, \text{Rank } k) = d(\Sigma, \text{Rank } k).$$

We now break Part (a) into two parts.

i. $d(\Sigma, C) \geq \sigma_{k+1}$ for all $C \in$ Rank $k$. To see this, let $C \in$ Rank $k$. For simplicity of notation, we will assume that first $k$ columns of

$C$ are linearly independent. Define an $(n - k - 1) \times n$ matrix $E$ in partitioned form, as

$$E = [\; 0 \quad I \;]$$

where $I$ is the $(n - k - 1) \times (n - k - 1)$ identity matrix. (If $n - k - 1 = 0$, $E$ is $0 \times n$, i.e., it won't appear in what follows.) Then, since $\text{rank}\, C = k$,

$$\text{rank} \begin{bmatrix} C \\ E \end{bmatrix} = k + (n - k - 1) = n - 1.$$

Thus there is an $x$, $\|x\|_2 = 1$, such that $\begin{bmatrix} C \\ E \end{bmatrix} x = 0$. Note that, since $Ex = 0$, $x_{k+2} = \cdots = x_n = 0$.

Now

$$\|\Sigma - C\|_2 = \max_{\|y\|_2 = 1} \|(\Sigma - C)\, y\|_2 \geq \|(\Sigma - C)\, x\|_2$$

$$= \|\Sigma x\|_2 = \left( \sigma_1^2 \, |x_1|^2 + \cdots + \sigma_{k+1}^2 \, |x_{k+1}|^2 \right)^{\frac{1}{2}}$$

$$\geq \sigma_{k+1} \|x\|_2 = \sigma_{k+1}.$$

Thus, since $C$ was chosen arbitrarily, (i) follows.

ii. $d(\Sigma, C) = \sigma_{k+1}$. To show this, by (i), we need to find only one $C \in \text{Rank}\, k$ such that $\|(\Sigma - C)\|_2 = \sigma_{k+1}$. For this $C$, set $C = diag\,(\sigma_1, \ldots, \sigma_k, 0, \ldots, 0)$. Then, as in the exercises, $\|(\Sigma - C)\|_2 = \sigma_{k+1}$.

Part b. Apply Corollary 7.1, and (a) of this theorem. ∎

Using the theorem, if $A$ is nonsingular, then by (a) the closest singular matrix to $A$ has rank $n - 1$ and the distance is $\sigma_n$. Also by (b)

$$\frac{1}{c_2(A)} = \frac{\sigma_n}{\sigma_1}$$

$$= \min \frac{\|A - B\|_2}{\|A\|_2}$$

where the minimum is over all singular matrices $B$. Thus

$$\frac{1}{c_2(A)} \|A\|_2 = \min \|A - B\|_2 .$$

This says that over all nonsingular matrices $A$, such that , say, $\|A\|_2 = c$, $c$ is a constant, the matrices which have the larger condition numbers are closer to being singular, and vice versa.

### 2. Moore-Penrose Pseudo-inverse of $A$

We know that the inverse exists for all nonsingular matrices. Actually, however, the notion of inverse has been extended to all matrices (even 0). We will show how.

The Moore-Penrose pseudo-inverse of an $m \times n$ matrix $A$ is an $n \times m$ matrix $X$ such that

   i. $AXA = A$.

  ii. $XAX = X$.

 iii. $AX$ is Hermitian.

 iv. $XA$ is Hermitian.

As we now show, each matrix has precisely one pseudo-inverse.

**Lemma 7.1** *Let $A$ be an $m \times n$ matrix. Then there is a unique $n \times m$ matrix $X$ which satisfies (i) through (iv).*

**Proof.** We prove two parts.
Part a. For the existence, set

$$X = V \operatorname{diag}\left(\sigma_1^{-1}, \ldots, \sigma_r^{-1}, 0, \ldots, 0\right) U^H.$$

Then $X$ satisfies (i) through (iv).
Part b. For the uniqueness, suppose $X$ and $Y$ are solutions to (i) through (iv). Then

$$
\begin{aligned}
X &= XAX = XX^H A^H = XX^H A^H Y^H A = XAXAY \\
&= XAY = XAYAY = A^H X^H A^H Y^H Y = A^H Y^H Y \\
&= YAY = Y.
\end{aligned}
$$

Thus $X = Y$. ∎

Since $A$ has precisely one pseudo-inverse, we can denote it by $A^+$. And since, if $A$ is nonsingular, $A^{-1}$ satisfies properties (i) through (iv), we have that $A^+ = A^{-1}$. So the pseudo-inverse extends the notion of the inverse to all matrices.

To see some use for this generalization, recall that if $A$ is nonsingular, then

$$Ax = b$$

has a solution $x = A^{-1}b$. We show an extended such result for all matrices.

**Corollary 7.2** *Let $Ax = b$ be a system of linear equations where $A$ is an $m \times n$ matrix and $b$ is $m \times 1$ vector. Then $x = A^+b$ is the least-squares solution to this equation that has the smallest 2-norm.*

**Proof.** To least-squares solve $Ax = b$, we least squares solve $\Sigma V^H x = U^H b$ or $\Sigma y = U^H b$ where $y = V^H x$. Now note that the least-squares solutions to $\Sigma y = U^H b$ are precisely the solutions to

$$\Sigma^+ \Sigma y = \Sigma^+ U^H b \qquad (7.8)$$

where

$$\Sigma^+ = \text{diag}\left(\sigma_1^{-1}, \ldots, \sigma_r^{-1}, 0, \ldots, 0\right).$$

For example, if $\Sigma = \begin{bmatrix} 3 & 0 & 0 \\ 0 & 2 & 0 \\ 0 & 0 & 0 \end{bmatrix}$ and $U^H b = \begin{bmatrix} 4 \\ 7 \\ 5 \end{bmatrix}$, then $\Sigma y = U^H b$ is

$$3y_1 = 4$$
$$2y_2 = 7$$
$$0 = 5,$$

while $\Sigma^+ \Sigma y = \Sigma^+ U^H b$ is

$$y_1 = \frac{4}{3}$$
$$y_2 = \frac{7}{2}$$
$$0 = 0.$$

Let $\hat{y}$ be the solution to (7.8) where all free variables are set to 0. Then $\Sigma^+ \Sigma \hat{y} = \hat{y}$ so (7.8) becomes

$$\hat{y} = \Sigma^+ U^H b.$$

Since $\hat{y} = V^H \hat{x}$ ($\hat{x}$ defined by $\hat{x} = V\hat{y}$)

$$V^H \hat{x} = \Sigma^+ U^H b$$

or

$$\hat{x} = V\Sigma^+ U^H b$$
$$= A^+ b.$$

Finally, since $\|\hat{y}\|_2$ is the smallest possible solution to (7.8), and $\|\hat{x}\|_2 = \|\hat{y}\|_2$, $\|\hat{x}\|_2$ is the smallest least-squares solutions to $Ax = b$, the desired result. ∎

There are many other uses of the psuedo-inverse.

### 3. Computing range and null space

To compute the range and null space for $A$, recall that

$$AV = U\Sigma.$$

By partitioned multiplication on the left and backward multiplying on the right, we have

$$Av_i = \sigma_i u_i$$

for $i = 1, \ldots, r$ and

$$Av_i = 0, \quad \text{otherwise.} \tag{7.9}$$

Both of our results will be obtained from (7.9). To see this, recall from Section 3 of Chapter 2 that, if $v_1, \ldots, v_n$ is a basis for $E^n$ then $Av_1, \ldots, Av_n$ form a spanning set for the range of $A$. Thus, the vectors $\alpha_1 u_1, \ldots, \alpha_r u_r$, and thus $u_1, \ldots, u_r$ span the range of $A$, i. e.,

$$R(A) = \text{span}\{u_1, \ldots, u_r\}$$

For $N(A)$, suppose $Ax = 0$. Then by writing $x = \alpha_1 v_1 + \cdots + \alpha_n v_n$, and noting (7.9), we see that

$$
\begin{aligned}
Ax &= \alpha_1 \left( \sigma_1 u_1 \right) + \cdots + \alpha_r \left( \sigma_r u_r \right) \\
&= \left( \alpha_1 \sigma_1 \right) u_1 + \cdots + \left( \alpha_r \sigma_r \right) u_r \\
&= 0.
\end{aligned}
$$

Since $u_1, \ldots, u_r$ are linearly independent, $\alpha_1 \sigma_1 = 0, \ldots, \alpha_r \sigma_r = 0$ or $\alpha_1 = 0, \ldots, \alpha_r = 0$. Thus, $x = \alpha_{r+1} v_{r+1} + \cdots + \alpha_n v_n$. And any vector in this form is in $N(A)$, so

$$N(A) = \text{span}\{v_{r+1}, \ldots, v_n\}.$$

(Numerically, this is a good way to compute $R(A)$ and $N(A)$.)

As given in Chapter 5, Section 4, if we set $U_1 = [u_1 \ldots u_r]$ and $V_2 = [v_{r+1} \ldots v_n]$, then $U_1 U_1^t$ and $V_2 V_2^t$ are orthogonal projection matrices onto $R(A)$ and $N(A)$, respectively.

**Example 7.5** *In Example 7.2 we showed for $A = \begin{bmatrix} 1 & 1 & 1 \\ 1 & 1 & 1 \end{bmatrix}$ that $U =$*

$\begin{bmatrix} \frac{1}{\sqrt{2}} & -\frac{1}{\sqrt{2}} \\ \frac{1}{\sqrt{2}} & \frac{1}{\sqrt{2}} \end{bmatrix}$, $V = \begin{bmatrix} \frac{1}{\sqrt{3}} & \frac{1}{\sqrt{2}} & \frac{1}{\sqrt{6}} \\ \frac{1}{\sqrt{3}} & -\frac{1}{\sqrt{2}} & \frac{1}{\sqrt{6}} \\ \frac{1}{\sqrt{3}} & 0 & -\frac{2}{\sqrt{6}} \end{bmatrix}$, *and* $\Sigma = \begin{bmatrix} \sqrt{6} & 0 & 0 \\ 0 & 0 & 0 \end{bmatrix}$.

*Thus, we have*

$$R(A) = \text{span}\{u_1\} = \text{span}\left\{ \begin{bmatrix} \frac{1}{\sqrt{2}} \\ \frac{1}{\sqrt{2}} \end{bmatrix} \right\}$$

*a line in $R^2$, and*

$$N\left(A\right) = \text{span}\left\{v_2, v_3\right\} = \text{span}\left\{\begin{bmatrix} \frac{1}{\sqrt{2}} \\ -\frac{1}{\sqrt{2}} \\ 0 \end{bmatrix}\begin{bmatrix} \frac{1}{\sqrt{6}} \\ \frac{1}{\sqrt{6}} \\ -\frac{2}{\sqrt{6}} \end{bmatrix}\right\},$$

*a plane in $R^3$.*

### 4. Computing numerical rank

Note that rank $A$ is not a continuous function of the entries of $A$. For example, let $A\left(\epsilon\right) = \begin{bmatrix} 1 & 1+\epsilon \\ 1 & 2 \end{bmatrix}$ and $r\left(\epsilon\right) = \text{rank}\, A\left(\epsilon\right)$. The graph of $r\left(\epsilon\right)$ is given in Figure 7.7. Thus $r$ is discontinuous at $\varepsilon = 1$.

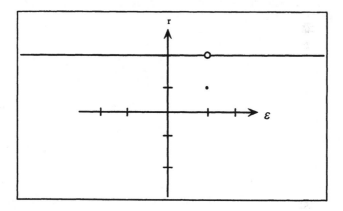

FIGURE 7.7.

Because of discontinuity, it can be difficult to compute rank. For example, MATLAB says it will 'approximate' the rank of a matrix.

The singular value decomposition,

$$A = U\Sigma V^H$$

is often used to estimate rank; a tolerance $\delta$ (estimating what singular values may actually be 0 but do not appear so due to rounding) is given. If

$$\sigma_1 \geq \cdots \geq \sigma_r > \delta \geq \sigma_{r+1} \geq \cdots \geq \sigma_s,$$

then the numerical rank is set equal to $r$. MATLAB uses

$$\delta = \text{tol} = \max\left(m, n\right) \cdot \|A\|_2 \cdot eps,$$

where our $eps = 2.2204 \times 10^{-16}$ (a MATLAB number which indicates computations are done in about 15 digits.)

As an example, if we define $Q = \begin{bmatrix} \frac{1}{\sqrt{2}} & \frac{1}{\sqrt{2}} \\ \frac{1}{\sqrt{2}} & -\frac{1}{\sqrt{2}} \end{bmatrix}$, $\Sigma = \text{diag}\left(100,\ 10^{-14}\right)$, and $A = Q\Sigma Q^t$, then MATLAB gives rank $A = 1$. Note that

$$\delta = 2 \times 100 \times 2.2204 \times 10^{-16}$$
$$= 4.4408 \times 10^{-14} > \sigma_2 = 10^{-14}.$$

So, rank $A$ was estimated at 1. (Here, we are looking at a rigged problem.)

## 5. Data compacting

Suppose, for simplicity, we have a $3 \times 3$ array of pixels which can be lit to form pictures. (Pixels could be in different colors, but we will use black and white.) We associate a $3 \times 3$ matrix $A$ with $a_{ij} = 1$ if the ij-th pixel is to be lit and 0's otherwise. An example is below.

$$\mathbf{L} \qquad A = \begin{bmatrix} 1 & 0 & 0 \\ 1 & 0 & 0 \\ 1 & 1 & 0 \end{bmatrix}$$

Note, to represent "$L$," or any figure in our $3 \times 3$ array of pixels, we need to store 9 entries in the array.

The singular value decomposition of $A$ can be written as

$$
\begin{aligned}
A &= \sigma_1 u_1 v_1^H + \sigma_2 u_2 v_2^H \\
&= 1.85 \begin{bmatrix} -.5 \\ -.5 \\ -.707 \end{bmatrix} [-.924, -.384, 0] + .715 \begin{bmatrix} -.5 \\ -.5 \\ -.707 \end{bmatrix} [-.384, -.924, 0] \\
&= \begin{bmatrix} .855 & .355 & 0 \\ .855 & .355 & 0 \\ 1.21 & .523 & 0 \end{bmatrix} + \begin{bmatrix} .137 & -.330 & 0 \\ .137 & -.330 & 0 \\ -.194 & .467 & 0 \end{bmatrix}.
\end{aligned}
$$

If we use a simple rounding rule on $\sigma_1 u_1 v_1^H$ that entries of .5 and above are 1's and that those less than .5 are 0's, $\sigma_1 u_1 v_1^H$ determines the picture. (It can be that on larger problems, $n \times n$ rather than $3 \times 3$, $\sigma_1 u_1 v_1^H + \sigma_2 u_2 v_2^H$ may be required to produce the picture, or we may need even more terms. However, it should be observed that $\sigma_1 \geq \sigma_2 \geq \cdots$ and $\|u_i\|_2 = 1, \|v_j\|_2 = 1$ for all $i$ and $j$, so we expect to add matrices of smaller size each time.)

Thus to keep $L$ we need only retain $\sigma_1$, $u_1$, and $v_1$. Counting entries which we need to form $L$, we have

$$1(\text{for } \sigma_1) + 3(\text{for } u^1) + 3(\text{for } v_1^H) = 7$$

Of course, this is a reduction of 2 from the original 9 entries we needed to keep. (Perhaps 22% would be a better view.) However in larger problems, the savings can be great.

## 6. Representations of linear transformations

The SVD gives an interesting view of the linear transformation $L(x) = Ax$, where $A$ is an $m \times n$ real matrix. To see this, set $A = U\Sigma V^H$. Using the columns of $U$ and $V$, define $Z = \{u_1, ..., u_m\}$ and $Y = \{v_1, ..., v_n\}$. Then $Z$ and $Y$ are orthonormal bases for the codomain $R^m$ and the domain $R^n$ of $L$, respectively. And, if $z = [x]_Z$ and $y = [x]_Y$, we have that

$$Uz = x \text{ and } Vy = x$$

are the change of coordinates from these bases, respectively, to the given vectors.

We now convert $L(x) = Ax$, so the domain is given in terms of the coordinates using $Y$ and the codomain given in terms of the coordinates using $Z$. To do this, change the coordinates of $Ax$ into those for the basis $Z$ by multiplying them by $U^H$. Thus we have

$$U^H Ax.$$

And we change the coordinates of $x$ to those for the basis $Y$ by replacing $x$ by $Vy$, getting

$$U^H AVy.$$

Thus,

$$U^H AVy = U^H U\Sigma V^H Vy = \Sigma y.$$

Hence, in terms of new coordinates in the domain and codomain, (See Figure 7.8.)

$$L(y) = \Sigma y.$$

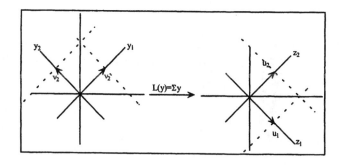

FIGURE 7.8.

An example may help.

**Example 7.6** *(Collapsing of space)*  Let $A = \begin{bmatrix} 1 & 2 \\ 1 & 2 \end{bmatrix}$. *An SVD of A is*

$$\begin{bmatrix} 0.7071 & -0.7071 \\ 0.7071 & 0.7071 \end{bmatrix} \begin{bmatrix} 3.1623 & 0 \\ 0 & 0 \end{bmatrix} \begin{bmatrix} 0.4472 & 0.8944 \\ -0.8944 & 0.4472 \end{bmatrix}.$$

*From this,* $Y = \{v_1, v_2\} = \left\{ \begin{bmatrix} 0.4472 \\ 0.8944 \end{bmatrix}, \begin{bmatrix} -0.8944 \\ 0.4472 \end{bmatrix} \right\}$ *and* $Z = \{u_1, u_2\} =$ $\left\{ \begin{bmatrix} 0.7071 \\ 0.7071 \end{bmatrix} \begin{bmatrix} -0.7071 \\ 0.7071 \end{bmatrix} \right\}$ *are bases for the domain and codomain, re- spectively. (See Figure 7.9.)*

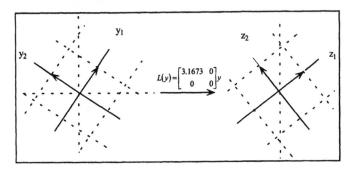

FIGURE 7.9.

*In terms of the coordinates of these bases, the transformation is described* as

$$\begin{bmatrix} z_1 \\ z_2 \end{bmatrix} = L(y) = \begin{bmatrix} 3.1623 & 0 \\ 0 & 0 \end{bmatrix} y$$
$$= \begin{bmatrix} 3.1623 y_1 \\ 0 \end{bmatrix}.$$

*Thus, the $y_2$-axis is collapsed (orthogonally projecting all points in $R^2$ onto the $y_1$-axis), and the $y_1$-axis is then stretched by 3.1623 and laid on the $z_1$-axis. (Note $L \begin{bmatrix} 1 \\ 0 \end{bmatrix} = 3.1626 \begin{bmatrix} 1 \\ 0 \end{bmatrix}$.)*

## 7.2.1   MATLAB (pinv, null, orth, and rank)

The computations discussed in this section can be done using MATLAB.

Use *pinv(A)* for the psuedo-inverse of $A$, *null(A)* for an orthonormal basis for the null space of $A$, and *orth(A)* for an orthonormal basis for the range of $A$.

# Exercises

1. What is $0^+$?

2. In proof of Lemma 7.1, Part b, tell why each statement is true.

3. Compute $A^+$ where $A$ is given below.

   (a) $A = \begin{bmatrix} 2 & 1 \\ 1 & 2 \end{bmatrix}$    (b) $A = \begin{bmatrix} 1 & 1 & 1 \\ 1 & 1 & 1 \end{bmatrix}$

   (c) $A = \begin{bmatrix} \sigma_1 & 0 & 0 \\ 0 & \sigma_2 & 0 \\ 0 & 0 & 0 \end{bmatrix}$    (d) $A = \begin{bmatrix} 1 & 0 \\ 0 & 1 \\ 0 & 1 \end{bmatrix}$

   where $\sigma_1$ and $\sigma_2$ are positive

4. Prove:

   (a) If $\sum = diag\,(\sigma_1, \ldots, \sigma_n)$ where $\sigma_1 \geq \cdots \geq \sigma_r > 0 = \sigma_{r+1} = \cdots = \sigma_n$, then $\sum^+ = diag\,(\sigma_1^{-1}, \ldots, \sigma_r^{-1}, 0, \ldots, 0)$.
   
   (b) If $A$ is nonsingular, then $A^+ = A^{-1}$.
   
   (c) If $A = U\Sigma V^H$, an SVD, then $A^+ = V\Sigma^+ U^H$.

5. How near are the following matrices to a singular matrix?

   (a) $A = \begin{bmatrix} 2 & 1 \\ 1 & 2 \end{bmatrix}$

   (b) $A = \begin{bmatrix} 1 & 1 & 0 \\ 1 & 0 & 1 \\ 0 & 1 & 1 \end{bmatrix}$

6. For each of the following matrices,

   (a) Find an orthonormal basis for $R(A)$ and $N(A)$, and
   
   (b) Find orthogonal projections on $R(A)$ and $N(A)$.

   $$A = \begin{bmatrix} 1 & 2 \\ 1 & 2 \end{bmatrix}, \qquad A = \begin{bmatrix} 2 & 0 & 0 \\ 0 & 1 & 1 \\ 0 & 1 & 1 \end{bmatrix}$$

7. Let $A = \begin{bmatrix} .501 & .499 \\ .499 & .501 \end{bmatrix}$. If $\delta$, as given in application 4, is .03, what is the numerical rank of $A$?

8. Use an SVD approach to represent $T$ in a $3 \times 3$ array, as done in application 5.

9. As in application 6, describe what $L$ does.

(a) $L(x) = \begin{bmatrix} 1 & 1 \\ 1 & 1 \end{bmatrix} x$

(b) $L(x) = \begin{bmatrix} 2 & -1 & -1 \\ 2 & 1 & 1 \end{bmatrix} x$

10. Let $A$ be a $2 \times 2$ matrix. Prove that in the Frobenius norm the closest unitary matrix to $A$ is $Q = UV^H$ where $A = U\Sigma V^H$ is a singular value decomposition of $A$. (Note that the proof can be extended to the $n \times n$ case.)

11. Prove $\|A\|_2^2 = \|A^t A\|_2$ and that $c_2(A^t A) = \|A^t A\|_2 \left\| A^{-1} (A^{-1})^t \right\|_2 = \|A\|_2^2 \|A^{-1}\|_2^2 = c_2(A)^2$. What does this say about solving normal equations numerically?

12. Prove that if $\Sigma = \text{diag}(\sigma, 0, \dots, 0)$, then $\|\Sigma\|_2 = |\sigma|$.

13. Prove that if $\sigma_1 > 0, \dots, \sigma_s > 0$ and $x_1, \dots, x_s$ vectors in $E^n$, then span $\{\sigma_1 x_1, \dots, \sigma_s x_s\} = \text{span}\{x_1, \dots, x_n\}$.

14. (MATLAB) Let $A = \begin{bmatrix} 1 & 1 & 1 & 1 \\ 1 & -1 & 1 & 0 \\ -1 & 0 & 0 & 1 \\ 0 & -1 & 0 & 1 \\ -1 & 1 & 0 & 1 \\ 1 & 0 & -1 & 1 \end{bmatrix}.$

(a) Find the distance from $A$ to the rank 1, rank 2, and rank 3 matrices.

(b) Find the pseudo-inverse of $A$ by using the singular value decomposition of $A$. Compare your results to that obtained using pinv($A$).

(c) Compute the range and null space of $A$, using an SVD and using the command orth and null.

(d) Compute rank $(A)$ and rank $(A^t)$.

(e) Solve $Ax = b$ where $b = [1, -1, 0, 1, 1, -1]^t$.

15. (MATLAB) Is there a matrix $A$ so that rank $A \neq$ rank $A^t$ in MAT-LAB?

16. (MATLAB) A 'house' is shown below.

$$\begin{bmatrix} 0 & 0 & 1 & 0 & 0 \\ 0 & 1 & 1 & 1 & 0 \\ 1 & 1 & 1 & 1 & 1 \\ 0 & 1 & 1 & 1 & 0 \\ 0 & 1 & 1 & 1 & 0 \end{bmatrix}$$

Using the expansion $A = \sigma_1 u_1 v_1^t + \cdots + \sigma_5 u_5 v_5^t$, show what the house looks like for

(a) $\sigma_1 u_1 v_1^t$.

(b) $\sigma_1 u_1 v_1^t + \sigma_2 u_2 v_2^t$.

(c) $\sigma_1 u_1 v_1^t + \sigma_2 u_2 v_2^t + \sigma_3 u_3 v_3^t$.

# 8

# LU and QR Decompositions

We have already seen that factoring matrices into simpler ones is important in developing and applying matrix theory. In this chapter we look at factoring an $m \times n$ matrix either as $LU$ (where $L$ is a lower triangular matrix and $U$ a row echelon form) or as $QR$ (where $Q$ is an orthogonal matrix and $R$ a row echelon form). Both factorizations involve a kind of Gaussian elimination approach and are highly used in numerical algorithms and software, such as MATLAB. Knowing this material also helps us understand the occasional warnings given with a MATLAB computation.

## 8.1  The LU Decomposition

Let $A$ be an $m \times n$ matrix. An $m \times m$ elementary matrix $E$, belonging to an elementary operation, is the matrix that produces by premultiplication, the elementary operation applied to $A$. Thus

$$EA = B$$

where $B$ is obtained by applying the elementary operation directly to $A$. For example if $m = 2$, we have the following.

$$R_1 \leftrightarrow R_2 \qquad E = \begin{bmatrix} 0 & 1 \\ 1 & 0 \end{bmatrix}$$

$$\alpha R \qquad E = \begin{bmatrix} \alpha & 0 \\ 0 & 1 \end{bmatrix}$$

$$\alpha R_1 + R_2 \qquad E = \begin{bmatrix} 1 & 0 \\ \alpha & 1 \end{bmatrix}.$$

Observe that each elementary matrix is nonsingular and its inverse reverses the elementary operation defining it. We will show this using the previous examples.

$$R_1 \leftrightarrow R_2 \quad E = \begin{bmatrix} 0 & 1 \\ 1 & 0 \end{bmatrix} \quad E^{-1} = \begin{bmatrix} 0 & 1 \\ 1 & 0 \end{bmatrix} \qquad R_1 \leftrightarrow R_2$$

$$\alpha R_1 \qquad E = \begin{bmatrix} \alpha & 0 \\ 0 & 1 \end{bmatrix} \quad E^{-1} = \begin{bmatrix} \alpha^{-1} & 0 \\ 0 & 1 \end{bmatrix} \qquad \alpha^{-1} R_1$$

$$\alpha R_1 + R_2 \quad E = \begin{bmatrix} 1 & 0 \\ \alpha & 1 \end{bmatrix} \quad E^{-1} = \begin{bmatrix} 1 & 0 \\ -\alpha & 1 \end{bmatrix} \qquad -\alpha R_1 + R_2.$$

Thus, if Gaussian elimination is applied to $A$ to obtain a row echelon form $U$, then elementary matrices corresponding to the elementary operations used, say, $E_1, \dots, E_r$, are such that

$$E_r \cdots E_1 A = U. \tag{8.1}$$

**Example 8.1** Let $A = \begin{bmatrix} 1 & -1 & 2 \\ 2 & 0 & 5 \\ 3 & 5 & 13 \end{bmatrix}$. Applying Gaussian elimination, we have

$$A \xrightarrow[\begin{subarray}{c} E_1 = \begin{bmatrix} 1 & 0 & 0 \\ -2 & 1 & 0 \\ 0 & 0 & 1 \end{bmatrix} \end{subarray}]{-2R_1 + R_2} \begin{bmatrix} 1 & -1 & 2 \\ 0 & 2 & 1 \\ 3 & 5 & 13 \end{bmatrix}$$

$$\xrightarrow[\begin{subarray}{c} E_2 = \begin{bmatrix} 1 & 0 & 0 \\ 0 & 1 & 0 \\ -3 & 0 & 1 \end{bmatrix} \end{subarray}]{-3R_1 + R_3} \begin{bmatrix} 1 & -1 & 2 \\ 0 & 2 & 1 \\ 0 & 8 & 7 \end{bmatrix}$$

$$\xrightarrow[\begin{subarray}{c} E_3 = \begin{bmatrix} 1 & 0 & 0 \\ 0 & 1 & 0 \\ 0 & -4 & 1 \end{bmatrix} \end{subarray}]{-4R_2 + R_3} \begin{bmatrix} 1 & -1 & 2 \\ 0 & 2 & 1 \\ 0 & 0 & 3 \end{bmatrix} = U.$$

Now

$$E_3 E_2 E_1 A = U.$$

If no interchange operations are used, then each $E_i$ is lower triangular, so $E_r \cdots E_1$ is lower triangular and so is $L = (E_r \cdots E_1)^{-1}$. Thus from (8.1),

$$A = LU.$$

If no scaling was applied (and there need not be) then, remarkably, $L$ can be computed easily from the multipliers used. (If $-\alpha R_i + R_j$ is applied, $\alpha$ is the multiplier.)    For example, to compute $L$, note that if

$$E_3 = \begin{bmatrix} 1 & 0 & 0 \\ 0 & 1 & 0 \\ 0 & -\gamma & 1 \end{bmatrix}, E_2 = \begin{bmatrix} 1 & 0 & 0 \\ 0 & 1 & 0 \\ -\beta & 0 & 1 \end{bmatrix}, E_1 = \begin{bmatrix} 1 & 0 & 0 \\ -\alpha & 1 & 0 \\ 0 & 0 & 1 \end{bmatrix}$$

then

$$L = E_1^{-1} E_2^{-1} E_3^{-1} = \begin{bmatrix} 1 & 0 & 0 \\ \alpha & 1 & 0 \\ 0 & 0 & 1 \end{bmatrix} \begin{bmatrix} 1 & 0 & 0 \\ 0 & 1 & 0 \\ \beta & 0 & 1 \end{bmatrix} \begin{bmatrix} 1 & 0 & 0 \\ 0 & 1 & 0 \\ 0 & \gamma & 1 \end{bmatrix}$$

$$= \begin{bmatrix} 1 & 0 & 0 \\ \alpha & 1 & 0 \\ \beta & \gamma & 1 \end{bmatrix}.$$

Thus $L$ can be computed by placing the multipliers used in their corresponding positions in $L$, so there is no need to keep track of the corresponding elementary matrices. (The order here, namely $E_1^{-1} E_2^{-1} E_3^{-1}$ is important. The computation of $E_3^{-1} E_2^{-1} E_1^{-1}$ cannot be done in the same way.) Thus, as shown in the following example, computing $LU$ is as efficient as finding $U$ by Gaussian elimination.

**Example 8.2** *Using the data from Example 8.1 and forming $L$ directly from the multipliers, we have*

$$L = \begin{bmatrix} 1 & 0 & 0 \\ 2 & 1 & 0 \\ 3 & 4 & 1 \end{bmatrix}.$$

*So*

$$A = LU$$

$$= \begin{bmatrix} 1 & 0 & 0 \\ 2 & 1 & 0 \\ 3 & 4 & 1 \end{bmatrix} \begin{bmatrix} 1 & -1 & 2 \\ 0 & 2 & 1 \\ 0 & 0 & 3 \end{bmatrix}.$$

Interchanges, however, may be required. We give two ways in which this can happen.

i. We obtain a form such as

$$\begin{bmatrix} \circledast & * & * & * & * \\ 0 & \circledast & * & * & * \\ 0 & 0 & 0 & * & * \\ 0 & 0 & \circledast & * & * \end{bmatrix}.$$

Now we would need to interchange rows 3 and 4 to continue toward a row echelon form.

ii. We have, say,

$$\begin{bmatrix} \circledast & * & * & * & * \\ 0 & 1 & * & * & * \\ 0 & 1000 & * & * & * \\ 0 & * & * & * & * \end{bmatrix}.$$

In numerical calculation, it is known that choosing large pivots, in general, leads to better results. Thus, at this step we would apply $R_2 \leftrightarrow R_3$, obtaining a larger pivot.

To see how to proceed when interchanges are used, we make an observation. Suppose

$$A = \begin{bmatrix} x_1 & y_1 & z_1 \\ x_2 & y_2 & z_2 \\ x_3 & y_3 & z_3 \end{bmatrix}.$$

We apply $\frac{-x_2}{x_1} R_1 + R_2$, $\frac{-x_3}{x_1} R_1 + R_3$ to get

$$\begin{bmatrix} 1 & 0 & 0 \\ 0 & 1 & 0 \\ \frac{-x_3}{x_1} & 0 & 1 \end{bmatrix} \begin{bmatrix} 1 & 0 & 0 \\ \frac{-x_2}{x_1} & 1 & 0 \\ 0 & 0 & 1 \end{bmatrix} A = \begin{bmatrix} x_1 & y_1 & z_1 \\ 0 & 0 & w_2 \\ 0 & v_3 & w_3 \end{bmatrix}. \qquad (8.2)$$

Now we would use $R_2 \leftrightarrow R_3$, whose corresponding elementary matrix is $C = \begin{bmatrix} 1 & 0 & 0 \\ 0 & 0 & 1 \\ 0 & 1 & 0 \end{bmatrix}$, which is not lower triangular. Note, however, if we apply $C$ first, we have

$$CA = \begin{bmatrix} x_1 & y_1 & z_1 \\ x_3 & y_3 & z_3 \\ x_2 & y_2 & z_2 \end{bmatrix}$$

and our corresponding elementary operations, eliminating $x_2$ first, would then be $\frac{-x_2}{x_1} R_1 + R_3$, $\frac{-x_3}{x_1} R_1 + R_2$. Using the corresponding elementary matrices, we would now have

$$\begin{bmatrix} 1 & 0 & 0 \\ \frac{-x_3}{x_1} & 1 & 0 \\ 0 & 0 & 1 \end{bmatrix} \begin{bmatrix} 1 & 0 & 0 \\ 0 & 1 & 0 \\ \frac{-x_2}{x_1} & 0 & 1 \end{bmatrix} CA = \begin{bmatrix} x_1 & y_1 & z_1 \\ 0 & v_3 & w_3 \\ 0 & 0 & w_2 \end{bmatrix}.$$

Comparing this result to that of (8.2), we can see that interchange operations, and their corresponding elementary matrices, can be moved up in the list of elementary operations done on $A$. To commute the elementary matrices, observe that applying $cR_i + R_j$ and $R_j \leftrightarrow R_k$ is the same as applying $R_j \leftrightarrow R_k$ and $cR_i + R_k$ $(i < j < k)$. Thus, we need only change the $c$'s in the row positions that might be in row $j$ or row $k$ according to the interchange. For example, for $\alpha R_1 + R_2$ and $R_2 \leftrightarrow R_3$ we have

$$\begin{bmatrix} 1 & 0 & 0 \\ 0 & 0 & 1 \\ 0 & 1 & 0 \end{bmatrix} \begin{bmatrix} 1 & 0 & 0 \\ \alpha & 1 & 0 \\ 0 & 0 & 1 \end{bmatrix} = \begin{bmatrix} 1 & 0 & 0 \\ 0 & 1 & 0 \\ \alpha & 0 & 1 \end{bmatrix} \begin{bmatrix} 1 & 0 & 0 \\ 0 & 0 & 1 \\ 0 & 1 & 0 \end{bmatrix}$$

or $R_2 \leftrightarrow R_3$ and $\alpha R_1 + R_3$.

So, if $P$, called a permutation matrix, is the product of elementary matrices corresponding to interchange operations (It accumulates all interchanges of rows into a single permutation of rows.), we see the following result.

**Theorem 8.1** *Let $A$ be an $m \times n$ matrix. Then there is a permutation matrix $P$ such that*

$$PA = LU$$

*where $L$ is lower triangular with 1's on the main diagonal and $U$ is an echelon form.*

**Proof.** In applying Gaussian elimination, commute the elementary matrices corresponding to the interchange operation so they are nearest $A$. Then we have

$$\hat{E}_r \cdots \hat{E}_{s+1} C_s \cdots C_1 A = U,$$

where $C_s, \ldots, C_1$ are the elementary matrices corresponding to interchange operations and $\hat{E}_r, \ldots, \hat{E}_{s+1}$ those corresponding to add operations. Set $P = C_s \cdots C_1$, a permutation matrix. Using this

$$PA = \left( \hat{E}_r \cdots \hat{E}_{s+1} \right)^{-1} U$$
$$= LU$$

where $L = \left( \hat{E}_r \cdots \hat{E}_{s+1} \right)^{-1}$ is lower triangular. Finally, since each $\hat{E}_i$ has 1's on the main diagonal, so does $\hat{E}_r \cdots \hat{E}_{s+1}$ and $\left( \hat{E}_r \cdots \hat{E}_{s+1} \right)^{-1}$. ∎

The form of Gaussian elimination we used to obtain $L$ and $U$ is called the *Doolittle method*. (In this method the main diagonal of $L$ consists of 1's.) If we scale rows to obtain pivots which are 1, the technique is called the *Crout method*. (This method produces pivots in $U$ which are 1's.)

**Example 8.3** *Let* $A = \begin{bmatrix} 1 & -1 & 2 \\ -2 & 2 & 1 \\ -3 & 1 & 3 \end{bmatrix}$. *Then*

$$A \quad \xrightarrow[\substack{E_1 = \begin{bmatrix} 1 & 0 & 0 \\ 2 & 1 & 0 \\ 0 & 0 & 1 \end{bmatrix}}]{2R_1 + R_2} \quad \begin{bmatrix} 1 & -1 & 2 \\ 0 & 0 & 5 \\ -3 & 1 & 3 \end{bmatrix}$$

$$\xrightarrow[\substack{E_2 = \begin{bmatrix} 1 & 0 & 0 \\ 0 & 1 & 0 \\ 3 & 0 & 1 \end{bmatrix}}]{3R_1 + R_3} \quad \begin{bmatrix} 1 & -1 & 2 \\ 0 & 0 & 5 \\ 0 & -2 & 9 \end{bmatrix}$$

$$\xrightarrow[\substack{C = \begin{bmatrix} 1 & 0 & 0 \\ 0 & 0 & 1 \\ 0 & 1 & 0 \end{bmatrix}}]{R_2 \leftrightarrow R_3} \quad \begin{bmatrix} 1 & -1 & 2 \\ 0 & -2 & 9 \\ 0 & 0 & 5 \end{bmatrix} = U.$$

*So*

$$C E_2 E_1 A = U.$$

*Now we need to move $C$ so it is next to $A$. Here*

$$C E_2 = \hat{E}_2 C$$

*where $\hat{E}_2$ is $3R_1 + R_2$. (Rows 2 and 3 were interchanged by $C$ so the elementary matrix is adjusted to show that.) And*

$$C E_1 = \hat{E}_1 C$$

*where $\hat{E}_1$ corresponds to $2R_1 + R_3$. (Rows 2 and 3 were interchanged by $C$, and $\hat{E}_1$ now needs to apply the addition to row 3.)*
*So we have*

$$\hat{E}_2 \hat{E}_1 C A = U$$

*and setting $P = C$,*

$$P A = L U$$

*where $L = \left( \hat{E}_2 \hat{E}_1 \right)^{-1}$ which can also be computed using multipliers, so*

$$L = \begin{bmatrix} 1 & 0 & 0 \\ -3 & 1 & 0 \\ -2 & 0 & 1 \end{bmatrix}.$$

We now show how the factorization can be used to solve a system of linear equation, say,

$$Ax = b.$$

To do this, factor $PA = LU$. Substitution yields

$$LUx = Pb.$$

i. We first find $L^{-1}Pb$. For this we solve

$$Ly = Pb$$

which can be solved by forward substitution. (Solve for $y_1$ first, then $y_2$, etc.) This gives $y = L^{-1}Pb$.

ii. Now we solve $Ux = L^{-1}Pb$. Knowing $y$, we can solve for $x$ by solving

$$Ux = y$$

by back substitution.

**Example 8.4** *Solve*

$$\begin{bmatrix} 1 & 1 \\ 1 & -1 \end{bmatrix} \begin{bmatrix} x_1 \\ x_2 \end{bmatrix} = \begin{bmatrix} 3 \\ -1 \end{bmatrix}.$$

*We factor $A = LU$ to get*

$$\begin{bmatrix} 1 & 1 \\ 1 & -1 \end{bmatrix} = \begin{bmatrix} 1 & 0 \\ 1 & 1 \end{bmatrix} \begin{bmatrix} 1 & 1 \\ 0 & -2 \end{bmatrix}.$$

*i. Solve $Ly = b$ by forward substitution. Here*

$$\begin{bmatrix} 1 & 0 \\ 1 & 1 \end{bmatrix} y = \begin{bmatrix} 3 \\ -1 \end{bmatrix}$$

*and we get $y = \begin{bmatrix} 3 \\ -4 \end{bmatrix}$.*

*ii. Now solve $Ux = y$ by back substitution. Here*

$$\begin{bmatrix} 1 & 1 \\ 0 & -2 \end{bmatrix} x = \begin{bmatrix} 3 \\ -4 \end{bmatrix}$$

*and we get*

$$x = \begin{bmatrix} 1 \\ 2 \end{bmatrix}.$$

## 8.1.1    Optional (Iterative Improvement in Solving $Ax = b$)

Let $A$ be a nonsingular matrix. We can solve

$$Ax = b \tag{8.3}$$

by Gaussian elimination applied to the augmented matrix $[A\,|\,b]$ or by using the $LU$ method. (Both use the same amount of arithmetic.) The $LU$ method has an advantage, however, when (8.3) needs to be solved for several different right sides. To show how this can occur, we describe the iterative improvement method which can be used to solve (8.3).

Suppose we numerically solve (8.3), obtaining $\widehat{x}$. Thus, we might expect some nonzero residual

$$r_1 = b - A\widehat{x}.$$

(For accuracy, the residual should be computed in double precision, i.e. using twice as many digits as normal.) We now try to improve our solution by solving

$$Ax = r_1 \tag{8.4}$$

for $a_1$ and adjusting the numerical solution to $\widehat{x} + a_1$. Note that if $a_1$ is the exact solution to (8.4), then

$$A\left(\widehat{x} + a_1\right) = A\widehat{x} + Aa_1 = b - r_1 + r_1 = b.$$

So $\widehat{x} + a_1$ would be the exact solution to (8.3). However, we may again expect some error, say, we get $\widehat{a}_1$ instead of $a_1$.

We compute the residual

$$r_2 = b - A\left(\widehat{x} + \widehat{a}_1\right).$$

If $r_2$ is not 0, then we would solve

$$Ax = r_2$$

for $a_2$ and adjust to $\widehat{x} + \widehat{a}_1 + \widehat{a}_2$.

It can be shown that, unless $c\left(A\right)$ is very large, the sequence $\widehat{x}, \widehat{x} + \widehat{a}_1, \widehat{x} + \widehat{a}_1 + \widehat{a}_2, \ldots$ converges to the solution to (8.3). And, usually, only a few iterations are required for desired results.

Note that in this process, we solve

$$Ax = b \tag{8.5}$$
$$Ax = r_1$$
$$\cdots$$

If $A$ is factored into $LU$, each solution can be computed by a forward, and then a backward substitution, which involves far fewer arithmetic operations than solving each equation in (8.5) by Gaussian elimination. So here, the $LU$ method has a distinct advantage.

An example follows.

**Example 8.5** *Let* $A = \begin{bmatrix} .982 & .573 \\ .402 & .321 \end{bmatrix}$ *and* $b = \begin{bmatrix} -.245 \\ .159 \end{bmatrix}$. *Solving* $Ax = b$ *in 3-digit arithmetic, by iterative improvement, we get* $L = \begin{bmatrix} 1 & 0 \\ .409 & 1 \end{bmatrix}$ *and* $U = \begin{bmatrix} .982 & .573 \\ 0 & .087 \end{bmatrix}$.

*On the first iteration, we have*

$$\hat{x} = \begin{bmatrix} -2 \\ 2.98 \end{bmatrix}.$$

*The second iteration gives*

$$\hat{x} + \hat{a}_1 = \begin{bmatrix} -2 \\ 3 \end{bmatrix}$$

*which is the exact solution.*

We should also add that there is a similar iterative improvement method for solving least-squares problems.

### 8.1.2  MATLAB (lu, [L, U] and [L, U, F])

MATLAB uses the $LU$ decomposition to solve $Ax = b$.

The MATLAB command for the $LU$ decomposition for any square matrix $A$ is $lu\,(A)$. To obtain $L$, $U$, and $P$ such that $PA = LU$, use $[L, U, F] = lu\,(A)$. Using $[L, U] = lu\,(A)$ produces $P^t L$ (not $L$) and $U$. MATLAB calls $P^t L$ a 'psychologically triangular matrix', i.e., $L$ with the rows permuted. An example follows.

$A = \begin{bmatrix} 1 & 1; & 2 & 3 \end{bmatrix}$;
$[L, U] = lu\,(A)$
ans: $L = \begin{bmatrix} .5 & 1 \\ 1 & 0 \end{bmatrix}$, $U = \begin{bmatrix} 2 & 3 \\ 0 & -.5 \end{bmatrix}$

Type in *help lu* for more information.

# Exercises

1. Let $A = \begin{bmatrix} a_1 \\ a_2 \\ a_3 \end{bmatrix}$ where $a_k$ is the $k$-th row of $A$. Find the $3 \times 3$ permutation matrix $P$ such that

(a) $PA = \begin{bmatrix} a_3 \\ a_1 \\ a_2 \end{bmatrix}$.  (b) $PA = \begin{bmatrix} a_2 \\ a_1 \\ a_3 \end{bmatrix}$.

(c) $PA = \begin{bmatrix} a_2 \\ a_3 \\ a_1 \end{bmatrix}$.  (d) $PA = \begin{bmatrix} a_3 \\ a_2 \\ a_1 \end{bmatrix}$.

2. Let $A = \begin{bmatrix} a_1 \\ a_2 \\ a_3 \end{bmatrix}$ where $a_k$ is the $k$-th row of $A$. Find the elementary matrix $E$ such that

(a) $EA = \begin{bmatrix} a_1 \\ a_2 \\ 3a_3 \end{bmatrix}$.

(b) $EA = \begin{bmatrix} a_1 \\ a_2 - 2a_1 \\ a_3 \end{bmatrix}$.

(c) $EA = \begin{bmatrix} a_1 \\ a_2 \\ a_3 - 3a_1 \end{bmatrix}$.

3. For the given matrix $A$, find the $LU$ decomposition.

(a) $A = \begin{bmatrix} 1 & 1 & 1 \\ 1 & 2 & 4 \\ 1 & 3 & 9 \end{bmatrix}$

(b) $A = \begin{bmatrix} 1 & 0 & -1 & 1 \\ 2 & 1 & 0 & -1 \\ 3 & 2 & -2 & 1 \end{bmatrix}$

4. For the given matrix $A$, find the $LU$ decomposition with partial pivoting.

(a) $A = \begin{bmatrix} 1 & -1 & 2 \\ -1 & 1 & -2 \\ 1 & 2 & 0 \end{bmatrix}$

(b) $A = \begin{bmatrix} 1 & 1 & 1 & 1 \\ 1 & 1 & 1 & 2 \\ 1 & 2 & 3 & 1 \\ 1 & 2 & 4 & 1 \end{bmatrix}$

5. Give an example of an elementary matrix for add operation which is not lower triangular.

6. Suppose we know that $A$ has an $LU$ decompositon. Solve for $L$ and $U$ by first finding the first row of $U$, then $l_{21}$, then the second row of $U$, etc.

(a) $A = \begin{bmatrix} 2 & 4 \\ 4 & 11 \end{bmatrix}$

(b) $A = \begin{bmatrix} 1 & 0 & 3 \\ 1 & 2 & 4 \\ 2 & 6 & 12 \end{bmatrix}$

7. Show that the two $3 \times 3$ elementary matrices corresponding to $2R_1 + R_2$ and $3R_2 + R_3$, don't commute.

8. Show that the two $3 \times 3$ elementary matrices corresponding to $\alpha R_1 + R_2$ and $\beta R_1 + R_3$ do commute. (Actually, for a given $i$, all elementary matrices corresponding to $\alpha_k R_i + R_k$ commute.)

9. Let $L_1$ and $L_2$ be two $3 \times 3$ lower triangular matrices with 1's on their main diagonals.

(a) Prove that $L_1 L_2$ has 1's on its main diagonal.

(b) Prove the result for $n \times n$ matrices.

10. Let $A$ be a nonsingular matrix. Suppose $LU$ and $\hat{L}\hat{U}$ are two $LU$ decompositons of $A$. Prove that $L = \hat{L}$ and $U = \hat{U}$.

11. Let $L = \begin{bmatrix} l_{11} & 0 & 0 \\ l_{21} & l_{22} & 0 \\ l_{31} & l_{32} & l_{33} \end{bmatrix}$ be a nonsingular matrix. By solving

$$LX = I,$$

find formulas for the entries of $L^{-1}$ in terms of those of $L$.

12. Let $E$ be an $m \times m$ elementary matrix corresponding to an elementary operation. Let $\hat{E}$ be the matrix obtained from the $m \times m$ identity matrix by applying the elementary operation to it. Prove that $\hat{E} = E$.

13. Let $E$ be a nonsingular $m \times m$ matrix. Let $A$ be an $m \times n$ matrix and $b$ an $m \times 1$ vector. Prove that $Ax = b$ and $EAx = Eb$ have the same solution set. (Loosely, this shows that most any kind of operation, e.g., $\alpha R_i + \beta R_j$ $(\beta \neq 0)$ can be applied to a $[A \mid b]$.)

14. (MATLAB) Let $A = \begin{bmatrix} 3 & 1 & -1 \\ -2 & 4 & 1 \\ 0 & 2 & 5 \end{bmatrix}$ and $b = \begin{bmatrix} 1 \\ -2 \\ 1 \end{bmatrix}$.

    (a) Find the $LU$ decomposition of $A$.
    (b) Solve $Ax = b$ by using this decomposition.
    (c) Solve $Ax = b$ by using the command $A \backslash b$.
    (d) Compare results.

## 8.2   The QR Decomposition

We can do with Householder matrices what we just did with elementary matrices. Although this work can be done with either real or complex numbers, we will do the work with real numbers, so we can give geometrical views.

Recall that a Householder matrix is defined by

$$H = I - 2u\,u^t$$

where $u$ is a vector such that $\|u\|_2 = 1$. Thus, if u is not of length 1, we would use the vector $\frac{u}{\|u\|_2}$ and obtain

$$H = I - \frac{2}{\|u\|_2^2}u\,u^t.$$

As shown in in Figure 8.1, what $H$ does is reflect (or invert) $R^n$ parallel to $u$ and through the subspace

$$W = \{y : (u, y) = 0\}.$$

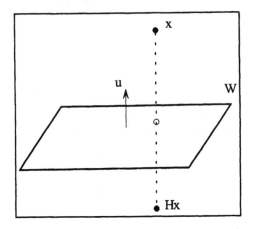

FIGURE 8.1.

For a given vector $x$, we now need to find a Householder matrix $H$ such that $Hx = \pm \|x\|_2 \, e_1$. (Recall that $H$ is orthogonal and orthogonal matrices don't change the length of vectors.) How such an $H$ can be determined is shown in the following example.

**Example 8.6** *Let $x$ be a nonzero vector in $R^2$. Notice in Figure 8.2 that two different choices, $\|x\|_2 \, e_1$ and $-\|x\|_2 \, e_1$, are possible.*

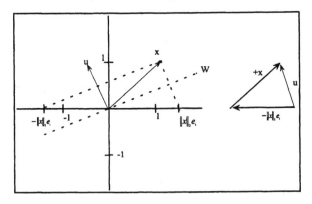

FIGURE 8.2.

1. For $Hx = \|x\|_2 \, e_1$, we use the vector $u = x - \|x\|_2 \, e_1$ *(equivalent to the arrow from $\|x\|_2 \, e_1$ to $x$).*

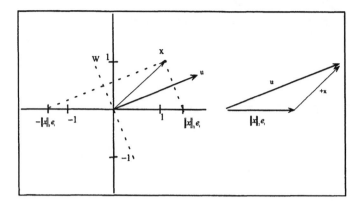

FIGURE 8.3.

2. For $Hx = -\|x\|_2 e_1$, we use the vector $u$, $u = x + \|x\|_2 e_1$ (equivalent to the arrow from $-\|x\|_2 e_1$ to $x$ in Figure 8.3).

Actually in software, the choice changes depending on $x$. For numerical reasons (to obtain better answers), whichever of $\|x\|_2 e_1$ or $-\|x\|_2 e_1$ is farthest from $x$, is chosen. For example, in our picture, $-\|x\|_2 e_1$ would be chosen. However, for our work, we will simply choose $\|x\|_2 e_1$.

To help recall the expression for $u$, observe in Figure 8.4 that

$$\frac{x + \|x\|_2 e_1}{2}$$

is the average of $x$ and $\|x\|_2 e_1$, and $x - \|x\|_2 e_1$ (a change of sign) provides the orthogonal vector. (Their dot product is 0.)

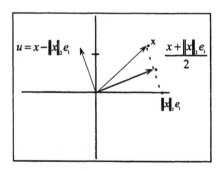

FIGURE 8.4.

From these remarks we have the following.

**Theorem 8.2** *Let $x$ be a nonzero $n \times 1$ vector. If $x \neq \|x\|_2 e_1$, set $u = x - \|x\|_2 e_1$. Then*

$$H = I - \frac{2}{\|u\|_2^2} u\, u^t$$

*and*

$$Hx = \|x\|_2\, e_1.$$

**Proof.** A direct calculation. ■

We show a numerical example below.

**Example 8.7** *Let* $x = \begin{bmatrix} 3 \\ 0 \\ 4 \end{bmatrix}$. *Then* $u = x - \|x\|_2\, e_1 = \begin{bmatrix} 3 \\ 0 \\ 4 \end{bmatrix} - \begin{bmatrix} 5 \\ 0 \\ 0 \end{bmatrix} = \begin{bmatrix} -2 \\ 0 \\ 4 \end{bmatrix}$. *Thus*

$$
\begin{aligned}
H &= I - \frac{2}{\|u\|_2^2} u\, u^t \\[4pt]
&= \begin{bmatrix} 1 & 0 & 0 \\ 0 & 1 & 0 \\ 0 & 0 & 1 \end{bmatrix} - \frac{2}{20} \begin{bmatrix} -2 \\ 0 \\ 4 \end{bmatrix} [-2 \ \ 0 \ \ 4] \\[4pt]
&= \begin{bmatrix} \frac{3}{5} & 0 & \frac{4}{5} \\ 0 & 1 & 0 \\ \frac{4}{5} & 0 & -\frac{3}{5} \end{bmatrix}
\end{aligned}
$$

*and* $Hx = \begin{bmatrix} 5 \\ 0 \\ 0 \end{bmatrix}$.

Now, to see how to do a Gaussian elimination process using Householder matrices, we let

$$A = \begin{bmatrix} x_1 & y_1 & z_1 \\ x_2 & y_2 & z_2 \\ x_3 & y_3 & z_3 \end{bmatrix}.$$

We find the Householder matrix $H_1$ such that

$$H_1 \begin{bmatrix} x_1 \\ x_2 \\ x_3 \end{bmatrix} = \begin{bmatrix} \ell_1 \\ 0 \\ 0 \end{bmatrix}$$

where $\ell_1 = \left\| \begin{bmatrix} x_1 \\ x_2 \\ x_3 \end{bmatrix} \right\|_2$. Then

$$H_1 A = \begin{bmatrix} \ell_1 & v_1 & w_1 \\ 0 & v_2 & w_2 \\ 0 & v_3 & w_3 \end{bmatrix}.$$

We now find the Householder matrix $H$ such that

$$H \begin{bmatrix} v_2 \\ v_3 \end{bmatrix} = \begin{bmatrix} \ell_2 \\ 0 \end{bmatrix}$$

where $\ell_2 = \left\| \begin{bmatrix} v_2 \\ v_3 \end{bmatrix} \right\|_2$. Set

$$H_2 = \begin{bmatrix} 1 & 0 \\ 0 & H \end{bmatrix}.$$

Then by partitioned multiplication,

$$H_2 H_1 A = H_2 \begin{bmatrix} \ell_1 & v_1 & w_1 \\ 0 & v_2 & w_2 \\ 0 & v_3 & w_3 \end{bmatrix} \tag{8.6}$$

$$= \begin{bmatrix} \ell_1 & v_1 & w_1 \\ 0 & \ell_2 & z_2 \\ 0 & 0 & z_3 \end{bmatrix}$$

which is a row echelon form.

So, if we let $R$ denote the row echelon form and set $Q = (H_2 H_1)^{-1}$, we have from (8.6) that

$$A = QR$$

where Q is an orthogonal matrix.

More generally, we have the following theorem.

**Theorem 8.3** *Let A be an $m \times n$ matrix. Then, there exists a sequence of Householder matrices $H_1, \ldots, H_s$, such that*

$$H_s \cdots H_1 A = R$$

*where R is a row echelon form. Thus, setting $Q = (H_s \cdots H_1)^{-1}$, $A = QR$.*

**Proof.** If $A = 0$, there is nothing to argue; thus, we assume $A \neq 0$. The proof is now given in steps.

Step 1. (Finding $H_1$) Let $b$ denote the first nonzero column of $A$. If $b = \|b\|_2 e_1$, the first row is staggered and we set $A_1 = A$. Otherwise, using

Theorem 8.2, let $H_1$ be the Householder matrix such that $H_1 b = ||b||_2\, e_1$. Then $H_1 A$ has its first row staggered. Set $A_1 = H_1 A$.

Step 2. (Finding $H_k$) Suppose $A_k$ has its first $k$ rows staggered. Then

$$A_k = \begin{bmatrix} C_1 & C_2 \\ 0 & B \end{bmatrix}$$

where $C_1$ has $k$ staggered rows. If $B$ has all 0 columns, we are through. Otherwise, let $b$ denote the first nonzero column of $B$. If $b = ||b||\, e_1$, the $(k+1)st$ row is staggered, and we set $A_{k+1} = A_k$. If $b \neq ||b||\, e_1$, let $H$ be the Householder transformation such that

$$Hb = ||b||_2\, e_1.$$

Setting $H_{k+1} = \begin{bmatrix} I & 0 \\ 0 & H \end{bmatrix}$, an $m \times m$ matrix, we have a Householder transformation such that if we set

$$H_{k+1} A_k = A_{k+1}$$

then $A_{k+1}$ has $k+1$ staggered rows.

Step 3. (Finding $Q$) Thus all rows can be staggered and a row echelon form $R$ achieved. Putting together, if $s$ Householder matrices were used, then $H_s \cdots H_1 A = R$ and $Q = (H_s \cdots H_1)^{-1}$. ∎

**Example 8.8** *Let* $A = \begin{bmatrix} 3 & 10 & 0 \\ 0 & 0 & 9 \\ 4 & 10 & -5 \end{bmatrix}$.

*Step 1. (Finding $H_1$) To find $H_1$, we use the first column of $A$. So,*

$x = \begin{bmatrix} 3 \\ 0 \\ 4 \end{bmatrix}$ *and by Example 8.7,* $H_1 = \begin{bmatrix} \frac{3}{5} & 0 & \frac{4}{5} \\ 0 & 1 & 0 \\ \frac{4}{5} & 0 & -\frac{3}{5} \end{bmatrix}$. *Now* $H_1 A = A_1 =$

$\begin{bmatrix} 5 & 14 & -4 \\ 0 & 0 & 9 \\ 0 & 2 & 3 \end{bmatrix}$.

*Step 2. (Finding $H_2$) For $H_2$ we use* $\begin{bmatrix} 0 \\ 2 \end{bmatrix}$. *We obtain a Householder matrix $H$ where* $H \begin{bmatrix} 0 \\ 2 \end{bmatrix} = \begin{bmatrix} 2 \\ 0 \end{bmatrix}$. *We get* $H = \begin{bmatrix} 0 & 1 \\ 1 & 0 \end{bmatrix}$, *and so* $H_2 =$

$\begin{bmatrix} 1 & 0 & 0 \\ 0 & 0 & 1 \\ 0 & 1 & 0 \end{bmatrix}$.

*Step 3. (Finding $Q$) Putting together, we have* $H_2 H_1 A = H_2 A_1 =$

$\begin{bmatrix} 5 & 14 & -4 \\ 0 & 2 & 3 \\ 0 & 0 & 9 \end{bmatrix} = R$, *the row echelon form.*

*Finally,*

$$H_2 H_1 = \begin{bmatrix} \frac{3}{5} & 0 & \frac{4}{5} \\ \frac{4}{5} & 0 & -\frac{3}{5} \\ 0 & 1 & 0 \end{bmatrix},$$

*so*

$$Q = (H_2 H_1)^t = \begin{bmatrix} \frac{3}{5} & \frac{4}{5} & 0 \\ 0 & 0 & 1 \\ \frac{4}{5} & -\frac{3}{5} & 0 \end{bmatrix}$$

*and $A = QR$.*

Recall, pivoting was used in the $LU$ decomposition to get large pivots. The same can be done in the $QR$ decomposition. For example, if

$$H_1 A = \begin{bmatrix} \ell_1 & v_1 & w_1 \\ 0 & v_2 & w_2 \\ 0 & v_3 & w_3 \end{bmatrix},$$

we can check the columns $\begin{bmatrix} v_2 \\ v_3 \end{bmatrix}$ and $\begin{bmatrix} w_2 \\ w_3 \end{bmatrix}$ to see which has the greater

length. If $\left\| \begin{bmatrix} w_2 \\ w_3 \end{bmatrix} \right\|_2 > \left\| \begin{bmatrix} v_2 \\ v_3 \end{bmatrix} \right\|_2$, then columns 2 and 3 of $H_1 A$ are inter-

changed. If $C$ is the elementary matrix corresponding to that elementary operation, then

$$C = \begin{bmatrix} 1 & 0 & 0 \\ 0 & 0 & 1 \\ 0 & 1 & 0 \end{bmatrix}, \text{ and } H_1 AC = \begin{bmatrix} \ell_1 & w_1 & v_1 \\ 0 & w_2 & v_2 \\ 0 & w_3 & v_3 \end{bmatrix}.$$

(When multiplying by $C$, we can use backward multiplication to see the result.) Now, we determine our next Householder matrix $H_2$ so that

$$H_2 H_1 AC = \begin{bmatrix} x_1 & z_1 & v_1 \\ 0 & \ell_2 & q_2 \\ 0 & 0 & q_3 \end{bmatrix} = R \qquad (8.7)$$

where $\ell_2 = \left\| \begin{bmatrix} w_2 \\ w_3 \end{bmatrix} \right\|_2$. (Had we not interchanged the columns, we would

have had a smaller pivot, namely $\left\| \begin{bmatrix} v_2 \\ v_3 \end{bmatrix} \right\|_2$.)

Thus, from (8.7) we have

$$AP = QR$$

where $Q = (H_2 H_1)^{-1}$ and $P = C$, a permutation matrix. (In general, $P$ is the product of the accumulated $C_i$'s.)

The $QR$ decomposition is often used in least-squares solving a system of linear equations, say,

$$Ax = b. \tag{8.8}$$

To find a least-squares solution $x$, we use the factorization $A = QR$ (or $QRF^t$) and substitute this into ( 8.8). We have

$$QRx = b.$$

As shown in Chapter 7, this is equivalent to finding least-squares solutions to

$$Rx = Q^t b. \tag{8.9}$$

Now we need to find vectors $\hat{x}$ so the left side is as close to the right side of (8.9) as possible. For example, if

$$R = \begin{bmatrix} r_{11} & r_{12} \\ 0 & r_{22} \\ 0 & 0 \end{bmatrix}$$

and $c = Q^t b$, we would find $x_i$'s so the left and right sides of

$$\begin{aligned} r_{11}x_1 + r_{12}x_2 &= c_1 \\ r_{22}x_2 &= c_2 \\ 0 &= c_3 \end{aligned}$$

are as close as we can make them.

We can't do anything about the last equation; however, we can get $x$'s so that the first two equations are satisfied. And any such $x$ will be a least-squares solution to (8.8).

**Example 8.9** *Find the least-squares solutions to*

$$\begin{aligned} 3x_1 - 3x_2 &= 1 \\ 5x_2 &= 1 \\ 4x_1 - 4x_2 &= 1. \end{aligned}$$

*Here,* $A = \begin{bmatrix} 3 & -3 \\ 0 & 5 \\ 4 & -4 \end{bmatrix}$ *and* $b = \begin{bmatrix} 1 \\ 1 \\ 1 \end{bmatrix}$. *Now, using Example 8.7,*

$$H_1 = \begin{bmatrix} \frac{3}{5} & 0 & \frac{4}{5} \\ 0 & 1 & 0 \\ \frac{4}{5} & 0 & -\frac{3}{5} \end{bmatrix} \text{ so } H_1 A = \begin{bmatrix} 5 & -5 \\ 0 & 5 \\ 0 & 0 \end{bmatrix} = R.$$

*Further,*

$$Q = H_1^{-1} = \begin{bmatrix} \frac{3}{5} & 0 & \frac{4}{5} \\ 0 & 1 & 0 \\ \frac{4}{5} & 0 & -\frac{3}{5} \end{bmatrix} \quad and \quad Q^t b = \begin{bmatrix} \frac{7}{5} \\ 1 \\ \frac{1}{5} \end{bmatrix}.$$

*So the equation we need to least-squares solve is $Rx = Q^t b$ or*

$$5x_1 - 5x_2 = \frac{7}{5}$$
$$5x_2 = 1$$
$$0 = \frac{1}{5}.$$

*The least-squares solution is*

$$x_2 = \frac{1}{5}$$
$$x_1 = \frac{12}{25}$$

*or $x = \begin{bmatrix} \frac{12}{25} \\ \frac{1}{5} \end{bmatrix}.$*

Note that, for this $x$, $\|Ax - b\|_2 = \frac{1}{5}$ since we could do nothing with the last equation.

In finding a $QR$ decomposition, we could also use Givens matrices. The basic idea for this approach is covered in the exercises.

## 8.2.1 Optional (QR Algorithm)

The numerical computation of $Q$ and $T$ of the real Schur form is usually done by the $QR$ algorithm. This algorithm is as important to eigenvalues and eigenvectors as Gaussian elimination is to systems of linear equations.

The algorithm sets $A_1 = A$ and iterative factors

$$A_k = Q_k R_k.$$

Then sets

$$A_{k+1} = R_k Q_k.$$

Under reasonable conditions $A_1, A_2, \ldots$ converges. Since

$$A_{k+1} = R_k Q_k$$
$$= Q_k^t Q_k R_k Q_k$$
$$= Q_k^t A_k Q_k.$$

we see that

$$A_{k+1} = Q_k^t \cdots Q_1^t A Q_1 \cdots Q_k$$

so, setting

$$Q = Q_1 \cdots Q_k$$
$$A_{k+1} = Q^t A Q$$

Thus $A_{k+1}$ is orthogonally similar to $A$. For sufficiently large $k$, $A_{k+1}$ is close to a block triangular matrix. Replacing those entries in $A_{k+1}$ below the blocks, which are sufficiently close to 0 by 0 yields the computed Schur form $T$ of $A$.

The $QR$ algorithm can be used to compute eigenvalues, and corresponding eigenvectors of $A$. (This is the best general such method.) We describe how this is done.

If $T$ is a real Schur form, its eigenvalues are the eigenvalues of the $1 \times 1$ or $2 \times 2$ matrices on the main diagonal, which are easy to compute. Thus, the eigenvalues of $T$ and hence $A$ are calculated.

Corresponding eigenvectors can be computed as follows. For an eigenvalue $\lambda$, solve

$$Ty = \lambda y, \quad y \neq 0$$

or

$$(T - \lambda I) y = 0$$

for corresponding eigenvectors. Then, for each such eigenvector, since

$$T = Q^t A Q$$
$$Q^t A Q y = \lambda y$$
$$A Q y = \lambda Q y,$$

$Qy$ is an eigenvector for $A$ corresponding to $\lambda$.

As we might expect, this algorithm has been improved (using Hessenberg matrices and an implicit shift, etc.)

It may be helpful to look at some data. We use the MATLAB program.

```
A = [1 2; 3 4]
for n = 1 : 5
    [Q, R] = qr (A);
    A = R * Q
end
```

Notice in the iterates, the 2,1-entries tend to 0, thus we get $T$.

1. $\begin{bmatrix} 5.2000 & 1.6000 \\ 0.6000 & -0.2000 \end{bmatrix}$

2.
$\begin{bmatrix} 5.3796 & -0.9562 \\ 0.0438 & -0.3796 \end{bmatrix}$

3.
$\begin{bmatrix} 5.3718 & 1.0030 \\ 0.0030 & -0.718 \end{bmatrix}$

4.
$\begin{bmatrix} 5.3723 & -0.998 \\ 0.0002 & -0.3723 \end{bmatrix}$

5.
$\begin{bmatrix} 5.3723 & 1.0000 \\ 0.0000 & -0.3723 \end{bmatrix}$

We can compare our result to the MATLAB result obtained by using the command

$$\text{shur}(A) = \begin{bmatrix} -0.3723 & -1.0000 \\ 0.0000 & 5.3723 \end{bmatrix}.$$

This is a little different; however, recall that MATLAB uses a more sophisticated $QR$-alogrithm (implicit shifts, etc.), and that $Q$ and $R$, in the $QR$ decomposition, are not unique.

## 8.2.2   MATLAB ($Ax = b$, QR, Householder, and Givens)

MATLAB uses the QR decomposition to least-squares solve $Ax = b$.

Comparing the $LU$ decomposition and the $QR$ decomposition, it is known that the latter requires about twice as much arithmetic to compute. And, to solve $Ax = b$, $LU$ with partial pivoting is known to be very satisfactory.

In comparing Householder and Givens, using Givens matrices requires about twice as much arithmetic as using Householder matrices.

The MATLAB commands for the $QR$ decomposition and Householder matrices follow.

1. $QR$ decomposition: The MATLAB command for the $QR$ decomposition of $A$ is $qr(A)$. To obtain the $Q$ and $R$, use $[Q, R] = qr(A)$. For the $QR$ decomposition with column pivoting, use $[Q, R, F] = qr(A)$. The $P$ here is as in $AF = QR$. An example follows.

   $A = [\, 1 \quad 2; \ 1 \quad 1 \,]$;
   $[Q, R, F] = qr(A)$
   ans: $Q = \begin{bmatrix} -0.8944 & -0.4472 \\ -0.4472 & 0.8944 \end{bmatrix}$, $R = \begin{bmatrix} -2.2361 & -1.3416 \\ 0 & 0.4472 \end{bmatrix}$
   and $P = \begin{bmatrix} 0 & 1 \\ 1 & 0 \end{bmatrix}$.

2. Householder matrix: Given a vector $x \neq 0$, the command $[H, r] = qr(x)$ provides a Householder matrix $H$ such that $Hx = \pm \|x\|_2 e_1$. The $r$ gives $\pm \|x\|_2 e_1$ as shown in the following example.

$x = [\ 1;\ \ 2\ ];$
$[H, r] = qr\,(x)$
ans: $H = \begin{bmatrix} -0.4472 & -0.8944 \\ -0.8944 & 0.4472 \end{bmatrix}$
$r = \begin{bmatrix} 2.2\overset{.}{3}61 \\ 0 \end{bmatrix}.$

3. **Givens matrix:** Given a vector $x \in R^2$, we can get a Givens matrix $G$ such that $Gx = \pm \|x\|_2\, e_1$ by using $[G, r]$ =planerot$(x)$. For example

$x = [\ 1;\ \ 2\ ];$
$[G, r] = planerot\,(x)$
ans: $G = \begin{bmatrix} 0.4472 & 0.8944 \\ -0.8944 & 0.4472 \end{bmatrix}$
$r = \begin{bmatrix} 2.2\overset{.}{3}61 \\ 0 \end{bmatrix}.$

Type in *help qr* for more information.

## Exercises

1. For the given $A$, find the $QR$ decomposition.

   (a) $A = \begin{bmatrix} 3 & 1 \\ 4 & 2 \end{bmatrix}$     (b) $A = \begin{bmatrix} 0 & 3 & 1 & -1 \\ 1 & 0 & 1 & 0 \\ 0 & 4 & 1 & 1 \end{bmatrix}$

2. For the given $A$, find the $QR$ decomposition with column pivoting.

   (a) $A = \begin{bmatrix} 1 & 3 \\ 1 & 4 \end{bmatrix}$     (b) $A = \begin{bmatrix} 1 & -1 & 0 & 3 \\ 1 & 0 & 4 & 0 \\ 1 & 1 & 0 & 4 \end{bmatrix}$

3. Given $x \neq -\|x\|_2\, e_1$, show, using sketches, how to find a Householder matrix $H$ such that

$$Hx = -\|x\|_2\, e_1.$$

4. Solve $\begin{bmatrix} 3 & 1 \\ 0 & 1 \\ 4 & 1 \end{bmatrix} x = \begin{bmatrix} 1 \\ 2 \\ 1 \end{bmatrix}$ using the $QR$ decomposition.

5. Solve $\begin{bmatrix} 1 & 0 \\ 1 & 4 \\ 1 & 3 \end{bmatrix} x = \begin{bmatrix} 2 \\ 1 \\ 1 \end{bmatrix}$ by using the $QR$ decomposition with column pivoting.

6. Let $u$ be a nonzero vector in $R^n$. Define $W = \{w : (w, u) = 0\}$. Prove that $W$ is a subspace of dimension $n - 1$.

7. Explain why a permutation matrix is orthogonal.

8. Let $H$ be a Householder matrix. Prove that

   (a) $H^t = H$.
   (b) $HH^t = H^tH = I$.
   (c) $H^2 = I$.
   (d) $\begin{bmatrix} I & 0 \\ 0 & H \end{bmatrix}$ is a Householder matrix.
   (e) $\det H = -1$.

9. Let $A$ be an $m \times n$ matrix having rank $r$.

   (a) Prove that $A$ can be factored

   $$A = Q_0 R_0$$

   where $Q_0$ is an $m \times r$ matrix with orthonormal columns and $R_0$ a row echelon form.

   (b) Prove that $Q_0$ is an orthonormal basis for span $\{a_1, \ldots, a_n\}$ where $a_i$ is the $i$-th column of $A$.

10. Prove that the $QR$ decomposition, with column pivoting assures that in $R$,

    $$|r_{11}| \geq |r_{22}| \geq \cdots$$

11. Using Exercise 10, prove that if rank $A = r$, then in partitioned form,

    $$R = \begin{bmatrix} R_{11} & R_{12} \\ 0 & 0 \end{bmatrix}$$

    where $R_{11}$ is a nonsingular $r \times r$ upper triangular matrix.

12. Let $H$ be a Householder matrix. Prove that $H$ has one eigenvalue which is $-1$, all others being 1.

13. Let $A$ be a nonsingular matrix. Prove that if $QR$ and $\hat{Q}\hat{R}$ are $QR$ decompositions of $A$, then $\hat{Q}Q$ is a diagonal matrix with main diagonal composed of 1's and $-1$'s.

14. Let $A$ be an $n \times n$ real matrix. Using the $QR$ decomposition, prove Hadamard's inequality.

    $$|\det A| \leq \|a_1\|_2 \cdots \|a_n\|_2$$

    Describe what this inequality says about the volume of a parallelepiped in $R^3$ determined from edges $a_1, a_2, a_3$. (A sketch can help support the description.)

15. Let $A = QR$, a $QR$ decomposition. Prove that $|r_{ii}| = $ distance from $a_i$ to span $\{a_1, \ldots, a_{i-1}\}$ where $a_1, \ldots, a_n$ are the columns of $A$. (So, the $r_{ii}$'s give some idea of how close the vectors are to being linearly independent.)

16. Two parts.

(a) Let $G = \begin{bmatrix} \cos\theta & -\sin\theta \\ \sin\theta & \cos\theta \end{bmatrix}$, a Givens matrix, and $x = \begin{bmatrix} a \\ b \end{bmatrix}$.
Show what trigonometric equation should be solved to find $\theta$ such that $GA = \begin{bmatrix} \|x\|_2 \\ 0 \end{bmatrix}$.

(b) Explain, for a $3 \times 3$ matrix $A$, how to find $Q$ and $R$ using Givens matrices. (Use as a guide the parallel result for Householder matrices shown in this section.)

17. Let $Q$ be an orthogonal matrix. Prove that $Q$ is a product of Givens matrices (plane rotations) if and only if $\det Q = 1$.

18. Let $x \in C^n$, $x \neq 0$. Choose $\theta$ such that $x_1 = |x_1| e^{i\theta}$. Define $u = x + e^{i\theta} \|x\|_2 e_1$ and

$$H = I - \frac{2}{u^H u} u u^H.$$

Prove that $H$ is unitary and $Hx = -e^{i\theta} \|x\|_2 e_1$. (This is the Householder matrix for complex numbers.)

19. (MATLAB) Factor the following matrices as $QR$.

(a) $A = \begin{bmatrix} -2 & 4 & 3 \\ 0 & 5 & 2 \\ 7 & -1 & 6 \end{bmatrix}$

(b) $A = \begin{bmatrix} 2 - 3i & i & 4 + 2i \\ 0 & 3 - 2i & -5 \\ -3 + 4i & 2 & 6i \end{bmatrix}$

20. (MATLAB) Let $A = \begin{bmatrix} 4 & -1 & 2 \\ 0 & 3 & 6 \\ -1 & 2 & 1 \end{bmatrix}$. Find

(a) The $QR$ decomposition of $A$.

(b) $orth\,(A)$.

Compare (b) and the $Q$ from (a) in light of exercise 9(b).

21. (MATLAB)  Let $W = \text{span}\left\{\begin{bmatrix} 1 \\ 1 \\ 1 \end{bmatrix}, \begin{bmatrix} 1 \\ 0 \\ 1 \end{bmatrix}\right\}$. Find the distance

from $\begin{bmatrix} 1 \\ 0 \\ 0 \end{bmatrix}$ to $W$

(a) Using the $QR$ decomposition and Exercise 15.

(b) Using *orth* to find the orthogonal projection matrix $P$ from $R^3$ to $W$, and computing $\|e_1 - Pe_1\|_2$.

(c) Using least-squares on $Ax = e_1$ where $A = \begin{bmatrix} 1 & 1 \\ 1 & 0 \\ 1 & 1 \end{bmatrix}$, to find $x$,

and computing $\|Ax - e_1\|_2$.

22. (MATLAB) Adjust the program in the MATLAB section and apply the $QR$ algorithm to $A = \begin{bmatrix} 0 & 1 & 0 \\ -1 & 0 & 0 \\ 1 & 1 & 2 \end{bmatrix}$.

# 9

# Properties of Eigenvalues and Eigenvectors

In this chapter we study how small changes in the entries of a matrix affect the eigenvalues and eigenvectors of that matrix. Such changes occur in modeling since the matrix in the model is often only an approximation of the actual one. Further, in numerical computations, usually the answer we get is (due to rounding) actually the exact answer to a matrix which is close to the given matrix. So, if matrices close to a given matrix have close eigenvalues and eigenvectors, we would have an ideal situation. Unfortunately, this is not always the case.

## 9.1 Continuity of Eigenvalues and Eigenvectors

We have seen that eigenvalues and eigenvectors are important to calculating $\lim_{k \to \infty} A^k$, solving systems of differential and difference equations, graphics, etc. Understanding about eigenvalues and eigenvectors also allows us to interact with software, such as MATLAB, knowing how to interpret answers.

In this section, we show that eigenvalues are continuous (in some sense), and, under certain hypotheses, so are corresponding eigenvectors.

Eigenvalues are roots of $\varphi(\lambda) = \det(A - \lambda I)$, the characteristic polynomial of $A$. To study these roots will require our obtaining formulas for the coefficients in $\varphi(\lambda)$. For this, let A be an $n \times n$ matrix and $i_1, \dots, i_r$ any $r$ integers between 1 and $n$ where $i_1 < \cdots < i_r$. Define

$$\Delta(i_1, \dots, i_r)$$

as the determinant of the submatrix found in rows $i_1, \ldots, i_r$ and columns $i_1, \ldots, i_r$ in $A$. For example, if

$$A = \begin{bmatrix} 1 & 2 & 3 \\ 4 & 5 & 6 \\ 7 & 8 & 9 \end{bmatrix}$$

then

$$\Delta(2) = \det[5] = 5,$$

$$\Delta(1,3) = \det \begin{bmatrix} 1 & 3 \\ 7 & 9 \end{bmatrix} = 9 - 21 = -12 \text{ and}$$

$$\Delta(1,2,3) = \det[A] = 0.$$

These $\Delta$'s can be used to calculate the coefficients in $\varphi(\lambda)$, as shown below.

**Lemma 9.1** *Let $A$ be an $n \times n$ matrix and $\varphi(\lambda) = c_n \lambda^n + c_{n-1} \lambda^{n-1} + \cdots + c_0$. Then*

$$c_n = (-1)^n \text{ and}$$
$$c_k = (-1)^k \sum \Delta(i_1, \ldots, i_{n-k})$$

*for all $k < n$, where the sum $\Sigma$ is over all $i_1, \ldots, i_{n-k}$ where $i_1 < \cdots < i_{n-k}$.*

**Proof.** We prove this result for a $3 \times 3$ matrix leaving the general argument as an exercise. For a $3 \times 3$ matrix $A$, with columns $a_1$, $a_2$, and $a_3$, observe that $\varphi(\lambda) = \det(A - \lambda I) = \det[a_1 - \lambda e_1, a_2 - \lambda e_2, a_3 - \lambda e_3]$. By using properties of the determinant, we have

$$\begin{aligned}
\varphi(\lambda) &= \det[-\lambda e_1, -\lambda e_2, -\lambda e_3] + (\det[a_1, -\lambda e_2, -\lambda e_3,] \\
&\quad + \det[-\lambda e_1, a_2, -\lambda e_3,] + \det[-\lambda e_1, -\lambda e_2, a_3]) \\
&\quad + \det[-\lambda e_1, a_2, a_3] + \det[a_1, -\lambda e_2, a_3] \\
&\quad + \det[a_1, a_2, -\lambda e_3] + \det[a_1, a_2, a_3] \\
&= -\lambda^3 + \left(\Delta(1)\lambda^2 + \Delta(2)\lambda^2 + \Delta(3)\lambda^2\right) \\
&\quad - (\Delta(2,3)\lambda + \Delta(1,3)\lambda + \Delta(1,2)\lambda) + \Delta(1,2,3) \\
&= -\lambda^3 + \sum \Delta(i)\lambda^2 - \sum \Delta(i_1, i_2)\lambda + \Delta(1,2,3),
\end{aligned}$$

which is the desired result. ■

An example follows.

**Example 9.1** *Let* $A = \begin{bmatrix} 1 & 2 & 3 \\ 4 & 5 & 6 \\ 7 & 8 & 9 \end{bmatrix}$ . *Then*

$$\Delta\left(1\right) = 1, \ \Delta\left(2\right) = 5, \ \Delta\left(3\right) = 9 \ so$$
$$c_2 = \left(-1\right)^2\left(1 + 5 + 9\right) = 15$$
$$\Delta\left(1, 2\right) = -3, \ \Delta\left(1, 3\right) = -12, \ \Delta\left(2, 3\right) = -3 \ so$$
$$c_1 = \left(-1\right)^3\left(-3 - 12 - 3\right) = 18$$
$$\Delta\left(1, 2, 3\right) = 0, \ so \ c_0 = 0. \ Thus,$$
$$\varphi\left(\lambda\right) = c_3\lambda^3 + c_2\lambda^2 + c_1\lambda + c_0$$
$$= -\lambda_3^3 + 15\lambda^2 + 18\lambda + 0.$$

Note that from Lemma 9.1 and Theorem 3.8

$$c_0 = \det A = \ \lambda_1 \cdots \lambda_n \ \text{and}$$
$$c_{n-1} = \left(-1\right)^{n-1} trace \ A = \left(-1\right)^{n-1}\left(\lambda_1 + \cdots + \lambda_n\right)$$

where $\lambda_1, \ldots, \lambda_n$ are the eigenvalues of $A$.

We now use this lemma to prove a type of continuity of eigenvalues result, the Continuous Dependence Theorem.

**Theorem 9.1** *Let* $A$ *be an* $n \times n$ *matrix with eignenvalues* $\lambda_1, \ldots, \lambda_n$. *Given* $\epsilon > 0$, *there is a* $\delta > 0$, *such that if* $B$ *is an* $n \times n$ *matrix and* $\|B - A\|_F < \delta$ *the eigenvalues of* $B$ *can be arranged, say,* $\beta_1, \ldots, \beta_n$, *such that*

$$|\lambda_i - \beta_i| < \epsilon \ for \ all \ i.$$

*(* $\|\cdot\|_F$ *can be replaced by any matrix norm.)*

**Proof.** We apply the following theorem from mathematical analysis: Let $p\left(t\right) = a_n t^n + \cdots + a_0$, have roots $\lambda_1, \ldots, \lambda_n$. Given $\epsilon > 0$, there is a $\delta_1 > 0$, such that if $q\left(t\right) = b_n t^n + \cdots + b_0$ and $|a_i - b_i| < \delta_1$ for all $i$, then the roots of $q(t)$ can be arranged, say, $\beta_1, \ldots, \beta_n$, such that $|\lambda_i - \beta_i| < \epsilon$ for all $i$.

To apply this result to our theorem, given $\epsilon > 0$, take $\delta > 0$ such that if $\|B - A\|_F < \delta$, then

$$\left|\sum \Delta_A\left(i_1, \ldots, i_k\right) - \sum \Delta_B\left(i_1, \ldots, i_k\right)\right| < \delta_1 \ \text{for all } k.$$

(Since the determinant is continuous, such a $\delta$ exists.) ∎

This theorem assures, in numerical calculations, that if $A$ is given and if we calculate the eigenvalues of a sequence of matrices $A_1, A_2, \ldots$, where $\lim_{k \to \infty} A_k = A$, then $A_1, A_2, \ldots$ have eigenvalues that tend to those of $A$.

A useful application of this theorem is Gershgorin's theorem, which gives a region in the complex plane that includes all eigenvalues of $A$. This theorem follows.

**Corollary 9.1** *Let $A$ be an $n \times n$ matrix. Consider the disks $D_i$ in the complex plane determined by graphing the inequalities, involving the variable $\lambda$,*

$$|\lambda - a_{ii}| \leq \sum_{\substack{k=1 \\ k \neq i}}^{n} |a_{ik}|$$

*$\left( \sum_{\substack{k=1 \\ k \neq i}}^{n} |a_{ik}| \text{ being the } i\text{-th off-diagonal absolute row sum} \right)$ for $i = 1, \ldots, n$.*

*(a) If $\hat{\lambda}$ is an eigenvalue of $A$, then $\hat{\lambda}$ is in $D_i$ for some $i$.*

*(b) If $K$ is a union of $m$ disks and $K$ is disjoint from all other disks, then $K$ contains $m$ eigenvalues of $A$. (See Figure 9.1.)*

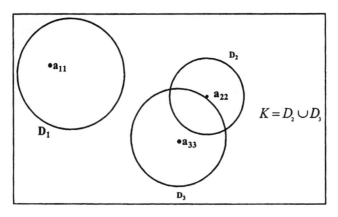

FIGURE 9.1.

**Proof.** There are two parts.
Part a.   Since $\hat{\lambda}$ is an eigenvalue of $A$, there is a eigenvector $x$ such that

$$Ax = \hat{\lambda}x. \tag{9.1}$$

Let $x_i$ denote the largest, in absolute value, entry in $x$. Then, equating the $i$-th entries in (9.1), we have

$$\sum_{k=1}^{n} a_{ik}x_k = \hat{\lambda}x_i.$$

Bringing the $a_{ii}x_i$ term to the right side yields

$$\sum_{\substack{k=1 \\ k \neq i}}^{n} a_{ik}x_k = \left(\hat{\lambda} - a_{ii}\right)x_i.$$

Taking absolute values, we have

$$\sum_{\substack{k=1 \\ k \neq i}}^{n} |a_{ik}|\,|x_k| \geq \left|\hat{\lambda} - a_{ii}\right||x_i|$$

and dividing by $|x_i|$,

$$\sum_{\substack{k=1 \\ k \neq i}}^{n} |a_{ik}|\frac{|x_k|}{|x_i|} \geq \left|\hat{\lambda} - a_{ii}\right|.$$

And since $\frac{|x_k|}{|x_i|} \leq 1$, for all $k$, we have

$$\sum_{\substack{k=1 \\ k \neq i}}^{n} |a_{ik}| \geq \left|\hat{\lambda} - a_{ii}\right|,$$

so $\hat{\lambda} \in D_i$.

Part b. Consider

$$B_t = diag\,(a_{11}, \dots, a_{nn}) + t\,(A - diag\,(a_{11}, \dots, a_{nn})).$$

Note that $B_0 = diag\,(a_{11}, \dots, a_{nn})$ and that $B_1 = A$.

Define $g(t) =$ the number of eigenvalues of $B_t$ in $K$. Note that the disks for $B_t$ lie in the corresponding disks for $A$.

Now suppose $g(t)$ is not continuous on $[0,1]$, say, not at $t_0$. Thus, there is a sequence $t_1, t_2, \dots$ converging to $t_0$ such that $g(t_k) \neq g(t_0)$ for all $k$. But this implies that the eigenvalues of $B_{t_k}$ can't approach those of $B_{t_0}$, providing a contradiction. (See Figure 9.1.) Thus $g(t)$ is constant, which says that $g(0) = g(1)$. Since $B_0$ has exactly $m$ eigenvalues in $K$, so does $B_1$, which is $A$. ∎

**Example 9.2** *Let* $A = \begin{bmatrix} 0 & 1 & 0 \\ -2 & 5 & 0 \\ -1 & 1 & 6 \end{bmatrix}$. *Then*

$$\begin{aligned} D_1 &: \quad |\lambda - 0| \leq |1| + |0| = 1 \\ D_2 &: \quad |\lambda - 5| \leq |-2| + |0| = 2 \\ D_3 &: \quad |\lambda - 6| \leq |-1| + |1| = 2. \end{aligned}$$

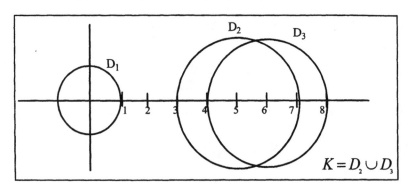

FIGURE 9.2.

*The graphs are in Figure 9.2.*

The eigenvalues of $A$ are $\lambda_1 = 0.4384, \lambda_2 = 4.5616$ *(to 5 digits), and* $\lambda_3 = 6$. *We see in the figure that these eigenvalues are covered by* $D_1$ *and* $D_2 \cup D_3$.

A corollary defining the radius of the disks by norms follows.

**Corollary 9.2** *Let* $A = D + B$ *where* $D = diag(d_1, \dots, d_n)$ *and* $B$ *an* $n \times n$ *matrix. Using any of the induced norms* $\|\cdot\|_1$, $\|\cdot\|_2$, *and* $\|\cdot\|_\infty$, *consider the disks* $D_i$ *defined by*

$$|\lambda - d_i| \leq \|B\|$$

*for* $i = 1, \dots, n$. *Then both (a) and (b) of Gershgorin's Theorem hold.*

**Proof.** The proof of this corollary, for induced matrix norms $\|\cdot\|_1$ and $\|\cdot\|_\infty$, is like that of Geshgorin's Theorem. For the induced matrix norm $\|\cdot\|_2$, the proof is more complicated. ∎

Note that Theorem 9.1 does not imply that eigenvalues are functions of the entries of their matrices. (As the entries of a matrix change, so do the multiplicities of the eigenvalues. A way to describe these eigenvalues as functions isn't known.)

For us to obtain a result of this type requires that there be no multiple eigenvalues. Thus to show eigenvalues as functions, we let $A$ be an $n \times n$ matrix with distinct eigenvalues $\lambda_1, \dots, \lambda_n$. Let $r$ be a radius which produces non-intersecting disks $D_1, \dots, D_n$ about these eigenvalues respectively.

Now we let the entries of $A$ vary, forming the matrix $B$. Then from results in function theory, if $B$ is sufficiently close to $A$ (say, $\|B - A\|_F < \epsilon$), then the eigenvalues $\beta_1, \dots, \beta_n$ of $B$ remain in the disks $D_1, \dots, D_n$. (See Figure 9.3.) Further, these eigenvalues are both continuous and differentiable.

Continuing, for each eigenvalue $\beta_i$, there is an eigenvector $x_i$, of length 1, that is continuous and differentiable.

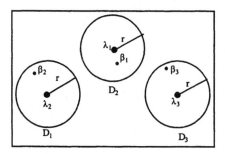

FIGURE 9.3.

## 9.1.1  *Optional (Eigenvectors and Multiple Eigenvalues)*

The eigenvalues of $A$ are continuously dependent on the entries of $A$ (as given in Theorem 9.1) even when those eigenvalues aren't described as functions. There is no such general result for eigenvectors.

Eigenvectors are continuous about matrices that have distinct eigenvalues. In the following we show an interesting example, a varient of one given by J. W. Givens, of what can happen when matrices are close to a matrix with multiple eigenvalues.

**Example 9.3** *Let*

$$A_\epsilon = \begin{bmatrix} \epsilon \cos \frac{2}{\epsilon} & -\epsilon \sin \frac{2}{\epsilon} \\ -\epsilon \sin \frac{2}{\epsilon} & -\epsilon \cos \frac{2}{\epsilon} \end{bmatrix}$$

*where $\epsilon$ is a positive scalar. This matrix has the form*

$$\begin{bmatrix} a & b \\ b & -a \end{bmatrix}$$

*whose eigenvalues are $\lambda = \pm\sqrt{a^2 + b^2}$. Thus the eigenvalues of $A_\epsilon$ are*

$$\lambda = \pm\epsilon.$$

*Taking $\lambda = \epsilon$, a corresponding eigenvector (when $\frac{2}{\epsilon}$ is not a multiple of $2\pi$) is*

$$x_\epsilon = \begin{bmatrix} \sin \frac{2}{\epsilon} \\ \cos \frac{2}{\epsilon} - 1 \end{bmatrix}.$$

*(This vector can be normalized, but for simplicity, we leave it as is.) Now, as $\epsilon \to 0$, the vectors $\begin{bmatrix} 1 \\ -1 \end{bmatrix}$ and $\begin{bmatrix} 0 \\ -2 \end{bmatrix}$ occur infinitely often. Thus, the eigenvectors 'wobble' and do not tend to any vector. So, even though the eigenvectors are continuous, they are very sensitive to change in the matrix.*

## Exercises

1. Compute the characteristic polynomial for

$$\begin{bmatrix} 1 & 2 & -1 \\ 2 & 3 & 1 \\ 3 & -1 & 0 \end{bmatrix}$$

by using Lemma 9.1.

2. Write out the general proof of Lemma 9.1.

3. The eigenvalues of $A = \begin{bmatrix} 1 & 1 & -1 & 1 \\ 2 & -1 & 1 & 2 \\ 3 & 0 & 2 & 1 \\ 1 & 5 & 2 & 4 \end{bmatrix}$ were calculated as, $\lambda_1 = 1.02, \lambda_2 = 3.14, \lambda_3 = 2.15., \lambda_4 = .73$. Is this correct? (Do not calculate the eigenvalues of $A$ and compare.)

4. Give one eigenvalue for

$$\begin{bmatrix} \pi & e & \sqrt{2} \\ 0 & 0 & 0 \\ i & \tan \delta & 200! \end{bmatrix}.$$

5. Apply Gershgorin's Theorem to

$$A = \begin{bmatrix} 3 & 1 & 1 \\ 1 & 2 & 0 \\ 2 & 1 & 4 \end{bmatrix}.$$

Is $A$ nonsingular?

6. Let $A$ be an $n \times n$ matrix.

   (a) Show that the eigenvalues of $A$ and $A^t$ are the same.

   (b) State Gershgorin's theorem for eigenvalues of $A$ in terms of column sums.

   (c) Determine the eigenvalues of $A = \begin{bmatrix} 2 & 1 & 1 \\ 0 & 5 & 0 \\ 0 & 0 & 6 \end{bmatrix}$, using Gerschgorin's Theorem. Check $A$ and $A^t$.

   (d) Estimate the eigenvalues of $\begin{bmatrix} .2 & .2 & .2 \\ .1 & .5 & .1 \\ .1 & .1 & 6 \end{bmatrix}$, using Gerschgorin's disks. Use both row and column sums.

(e) Compute the eigenvalues of the matrix in (d) using MATLAB. Plot these in the disks determined by (d).

7. Apply Gershgorin's theorem to $A = \begin{bmatrix} 1 & 1 \\ 16 & 1 \end{bmatrix}$. Note that one of the disks contains no eigenvalues. Does this contradict the theorem?

8. Prove Corollary 9.2 for $\|\cdot\|_\infty$.

## 9.2  Perturbation of Eigenvalues and Eigenvectors

To perturb means to change slightly. Thus, in this section, we study how much eigenvalues and corresponding eigenvectors of an $n \times n$ matrix, say, $A$, change under small perturbations of the entries of $A$ obtaining, say, $A + E$. We should recall from calculus that even if a function (like eigenvalue or eigenvector functions) is continuous, small changes in the variables can yield huge changes in the corresponding functional values.

Our first theorem is an eigenvalue result.

**Theorem 9.2** *Let $A$ be an $n \times n$ diagonalizable matrix, say, $A = PDP^{-1}$. Let $E$ be an $n \times n$ matrix and $\|\cdot\|$ the induced matrix norm $\|\cdot\|_1$, $\|\cdot\|_2$, or $\|\cdot\|_\infty$.*

(a) *If $\lambda$ is an eigenvalue of $A + E$,*

$$|\lambda - \lambda_i| \leq c(P)\|E\|$$

*for some eigenvalue $\lambda_i$ of $A$.*

(b) *If $K$ is the union of $m$ of the disks described in (a), and $K$ is disjoint from all other disks, then $K$ contains $m$ eigenvalues of $A$.*

**Proof.** Since the proof of (b) is as that in Gershgorin's Theorem, we only prove part (a).

By hypothesis,

$$P^{-1}AF = D$$

where $D = diag(\lambda_1, \dots, \lambda_n)$ and $\lambda_1, \dots, \lambda_n$ the eigenvalues of $A$. Thus

$$P^{-1}(A + E)P = D + P^{-1}EP.$$

Now, if $\lambda$ is an eigenvalue of $A + E$, using similarity, it is an eigenvalue of $D + P^{-1}EF$. Applying the Corollary 9.2 to $D + P^{-1}EF$ yields that, for some $\lambda_i$,

$$\begin{aligned}|\lambda - \lambda_i| &\leq \|P^{-1}EF\| \\ &\leq \|P^{-1}\|\,\|P\|\,\|E\| \\ &= c(P)\|E\|,\end{aligned}$$

the desired result. ∎

We can think of $c(P)$ here as somewhat like a derivative. Setting $\lambda = \lambda_i(A + E)$ and $\lambda_i = \lambda_i(A)$, we have

$$|\lambda_i(A + E) - \lambda_i(A)| \le c(P)\|E\|,$$

somewhat similar to the Mean Value Theorem that we studied in calculus.

Note that if we change the entries in $A$ a bit, say, by adding $E$ where $\|E\| < .01$, then we can't assume the eigenvalues of $A + E$ are within .01 of those of $A$. This would be true if $A$ is normal since in this case we would have $P$ unitary and then using the induced matrix norm, $\|\cdot\|_2$, $c(P) = 1$. However, if $c(P)$ is larger, the eigenvalues of $A + E$ may be farther away from those of $A$ than $\|E\|$.

**Example 9.4** *Let* $A = \begin{bmatrix} 1 & 1 \\ 0 & 2 \end{bmatrix}$. *Then* $\lambda_1 = 1$, $\lambda_2 = 2$, $P = \begin{bmatrix} 1 & 1 \\ 0 & 1 \end{bmatrix}$, $P^{-1} = \begin{bmatrix} 1 & -1 \\ 0 & 1 \end{bmatrix}$ *and* $c_\infty(P) = 2^2 = 4$. *Now let* $E = \begin{bmatrix} 0.01 & 0.02 \\ 0.01 & 0.03 \end{bmatrix}$, *so* $A + E$ *is a change in* $A$ *of* .04 ($\|E\|_\infty = .04$). *What we know now is that the eigenvalues of* $A + E$ *are within* $c_\infty(P)\|E\|_\infty = .16$ *of those of* $A$. *Calculating the eigenvalues of* $A + E$, *we have* $\hat{\lambda}_1 = 1.0001$, $\hat{\lambda}_2 = 2.0399$, *well within our bound.*

We now give another perturbation result. This result, concerning both eigenvalues and eigenvectors, is obtained by differentiation of the eigenvalue and eigenvector functions which we described in the previous section. So, we assume that the eigenvalues $\lambda_1, \ldots, \lambda_n$ of $A$ are distinct.

For this work, we let $y_1, \ldots, y_n$ be left eigenvectors of $A$, corresponding to $\lambda_1, \ldots, \lambda_n$, respectively. We assume these eigenvectors have been normalized so that their lengths are 1.

In addition, let $E$ be an $n \times n$ matrix, $\|E\|_2 = 1$, and $B = A + tE$ where $t$ is a real variable. We take $t$ sufficiently small so that the eigenvalues of $B$ remain in the disks described in the previous section. It can be shown from function theory, that for each $i$, there is a differentiable eigenvector $x_i(t)$, $x_i(0) = x_i$, and a differentiable eigenvalue $\lambda_i(t)$, $\lambda_i(0) = \lambda_i$, such that for sufficiently small $t$,

$$(A + tE)x_i(t) = \lambda_i(t)x_i(t)$$

where $\|x_i(t)\|_2 = 1$. (Note $\lambda_i(t) = \beta_i$ of the previous section.)

The idea now is to compute $\lambda_i'(0)$, to see how small changes in $t$ affect $\lambda_i(t)$.

**Theorem 9.3** *Using the above notations,*

(a) $\lambda_i'(0) = \frac{y_i E x_i}{y_i x_i}$ and

(b) If $x_i(t) = \delta_1(t) x_1 + \cdots + \delta_n(t) x_n$ $(\delta_i(t) = 1)$, then

$$\delta_j'(0) = \frac{y_j E x_i}{(\lambda_i - \lambda_j) y_j x_j}$$

for all $j \neq i$.

**Proof.** We prove both parts.
Part a. Expanding

$$(A + tE) x_i(t) = \lambda_i(t) x_i(t)$$

we have

$$A x_i(t) + t E x_i(t) = \lambda_i(t) x_i(t).$$

Differentiating

$$A x_i'(t) + E x_i(t) + t E x_i'(t) = \lambda_i'(t) x_i(t) + \lambda_i(t) x_i'(t).$$

Setting $t = 0$ yields

$$A x_i'(0) + E x_i = \lambda_i'(0) x_i(0) + \lambda_i(0) x_i'(0).$$

Rearranging leads to

$$(A - \lambda_i I) x_i'(0) = \lambda_i'(0) x_i - E x_i. \qquad (9.2)$$

Multiplying through by $y_i$, we have

$$0 = \lambda_i'(0) y_i x_i - y_i E x_i.$$

Thus

$$\lambda_i'(0) = \frac{y_i E x_i}{y_i x_i}.$$

Part b. Now, note that for any $j$,

$$y_j x_i(t) = \delta_j(t) y_j x_j.$$

So

$$\delta_j(t) = \frac{y_j x_i(t)}{y_j x_j}.$$

Since $x_i(t)$ is differentiable, so is $\delta_j(t)$.

Multiplying (9.2) through by $y_j$, using the Principle of Biothogonality, and that

$$x_i' (0) = \delta_1' (0) x_1 + \cdots + \delta_n' (0) x_n$$

yields

$$(\lambda_j - \lambda_i) y_j \delta_j' (0) x_j = -y_j E x_i \text{ when } j \neq i.$$

So

$$\delta_j' (0) = \frac{y_j E x_i}{(\lambda_i - \lambda_j) y_j x_j},$$

the result desired. ∎

Note that

$$|\lambda_i' (0)| = \frac{|y_i E x_i|}{|y_i x_i|}.$$

Then by the Cauchy-Schwarz inequality,

$$|\lambda_i' (0)| \leq \frac{\|y_i\|_2 \, \|E x_i\|_2}{|y_i x_i|}$$

$$\leq \frac{\|y_i\|_2 \, \|E\|_2 \, \|x_i\|_2}{|y_i x_i|}$$

$$\leq \frac{1}{s_i} \tag{9.3}$$

where

$$s_i = |y_i x_i|.$$

(Observe that since $|y_i x_i| \leq \|x_i\|_2 \, \|y_i\|_2 = 1$, $0 < s_i \leq 1$.)

We call $\frac{1}{s_i}$ the *condition number* for $\lambda_i$. Using (9.3), we interpret $\frac{1}{s_i}$ like the derivative in calculus, if $s_i$ is close to 0, $\lambda_i$ is ill-conditioned. (Small changes in $A$ can lead to much larger changes in $\lambda_i$.) And if $s_i$ is close to 1, $\lambda_i$ is well conditioned. (Small changes in $A$ lead to small changes in $\lambda_i$.)

Geometry can help us see when $\frac{1}{s_i}$ is large. ($\lambda_i$ is ill-conditioned.) Note from Figure 9.4 that

$$|\cos \theta| = |y_i x_i|$$

$$= s_i.$$

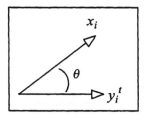

FIGURE 9.4.

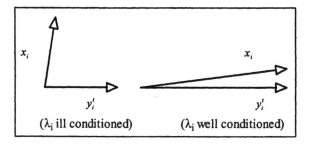

FIGURE 9.5.

So $s_i$ is the absolute value of the cosine of the angle between $x_i$ and $y_i^t$. Thus if the left and right eigenvectors (left eigenvector transposed) of $\lambda_i$ are nearly orthogonal ($s_i$ is near 0), then $\frac{1}{s_i}$ is large, and the condition number is large. If the left and right eigenvectors of $\lambda_i$ are nearly parallel ($s_i$ is near 1), then $\frac{1}{s_i}$ is near 1, 1 being the best possible condition number.

Since a normal matrix (This includes symmetric and Hermitian matrices.) is orthogonally diagonalizable, as given in the exercises, $x_i = y_i^t$ for all $i$ and so its eigenvalues are well conditioned. (This sometimes prompts the remark that matrices which have ill-conditioned eigenvalues are non-normal.) (See Figure 9.5.)

Similar to our analysis of eigenvalues, the numbers

$$\frac{1}{|\lambda_1 - \lambda_i|\, s_i}, \; \frac{1}{|\lambda_2 - \lambda_i|\, s_i}, \; \cdots \; , \frac{1}{|\lambda_n - \lambda_i|\, s_i} \tag{9.4}$$

(where the expression $\frac{1}{|\lambda_i - \lambda_i|\, s_i}$ is omitted) indicate the sensitivity of the coefficients of $x_i\,(t)$ to small changes in $t$ in $A + tE$. Thus, if $A$ has distinct, well-separated eigenvalues and $\frac{1}{s_i}$ is not too large, then the eigenvectors $x_i\,(t)$ have well-conditioned coefficients.

**Example 9.5** *Let* $A = \begin{bmatrix} 3 & 2 \\ 2 & 3 \end{bmatrix}$. *Then* $\lambda_1 = 5, \lambda_2 = 1$ *with corresponding right eigenvectors, normalized,* $x_1 = \begin{bmatrix} \frac{1}{\sqrt{2}} \\ \frac{1}{\sqrt{2}} \end{bmatrix}$, $x_2 = \begin{bmatrix} \frac{1}{\sqrt{2}} \\ -\frac{1}{\sqrt{2}} \end{bmatrix}$ *and left eigenvectors* $y_1 = \left[\frac{1}{\sqrt{2}}, \frac{1}{\sqrt{2}}\right]$, $y_2 = \left[\frac{1}{\sqrt{2}}, -\frac{1}{\sqrt{2}}\right]$.

(a) *The condition numbers for the eigenvalues are*

    i. $\lambda_1 = 5; \frac{1}{s_1} = \frac{1}{|y_1 x_1|} = 1.$

    ii. $\lambda_2 = 2; \frac{1}{s_2} = \frac{1}{|y_2 x_2|} = 1.$

*Thus for a small t, the eigenvalues of $A + tE$ will differ from those of A by about t, at most. (Note that A is symmetric.)*

(b) *For the condition of eigenvectors, we look at two parts. Note that*

    i. $\frac{1}{s_1} = 1$ *and* $\frac{1}{s_2} = 1.$

    ii. $\frac{1}{|\lambda_1 - \lambda_2|} = \frac{1}{3}.$

*So the eigenvectors have well-conditioned coefficients.*

### 9.2.1 Optional (Pictures of Eigenvalue and Eigenvector Sensitivity)

In this optional, we show the sensitivity of eigenvalues and eigenvectors in terms of pictures. To do this, we need some preliminary work.

It is known that if a matrix $A$ has an ill-conditioned eigenvalue ($\frac{1}{s_i}$ is large for some $i$), then $A$ is close to a matrix having multiple eigenvalues. (The converse is not true.) So if a matrix has close eigenvalues, it is a signal that eigenvalues and eigenvectors could be ill conditioned.

We now look at two examples showing this, one for eigenvalues and the other for eigenvectors.

**Example 9.6** *For eigenvalues we let* $A = \begin{bmatrix} 1 & b \\ c & 1 \end{bmatrix}$. *Then* $\lambda_1 = 1 + \sqrt{bc}$ *and* $\lambda_2 = 1 - \sqrt{bc}$ *give the eigenvalues of A. As a function of b and c, the graph of $\lambda_1$ is shown in Figure 9.6. We use the Frobenius norm on $R^{2\times2}$ so the matching* $\begin{bmatrix} 1 & b \\ c & 1 \end{bmatrix} \leftrightarrow (b,c)^t$ *preserves distance. Note that near the b and c axes (b or c is small), the partial derivatives of $\lambda$ are very large. So we know that small changes in b or c (near the axes) can cause much larger changes in $\lambda_1$. For example, for* $A = \begin{bmatrix} 1 & 0 \\ 1000 & 1 \end{bmatrix}$, *the eigenvalues*

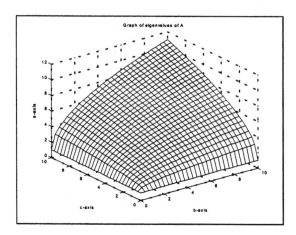

FIGURE 9.6.

are $\lambda_1 = 1.0000$, $\lambda_2 = 1.0000$. *But for* $\hat{A} = \begin{bmatrix} 1 & .0001 \\ 1000 & 1 \end{bmatrix}$, *we have*
$\lambda_1 = 1.3162$, $\lambda_2 = 0.6838$. *So a change of .0001 in A* ($\|E\|_2 = .0001$), *changed* $\lambda_1$ *from 1.000 to 1.3162 and* $\lambda_2$ *from 1.0000 to 0.6838. (This was somewhat predictable since the eigenvalues were close. Close eigenvalues are a red flag.)*

*Observe also that when* $b = c$, $\lambda_1$ *doesn't change much for changes in, say, b. As we know, this is true for symmetric matrices in general (even when eigenvalues are close).*

**Example 9.7** *For eigenvectors, we let* $A = \begin{bmatrix} 1 & 0 \\ 0 & 1.1 \end{bmatrix}$ *and* $A(t) = \begin{bmatrix} 1 & t \\ t & 1.1 \end{bmatrix}$,
*so* $E = \begin{bmatrix} 0 & 1 \\ 1 & 0 \end{bmatrix}$. *The eigenvalues of* $A(t)$ *are given by*

$$\lambda_1 = \frac{2.1 + \sqrt{.01 + 4t^2}}{2} \qquad \lambda_2 = \frac{2.1 - \sqrt{.01 + 4t^2}}{2}.$$

*The corresponding eigenvector, normalized to length 1, for* $\lambda_1$ *is*

$$x = \begin{bmatrix} \frac{2t}{d} \\ \frac{.1 + \sqrt{.01 + 4t^2}}{d} \end{bmatrix}$$

*where* $d = \left( 8t^2 + .02 + .2\sqrt{.01 + 4t^2} \right)^{\frac{1}{2}}$.

*We let* $x = t$, $y = \frac{2t}{d}$, $z = \frac{.1 + \sqrt{.01 + 4t^2}}{d}$ *and graph* $(x, y, z)^t$ *for* $-1 \le t \le 1$. *In the graph, the eigenvector is* $(y, z)^t$. *Observe in Figure 9.7 that about* $t = 0$ *the eigenvector shows a lot of change for small changes in t. This is*

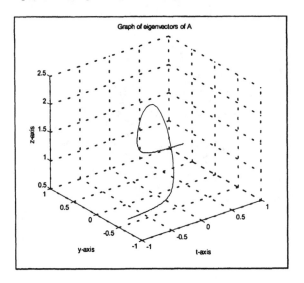

FIGURE 9.7.

*confirmed numerically by computing the eigenvector for several values of t.*

| t | −.1 | 0 | .1 |
|---|---|---|---|
| $\begin{bmatrix} x \\ y \end{bmatrix}$ | $\begin{bmatrix} -0.5257 \\ 0.8507 \end{bmatrix}$ | $\begin{bmatrix} 0 \\ 1 \end{bmatrix}$ | $\begin{bmatrix} 0.5259 \\ 0.8507 \end{bmatrix}$ |

*(Note that A has close eigenvalues when t is small, a red flag that indicates eigenvectors could be ill conditioned.)*

## 9.2.2    MATLAB (Condeig)

MATLAB provides condition numbers for eigenvalues. The command is *condeig.* An example follows.

$A = \begin{bmatrix} 1 & 0 & 1 ; & 2 & 7 & 3 ; & 7 & 2 & 0 \end{bmatrix}$;
$\text{eig}(A)$

$$\text{ans} = \begin{bmatrix} 2.5958 \\ -2.5071 \\ 7.9475 \end{bmatrix}$$

$\text{condeig}(A)$

$$\text{ans} = \begin{bmatrix} 1.5766 \\ 1.4399 \\ 1.1994 \end{bmatrix}$$

If we want eigenvectors (recorded as columns in a matrix), eigenvalues, and their condition numbers, we use

$[V, D, s] = condeig\,(A)$

## 1. Code for eigenvalue picture

```
b = linspace(0, 10, 30);
c = linspace(0, 10, 30);
[b, c] = meshgrid(b, c);
s = 1+sqrt(b. * c);
mesh(b, c, s)
grid on
```

## 2. Code for eigenvector pictures

```
t = linspace(-1, 1, 50);
d = 8 * (t. ∧ 2) + .02 + .2*sqrt(.01 + 4 * (t. ∧ 2));
y = 2 * t./d;
z = .1+sqrt(.01 + 4 * (t. ∧ 2)))./d;
plot3(t, y, z)
grid on
```

For more information, type in *help condeig*. Also for the graphs, type in *help plot3*.

# Exercises

1. Let $A = \begin{bmatrix} 2 & 1 \\ 1 & 2 \end{bmatrix}$.

    (a) Factor $A = PDF^{-1}$ and compute $c_\infty (P)$.

    (b) If $E = \begin{bmatrix} .1 & .1 \\ .1 & .1 \end{bmatrix}$, find $c_\infty (P) \|E\|_\infty$.

    (c) Plot the eigenvalues of $A$ in $R^2$. Draw circles of radius $c_\infty (P) \|E\|_\infty$ about the eigenvalues.

    (d) Find, and plot in (c), the eigenvalues of $A + E$.

2. Repeat Exercise 1 for $A = \begin{bmatrix} 1 & 0 \\ 1000 & 1.1 \end{bmatrix}$ and $E = \begin{bmatrix} .1 & .1 \\ .1 & .1 \end{bmatrix}$.

3. Let $A = \begin{bmatrix} 3 & 1 \\ 1 & 3 \end{bmatrix}$.

    (a) Using $E = \begin{bmatrix} 1 & 0 \\ 0 & 0 \end{bmatrix}$, compute $\lambda_i' (0)$, for $i = 1, 2$, using Theorem 9.3 (a).

    (b) Repeat (a) for $E = \begin{bmatrix} 0 & 0 \\ 1 & 0 \end{bmatrix}$.

4. Repeat Exercise 3 for $A = \begin{bmatrix} 1 & 1000 \\ 0 & 4 \end{bmatrix}$.

5. For the matrix given in Exercise 3,

   (a) Find $\frac{1}{s_1}$ for $\lambda_1$. Explain what this means in terms (9.3).

   (b) Find $|\lambda_1 - \lambda_2|$. Using (9.4), explain what $\frac{1}{|\lambda_1-\lambda_2|s_1}$ means in terms of the condition of the coefficients of $x_1$.

   (c) Repeat (a) for $\lambda_2$.

   (d) Repeat (b) for $x_2$.

6. Repeat Exercise 5 for the matrix in Example 4.

7. Let $A$ be an $n \times n$ normal matrix with distinct eigenvalues. Prove that if $y_i$ is a left eigenvector, for the eigenvalue $\lambda_i$ of $A$, then $y_i^t$ is a right eigenvector belonging to that eigenvalue.

8. (MATLAB) Let

$$A = \begin{bmatrix} 2 & 1 & -1 \\ 3 & -1 & -2 \\ 2 & -5 & 3 \end{bmatrix}.$$

   (a) Find the eigenvalues of $A$.

   (b) Find $\frac{1}{s_1}$ for $i = 1$, 2, and 3.

   (c) Make some conclusion from (b).

9. (MATLAB) Find a $3 \times 3$ matrix, with $\frac{1}{s_i} > 100$, for some $i$. (Use theory to see where to look.)

# 10
# Hermitian and Positive Definite Matrices

As we will see, Hermitian matrices (In the real case we are talking about symmetric matrices.) arise in mathematical models of mechanical systems, in Hermitian forms, and in optimization. (There are many other areas as well.) In this chapter we look at several results about Hermitian matrices which are useful in these areas.

## 10.1  Positive Definite Matrices

As we have seen, a Hermitian can be diagonalized by a unitary matrix. By using a special class of Hermitian matrices, the positive definite Hermitian matrices, we show in this section how two matrices can be simultaneously diagonalized in a special way.

A Hermitian matrix $A$ is *positive definite* if all of the eigenvalues of $A$ are positive. And if $A$ has all nonnegative eigenvalues, we use the words *positive semidefinite*. (If $-A$ is positive definite, we call $A$ *negative definite* and if $-A$ is positive semidefinite, we call $A$ *negative semidefinite*.) Positive definite matrices can be factored in a special way.

**Lemma 10.1** *Let $A$ be an $n \times n$ matrix. Then $A$ is Hermitian and positive definite if and only if there is an $n \times n$ nonsingular matrix $R$ such that*

$$A = RR^H.$$

*(If $A$ is symmetric, $R$ is real, and we have $A = RR^t$.)*

**Proof.** We prove the biconditional in two parts.

Part a. If $A$ is Hermitian and positive definite, we can factor

$$A = UDU^H$$

where $U$ is a unitary matrix and $D = \text{diag}\,(\lambda_1, \ldots, \lambda_n)$ where each $\lambda_i > 0$. Define $D^{\frac{1}{2}} = \text{diag}\,(\sqrt{\lambda_1}, \ldots, \sqrt{\lambda_n})$ and set $R = UD^{\frac{1}{2}}$, a nonsingular matrix. Then

$$A = RR^H.$$

Part b. If $A = RR^H$, then $A$ is clearly Hermitian. To show that $A$ is positive definite, we proceed as follows.

Suppose $\lambda$ is an eigenvalue of $A$. Then $Ax = \lambda x$ for some eigenvector $x$. Since $A = RR^H$,

$$RR^H x = \lambda x.$$

Thus

$$x^H RR^H x = \lambda x^H x$$

$$\left\| R^H x \right\|_2^2 = \lambda \left\| x \right\|_2^2.$$

Since $R$ is nonsingular, so is $R^H$ ($\det R^H = \overline{\det R} \neq 0$) and thus $R^H x \neq 0$. Hence $\left\| R^H x \right\|_2^2 > 0$. Since $\left\| x \right\|_2^2 > 0$, it follows that $\lambda > 0$. As $\lambda$ was an arbitrarily chosen eigenvalue, $A$ is positive definite. ∎

**Example 10.1** *Let* $A = \begin{bmatrix} 3 & 1 \\ 1 & 3 \end{bmatrix}$, *a symmetric matrix. The eigenvalues and corresponding eigenvectors (of length 1) are given by* $\lambda_1 = 4$, $\lambda_2 = 2$, *and* $u_1 = \begin{bmatrix} \frac{\sqrt{2}}{2} \\ \frac{\sqrt{2}}{2} \end{bmatrix}$, $u_2 = \begin{bmatrix} -\frac{\sqrt{2}}{2} \\ \frac{\sqrt{2}}{2} \end{bmatrix}$. *Thus, $A$ is positive definite. Now*

$$A = UDU^t = \begin{bmatrix} \frac{\sqrt{2}}{2} & -\frac{\sqrt{2}}{2} \\ \frac{\sqrt{2}}{2} & \frac{\sqrt{2}}{2} \end{bmatrix} \begin{bmatrix} 4 & 0 \\ 0 & 2 \end{bmatrix} \begin{bmatrix} \frac{\sqrt{2}}{2} & \frac{\sqrt{2}}{2} \\ -\frac{\sqrt{2}}{2} & \frac{\sqrt{2}}{2} \end{bmatrix}$$

$$= \left( \begin{bmatrix} \frac{\sqrt{2}}{2} & -\frac{\sqrt{2}}{2} \\ \frac{\sqrt{2}}{2} & \frac{\sqrt{2}}{2} \end{bmatrix} \begin{bmatrix} 2 & 0 \\ 0 & \sqrt{2} \end{bmatrix} \right) \left( \begin{bmatrix} 2 & 0 \\ 0 & \sqrt{2} \end{bmatrix} \begin{bmatrix} \frac{\sqrt{2}}{2} & \frac{\sqrt{2}}{2} \\ -\frac{\sqrt{2}}{2} & \frac{\sqrt{2}}{2} \end{bmatrix} \right)$$

$$= \begin{bmatrix} \sqrt{2} & -1 \\ \sqrt{2} & 1 \end{bmatrix} \begin{bmatrix} \sqrt{2} & \sqrt{2} \\ -1 & 1 \end{bmatrix}$$

$$= RR^t$$

*where* $R = \begin{bmatrix} \sqrt{2} & -1 \\ \sqrt{2} & 1 \end{bmatrix}$.

Adjusting the proof slightly, we can show that an $n \times n$ matrix $A$ is Hermitian and positive semidefinite if and only if $A = RR^H$ for some $n \times n$ matrix $R$.

Note that if $S = R^H$, then

$$A = S^H S$$

so which matrix in the factorization has the superscript $H$ doesn't matter.

From this lemma we can produce a simpler factorization, the Choleski's decomposition.

**Corollary 10.1** *Let $A$ be an $n \times n$ matrix which is Hermitian and positive definite. Then*

$$A = T^H T$$

*where $T$ is an upper triangular matrix.*

**Proof.** Using Lemma 10.1, factor

$$A = S^H S.$$

Now factor $S = UT$, where $U$ is unitary and $T$ upper triangular. ($UT$ found from the Gram-Schmidt process, i.e. the $QR$ factorization.) By substitution

$$A = (UT)^H (UT)$$
$$= T^H T,$$

which is Choleski's decomposition. ∎

Note that the lemma (or corollary) also implies that if $A$ is Hermitian and positive definite, and $B$ is a nonsingular matrix, then $B^H AB$ is Hermitian and positive definite. (To see this, factor $A = R^H R$ and substitute to get $B^H AB = B^H R^H RB = (RB)^H (RB)$.)

We can now show how two Hermitian matrices can be simultaneously factored into diagonal matrices.

**Theorem 10.1** *Let $A$ and $B$ be $n \times n$ Hermitian matrices with $B$ positive definite. Then there is an $n \times n$ nonsingular matrix $P$ such that*

$$P^H BP = I \text{ and } P^H AP = D$$

*where $D = \text{diag}(\lambda_1, \dots, \lambda_n)$. Further, if $A$ is also positive definite, $D$ has a positive main diagonal.*

*Finally, if $A$ and $B$ are real, so is $P$.*

**Proof.** The proof outlines the method to find $P$.

Step 1. (Find $R$.) As in Lemma 10.1, find $R$ such that

$$B = RR^H.$$

Note that $R^{-1}B\left(R^H\right)^{-1} = I$.
   Step 2. (Find $U$.) Set

$$C = R^{-1}A\left(R^H\right)^{-1}$$

a Hermitian matrix. Thus, we can find a unitary matrix $U$ such that

$$U^H CU = D, \quad \text{or}$$
$$U^H R^{-1}A\left(R^H\right)^{-1}U = D.$$

Step 3. (Find $P$.) Set

$$P = \left(R^H\right)^{-1}U.$$

Then

$$P^H BF = U^H R^{-1}\left(RR^H\right)\left(R^H\right)^{-1}U = I$$

and

$$P^H AF = U^H R^{-1}A\left(R^H\right)^{-1}U = U^H CU = D.$$

Thus, $P$ has the required properties. ∎

Finally, if $A$ is positive definite, so is $P^H AP$, so $D$ has a positive main diagonal.

**Example 10.2** *Let* $A = \begin{bmatrix} 2 & -1 \\ -1 & 2 \end{bmatrix}$ *and* $B = \begin{bmatrix} 4 & 0 \\ 0 & 4 \end{bmatrix}$. *We follow the steps of the proof of the theorem.*
   *Step 1. (Finding $R$) Factoring*

$$B = RR^t = \begin{bmatrix} 2 & 0 \\ 0 & 2 \end{bmatrix}\begin{bmatrix} 2 & 0 \\ 0 & 2 \end{bmatrix},$$

*so* $R = \begin{bmatrix} 2 & 0 \\ 0 & 2 \end{bmatrix}$.
   *Step 2. (Finding $U$) Set*

$$C = R^{-1}A\left(R^t\right)^{-1}$$
$$= \begin{bmatrix} \frac{1}{2} & 0 \\ 0 & \frac{1}{2} \end{bmatrix}\begin{bmatrix} 2 & -1 \\ -1 & 2 \end{bmatrix}\begin{bmatrix} \frac{1}{2} & 0 \\ 0 & \frac{1}{2} \end{bmatrix}$$
$$= \begin{bmatrix} \frac{1}{2} & -\frac{1}{4} \\ -\frac{1}{4} & \frac{1}{2} \end{bmatrix}.$$

*Orthogonally diagonalizing $C$, we have*

$$U = \begin{bmatrix} \frac{1}{\sqrt{2}} & -\frac{1}{\sqrt{2}} \\ \frac{1}{\sqrt{2}} & \frac{1}{\sqrt{2}} \end{bmatrix} \text{ and } D = \begin{bmatrix} \frac{1}{4} & 0 \\ 0 & \frac{3}{4} \end{bmatrix}.$$

*Step 3.  (Finding $P$)  Set*

$$P = \left(R^t\right)^{-1} U = \begin{bmatrix} \frac{1}{2} & 0 \\ 0 & \frac{1}{2} \end{bmatrix} \begin{bmatrix} \frac{1}{\sqrt{2}} & -\frac{1}{\sqrt{2}} \\ \frac{1}{\sqrt{2}} & \frac{1}{\sqrt{2}} \end{bmatrix}$$

$$= \begin{bmatrix} \frac{\sqrt{2}}{4} & -\frac{\sqrt{2}}{4} \\ \frac{\sqrt{2}}{4} & \frac{\sqrt{2}}{4} \end{bmatrix}.$$

Simultaneous diagonalization of positive definite Hermitian matrices arises in simplifying quadratic forms, say, $q_1(x) = x^t A x$ and $q_2(x) = x^t B x$ (which represent, say, kinetic and potential energies). It is also sometimes used to solve systems of differential equations of the form

$$Mx''(t) + Kx(t) = 0 \tag{10.1}$$

as in the spring-mass, building, etc. problems. The simultaneous reduction here (using $y(t) = P^{-1}x(t)$) yields

$$y''(t) + Dy(t) = 0.$$

These equations are easily solved for $y(t)$ and then

$$x(t) = Py(t)$$
$$= y_1(t)p_1 + \cdots + y_n(t)p_n$$

where $y(t) = (y_1(t), \ldots, y_n(t))$ and $p_i$ is the $i$-th column of $P$. We will show such an approach in Optional. Now, however, we want to observe that (10.1) is equivalent to

$$x''(t) + M^{-1}Kx(t) = 0.$$

If $M^{-1}K$ is diagonalizable, we can solve this problem as described in Chapter 4.

We need the following.

**Corollary 10.2** *Let $A$ and $B$ be $n \times n$ Hermitian matrices with $B$ positive definite. Then $B^{-1}A$ is diagonalizable. And if $A$ is positive definite, $B^{-1}A$ has positive eigenvalues.*

**Proof.** From the theorem, there is a nonsingular matrix $P$ such that

$$P^H B P = I \tag{10.2}$$

$$P^H AF = D. \tag{10.3}$$

Taking the inverse of the matrices in (10.2) and multiplying sides to those of the equation in (10.3), we have

$$P^{-1}B^{-1}\left(P^H\right)^{-1}P^H AF = ID$$

or

$$P^{-1}B^{-1}AF = D.$$

Thus, $B^{-1}A$ is diagonalizable.

Finally, if $A$ is positive definite, then so is $P^H AF$ and thus by (10.3), $D$ has a positive diagonal. And using similarity, $B^{-1}A$ has positive eigenvalues. ∎

In Chapter 4, we derived a formula for the solutions to

$$x''(t) + Ax(t) = 0 \tag{10.4}$$

where $A$ is a $2 \times 2$ diagonalizable matrix with positive eigenvalues.

The formula is

$$x(t) = \left(\alpha_1 \cos \sqrt{\lambda_1}t + \beta_1 \sin \sqrt{\lambda_1}t\right) p_1 \tag{10.5}$$
$$+ \left(\alpha_2 \cos \sqrt{\lambda_2}t + \beta_2 \sin \sqrt{\lambda_2}t\right) p_2$$

where $\lambda_1$ and $\lambda_2$ are the eigenvalues of $A$ having corresponding eigenvectors $p_1$ and $p_2$, respectively.

An example applying this formula follows.

**Example 10.3** *We solve the spring-mass problem given by the equation*

$$\begin{bmatrix} 1 & 0 \\ 0 & \frac{1}{4} \end{bmatrix} x'' + \begin{bmatrix} 4 & -1 \\ -1 & 1 \end{bmatrix} x = 0.$$

*This system is equivalent to*

$$x'' + \begin{bmatrix} 1 & 0 \\ 0 & 4 \end{bmatrix} \begin{bmatrix} 4 & -1 \\ -1 & 1 \end{bmatrix} x = 0.$$

*Now* $\begin{bmatrix} 1 & 0 \\ 0 & 4 \end{bmatrix} \begin{bmatrix} 4 & -1 \\ -1 & 1 \end{bmatrix} = \begin{bmatrix} 4 & -1 \\ -4 & 4 \end{bmatrix}$ *is diagonalizable with* $D =$ $\begin{bmatrix} 6 & 0 \\ 0 & 2 \end{bmatrix}$ *and* $P = \begin{bmatrix} 1 & 1 \\ -2 & 2 \end{bmatrix}$. *Thus*

$$x(t) = \left(\alpha_1 \cos \sqrt{6}t + \beta_1 \sin \sqrt{6}t\right) \begin{bmatrix} 1 \\ -2 \end{bmatrix} + \left(\alpha_2 \cos \sqrt{2}t + \beta_2 \sin \sqrt{2}t\right) \begin{bmatrix} 1 \\ 2 \end{bmatrix}.$$

*If the problem has initial conditions, say,*

$$x(0) = \begin{bmatrix} 5 \\ 2 \end{bmatrix}$$

$$x'(0) = \begin{bmatrix} 8 \\ -8 \end{bmatrix},$$

*we compute coefficients to satisfy these. From*

$$x(0) = \alpha_1 \begin{bmatrix} 1 \\ -2 \end{bmatrix} + \alpha_2 \begin{bmatrix} 1 \\ 2 \end{bmatrix},$$

*we get $\alpha_1 = 2$, $\alpha_2 = 3$. And from*

$$x'(0) = \sqrt{6}\beta_1 \begin{bmatrix} 1 \\ -2 \end{bmatrix} + \sqrt{2}\beta_2 \begin{bmatrix} 1 \\ 2 \end{bmatrix}.$$

*We find $\beta_1 = \sqrt{6}$ and $\beta_2 = \sqrt{2}$. So*

$$x(t) = \left(2\cos\sqrt{6} + \sqrt{6}\sin\sqrt{6}t\right) \begin{bmatrix} 1 \\ -2 \end{bmatrix} + \left(3\cos\sqrt{2}t + \sqrt{2}\sin\sqrt{2}t\right) \begin{bmatrix} 1 \\ 2 \end{bmatrix}.$$

### 10.1.1   *Optional (Solving the Motion of a Building Problem)*

A building as diagramed in Figure 10.1, can show some motion if displaced from vertical.

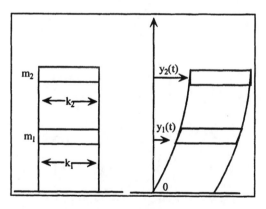

FIGURE 10.1.

The mathematical model for this building was derived in Chapter 4 as

$$M\frac{d^2}{dt^2}y(t) + Ky(t) = 0$$

where $M = \begin{bmatrix} m_1 & 0 \\ 0 & m_2 \end{bmatrix}$ and $K = \begin{bmatrix} k_1 + k_2 & -k_2 \\ -k_2 & k_2 \end{bmatrix}$. It can be shown that $M$ and $K$ are symmetric and positive definite.

We now demonstrate how to use Theorem 10.1 to solve this equation. A particular example follows.

**Example 10.4** *Consider the building drawn in Figure 10.2. Then $M =$*

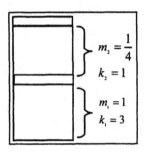

$$m_2 = \frac{1}{4}$$
$$k_2 = 1$$
$$m_1 = 1$$
$$k_1 = 3$$

FIGURE 10.2.

$\begin{bmatrix} 1 & 0 \\ 0 & \frac{1}{4} \end{bmatrix}$ and $K = \begin{bmatrix} 4 & -1 \\ -1 & 1 \end{bmatrix}$. *Using MATLAB and the algorithm in the MATLAB section, we found*

$$P = \begin{bmatrix} -0.7071 & -0.7070 \\ 1.4142 & -1.4142 \end{bmatrix}$$

*and*

$$D = \begin{bmatrix} 6 & 0 \\ 0 & 2 \end{bmatrix}.$$

*Thus using (10.5),*

$$
\begin{aligned}
x(t) &= \left( \alpha_1 \cos \sqrt{\lambda_1} t + \beta_1 \sin \sqrt{\lambda_1} t \right) p_1 \\
&+ \left( \alpha_2 \cos \sqrt{\lambda_2} t + \beta_2 \sin \sqrt{\lambda_2 2} t \right) p_2 \\
&= \left( \alpha_1 \cos \left( \sqrt{2} t \right) + \beta_1 \sin \left( \sqrt{6} t \right) \right) \begin{bmatrix} -0.7071 \\ 1.4142 \end{bmatrix} \\
&+ \left( \alpha_2 \cos \left( \sqrt{2} t \right) + \beta_2 \sin \left( \sqrt{2} t \right) \right) \begin{bmatrix} -0.7071 \\ -1.4142 \end{bmatrix}.
\end{aligned}
$$

*Now suppose at $t = 0$ the building is erect, so $x_1(0) = 0$ and $x_2(0) = 0$. With a gust of wind, we have $x_1'(0) = 1$ and $x_2'(0) = 1$. Plugging in $t = 0$, we have*

$$x(0) = \alpha_1 p_1 + \alpha_2 p_2$$

*or*

$$0 = \alpha_1 \rho_1 + \alpha_2 \rho_2.$$

*Solving yields* $\alpha_1 = \alpha_2 = 0$.
  *Now*

$$x'(0) = \beta_1 \sqrt{\lambda_1} p_1 + \beta_2 \sqrt{\lambda_2} p_2.$$

*Solving*

$$\begin{bmatrix} 1 \\ 1 \end{bmatrix} = \beta_1 \sqrt{\lambda_1} p_1 + \beta_2 \sqrt{\lambda_2} p_2,$$

*we get* $\beta_1 = -.01443$ *and* $\beta_2 = -0.7500$.
  *Thus, the subsequent motion of the building is*

$$
\begin{aligned}
x(t) &= \beta_1 \sqrt{\lambda_1} p_1 + \sqrt{\beta_2} p_2 \\
&= -0.1443 \sin\left(\sqrt{6}t\right) \begin{bmatrix} -0.7071 \\ 1.4142 \end{bmatrix} \\
&\quad - 0.7500 \sin\left(\sqrt{2}t\right) \begin{bmatrix} -0.7071 \\ -1.4142 \end{bmatrix}.
\end{aligned}
$$

*A graph depicting the building when* $t = 1$ *can be obtained by calculating*

$$x(1) = \begin{bmatrix} 0.5890 \\ 0.9175 \end{bmatrix}.$$

*Thus we have the shape shown in Figure 10.3.*

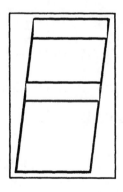

FIGURE 10.3.

## 10.1.2  *MATLAB (Code for Computing P)*

The commands for finding $P$, such that $P^t BF = I$ and $P^t AF = D$, as described in Theorem 10.1, follow.

$L = \text{chol}(B)$;                    % Gives the Cholesky decomposition
of $B$ as $LL^t$ where $L$ is lower
triangular.

$C = inv(L) * A * inv(L')$;
$[Q, T] = schur(C)$;
$P = inv(L') * Q$
$D = P' * A * P$

(MATLAB calculations may not agree entrywise with hand calculations since the factorizations $RR^t$ and $QDQ^t$ are not unique.)

See the exercises for a few problems on which to use this algorithm.

## Exercises

1. Factor $A$ as $RR^H$.

   (a) $\begin{bmatrix} 5 & 2 \\ 2 & 5 \end{bmatrix}$   (b) $\begin{bmatrix} 2 & 1 & 1 \\ 1 & 2 & 1 \\ 1 & 1 & 3 \end{bmatrix}$

2. Prove that $A$ is Hermitian and positive semidefinite if and only if $A = RR^H$ for some matrix $R$.

3. Using Exercise 2, factor $\begin{bmatrix} 1 & 1 \\ 1 & 1 \end{bmatrix}$ as $RR^H$.

4. Prove that if $A$ is an $n \times n$ positive semidefinite Hermitian matrix, then so is $B^H AB$ where $B$ is an $n \times n$ matrix.

5. Find a matrix $P$ that diagonalized "both"

$$A = \begin{bmatrix} 1 & 1 \\ 1 & 2 \end{bmatrix}, \quad B = \begin{bmatrix} 1 & 2 \\ 2 & 4 \end{bmatrix}.$$

   Is $P$ an orthogonal matrix? Are the columns of $P$ orthogonal?

6. Solve $\begin{bmatrix} 1 & 1 \\ 1 & 2 \end{bmatrix} x'' + \begin{bmatrix} 1 & 2 \\ 2 & 4 \end{bmatrix} x = 0$ where $x(0) = \begin{bmatrix} 0 \\ 0 \end{bmatrix}$ and $x'(0) = \begin{bmatrix} 2 \\ 3 \end{bmatrix}$. (Use the results of Exercise 5.)

7. Solve $\begin{bmatrix} 2 & 0 \\ 0 & 4 \end{bmatrix} x'' + \begin{bmatrix} 1 & 2 \\ 2 & 4 \end{bmatrix} x = 0$ using Corollary 10.2.

8. Solve the spring-mass problem in Figure 10.4 using Corollary 10.2.

9. Two parts.

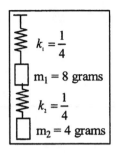

FIGURE 10.4.

(a) Prove that $\lambda_1, \ldots, \lambda_n$ (called generalized eigenvalues) of Theorem 10.1 can be found by solving $\det(\lambda B - A) = 0$.

(b) Prove that the columns $p_i$ of $P$ in Theorem 10.1, called generalized eigenvectors, can be computed by solving

$$(\lambda_i B - A) x = 0$$

for $m_i$ linearly independent vectors, where $\lambda_i$ has multiplicity $m_i$, and then by applying Gram-Schmidt to these using the inner product $(x, y) = y^H B x$.

10. (MATLAB) Let $A = \begin{bmatrix} 2 & 1 & 0 \\ 1 & 3 & -1 \\ 0 & -1 & 2 \end{bmatrix}$ and $B = \begin{bmatrix} 1 & 2 & 1 \\ 2 & 1 & -1 \\ 1 & -1 & 0 \end{bmatrix}$.

Use the algorithm in Optional to find the matrix $P$ described in Theorem 10.1.

11. (MATLAB) Solve the building problem for the building in Figure 10.5 where initially $x_1(0) = 1$, $x_1'(0) = 0$, $x_2(0) = 1$, $x_2'(0) = 0$. Draw the building when $t = 5$.

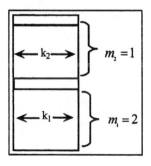

FIGURE 10.5.

## 10.2 Special Eigenvalue Results on Hermitian Matrices

In this section, we look at several eigenvalue results about Hermitian matrices. To do this, we define a special function called a Hermitian form.

Let $A$ be an $n \times n$ Hermitian matrix. Define a *Hermitian form* $h$ as

$$h : C^n \to C$$

where

$$h(x) = x^H A x.$$

If $\alpha = x^H A x$, then $\bar{\alpha} = \alpha^H = \left(x^H A x\right)^H = x^H A x = \alpha$. Thus the value of a Hermitian form is always real. If $A$ is symmetric and we let $q$ denote $h$ restricted to $R^n$, that is

$$q : R^n \to R$$

$$q(x) = x^t A x,$$

we call $q$ a *quadratic form*.

Hermitian and quadratic forms arise in representations of potential and kinetic energy in a system. Using Lagrange's equation and energy expressions, mathematical models of the system can be derived. In addition, these forms are used to develop numerical methods for computing eigenvalues, as well as in solving optimization problems for functions of several variables.

We first give a description of Hermitian forms, obtaining some view of the shapes of their graphs. To do this, let $A$ be an $n \times n$ Hermitian matrix. Then we can factor $A$ as

$$A = U D U^H$$

for some unitary matrix $U = [u_1 \ldots u_n]$ where $u_k$ is the $k$-th column of $U$, and diagonal matrix $D = diag\left(\lambda_1, \ldots, \lambda_n\right)$. We assume these eigenvalues are arranged in $D$ so that

$$\lambda_1 \geq \lambda_2 \geq \ldots \geq \lambda_n. \tag{10.6}$$

Now

$$h(x) = x^H A x = x^H \left(U D U^H\right) x = \left(U^H x\right)^H D \left(U^H x\right).$$

Setting $y = U^H x$, and defining $h_Y(y) = y^H D y$, we have

$$h(x) = y^H D y \tag{10.7}$$
$$= \lambda_1 |y_1|^2 + \cdots + \lambda_n |y_n|^2$$
$$= h_Y(y).$$

To see what this equation means geometrically in $R^n$, set $Y = \{u_1, \ldots, u_n\}$. Then the equation

$$x = Uy$$

converts the $y$ coordinates of $x$ into $x$. Thus

$$q(x) = q_Y(y)$$

says that $q(x)$ can be graphed by graphing $q_Y(y)$ in the $Y$-coordinate system.

We show an example.

**Example 10.5** *Let* $A = \begin{bmatrix} 3 & 1 \\ 1 & 3 \end{bmatrix}$.

*Then* $D = \begin{bmatrix} 4 & 0 \\ 0 & 2 \end{bmatrix}, U = \begin{bmatrix} \frac{1}{\sqrt{2}} & -\frac{1}{\sqrt{2}} \\ \frac{1}{\sqrt{2}} & \frac{1}{\sqrt{2}} \end{bmatrix}$. *So*

$$Y = \left\{ \begin{bmatrix} \frac{1}{\sqrt{2}} \\ \frac{1}{\sqrt{2}} \end{bmatrix}, \begin{bmatrix} -\frac{1}{\sqrt{2}} \\ \frac{1}{\sqrt{2}} \end{bmatrix} \right\}.$$

*The graph of*

$$q(x) = x^t A x = 3x_1^2 + 2x_1 x_2 + 3x_2^2,$$

*is in Figure 10.6. The graph of*

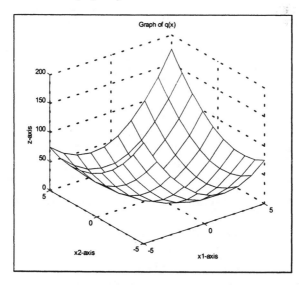

FIGURE 10.6.

$$q_Y(y) = y^t D y = 4y_1^2 + 2y_2^2,$$

*using the Y-coordinate system is in Figure 10.7.*

*Note that both graphs are identical when the axes of the Y-coordinate systems are rotated $-\frac{\pi}{4}$ radian, showing them relative to the axes in $R^2$.*

*The use of $h_Y(y)$ to derive information about $h(x)$ can be seen in the following theorem.*

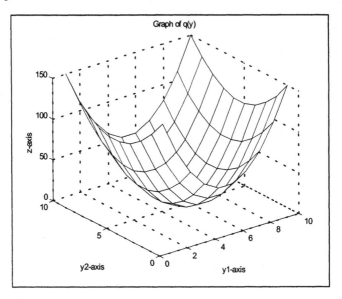

FIGURE 10.7.

**Theorem 10.2** *Let $h(x) = x^H A x$ be a Hermitian form. Then $A$ is a positive definite Hermitian matrix if and only if $h(x) > 0$, for all $x$, except at $x = 0$.*

**Proof.** Follows from (10.7). ∎

We now give a sequence of results about the eigenvalues of Hermitian matrices. In Rayleigh's Principle, we show how the smallest and largest eigenvalues of a Hermitian matrix can be found from Hermitian forms.

**Theorem 10.3** *Let $h(x) = x^H A x$ be a Hermitian form. Then*

$$\max_{\|x\|_2 = 1} h(x) = \lambda_1, \quad \min_{\|x\|_2 = 1} h(x) = \lambda_n$$

*where $\lambda_1$ and $\lambda_n$ are the largest and smallest, respectively, eigenvalues of $A$. Further, the maximum and minimum values of $h$ are achieved at $u_1$ and $u_n$, the eigenvectors of length 1 corresponding to $\lambda_1$ and $\lambda_n$ respectively.*

**Proof.** We prove the maximum result.

Part a. We show that if $\|x\|_2 = 1$, then $h(x) \leq \lambda_1$. To see this, by (10.7),

$$h(x) = h_Y(y),$$

$x$ and $y$ related by $x = Uy$. Thus

$$
\begin{aligned}
h(x) &= y^H D y \\
&= \lambda_1 |y_1|^2 + \cdots + \lambda_n |y_n|^2 \\
&\leq \lambda_1 \left( |y_1|^2 + \cdots + |y_n|^2 \right) \\
&= \lambda_1 \|y\|_2^2.
\end{aligned}
$$

Since $\|x\|_2 = 1$ and $U$ is unitary, $\|y\|_2 = \|Ux\|_2 = \|x\|_2 = 1$. So

$$h(x) \leq \lambda_1.$$

Part b. We show there is an $x$, $\|x\|_2 = 1$, such that $h(x) = \lambda_1$. For this, let $y = e_1$. Then $x = Ue_1$ ($x = u_1$) and

$$h(x) = h_Y(y) = \lambda_1,$$

the desired result ■

A result extending Rayleigh's Principle, namely Courant's Minimax Theorem, shows how each eigenvalue of a Hermitian matrix can be found using expressions like those of Rayleigh. This work is rather intricate, a bit more than what is intended in this text. However, we will state a useful consequence of Courant's work, the Inclusion Principle, without proof.

**Theorem 10.4** *Let $A$ be an $n \times n$ Hermitian matrix and $B$ the $(n-1) \times (n-1)$ submatrix of $A$ obtained by deleting its last row and last column. If the eigenvalues of $A$ and $B$ are indexed such that, $\lambda_1 \geq \cdots \geq \lambda_n$ and $\beta_1 \geq \cdots \geq \beta_{n-1}$, respectively, then*

$$\lambda_1 \geq \beta_1 \geq \lambda_2 \geq \beta_2 \geq \cdots \geq \beta_{n-1} \geq \lambda_n.$$

An example demonstrating the Inclusion Principle follows.

**Example 10.6** *Let $A = \begin{bmatrix} 2 & 1 & 3 \\ 1 & 2 & 3 \\ 3 & 3 & 0 \end{bmatrix}$. Then, the eigenvalues of $A$ are given by $\lambda_1 = 6$, $\lambda_2 = 1$, $\lambda_3 = -3$, while the eigenvalues of $B$ are given by $\beta_1 = 3$, $\beta_2 = 1$. Observe that*

$$\lambda_1 \geq \beta_1 \geq \lambda_2 \geq \beta_2 \geq \lambda_3.$$

Using the Inclusion Principle, we can give a test for positive definite Hermitian matrices.

**Theorem 10.5** *Let $A$ be an $n \times n$ Hermitian matrix. Let $A_k$ be the submatrix in the first $k$ rows and columns of $A$. Then $A$ is positive definite if and only if $\det A_k > 0$ for all $k$.*

**Proof.** We prove both parts of this biconditional.

Part a. We show that if $A$ is positive definite, then $\det A_k > 0$ for all $k$. For this, let $x = (x_1, \dots, x_k, 0, \dots, 0) \in C^n$. Then, since $A$ is positive definite

$$(\bar{x}_1, \dots, \bar{x}_k) A_k \begin{bmatrix} x_1 \\ \cdots \\ x_k \end{bmatrix} = x^H A x \geq 0.$$

Since this holds for all such $x$'s, with equality only when $x = 0$, $A_k$ is positive definite, and thus all its eigenvalues are positive. Since $\det A_k$ is the product of the eigenvalues of $A_k$, $\det A_k > 0$.

Part b. We show that if $\det A_k > 0$ for all $k$, then $A$ is positive definite. Here, we use induction on $n$. If $A$ is $1 \times 1$, then the result is obvious. Thus, suppose the result holds for all $n$, $n < k$. Now let $A$ be an $k \times k$ Hermitian matrix satisfying the hypothesis of this part.

Since $A_{k-1}$ is Hermitian and satisfies the hypothesis of this part, we have by the induction hypothesis that $A_{k-1}$ is positive definite and thus its eigenvalues, say, $\beta_1, \dots, \beta_{k-1}$ are all positive. Now, by the Inclusion Principle, if $\lambda_1, \dots, \lambda_k$ are the eigenvalues of $A$, then using the notation of (10.6),

$$\lambda_1 \geq \beta_1 \geq \cdots \geq \beta_{k-1} \geq \lambda_k.$$

Thus, $\lambda_{k-1}, \dots, \lambda_1$ are positive. Since $\det A > 0$ and $\det A = \lambda_1 \cdots \lambda_{k-1} \lambda_k$, it follows that $\lambda_k > 0$ as well. Thus, $A$ is positive definite. ∎

We demonstrate the theorem with an example.

**Example 10.7** *Let $K = \begin{bmatrix} k_1 + k_2 & -k_2 \\ -k_2 & k_2 \end{bmatrix}$ where $k_1 > 0$ and $k_2 > 0$. Then $\det K_1 = k_1 + k_2 > 0$ and $\det K = k_1 k_2 > 0$. Thus, $K$ is positive definite.*

The final result of this section is an interesting result about linear transformations. For this result, we work with real numbers.

Recall that if $A$ is symmetric, then we can factor $A$ as

$$A = QDQ^t$$

where $Q = [q_1 \dots q_n]$ is orthogonal and $D = \text{diag}(\lambda_1, \dots, \lambda_n)$. And

$$x^t A x = c$$

can be graphed by graphing

$$y^t D y = c,$$

where $Qy = x$, in the coordinates determined by the basis $Y = \{q_1, \ldots, q_n\}$. The graph in these coordinates is

$$\lambda_1 y_1^2 + \cdots + \lambda_n y_n^2 = c.$$

If $\lambda_1 > 0, \ldots, \lambda_n > 0$ and $c > 0$, the graph of this equation is an ellipsoid. Thus, if $A$ is positive definite symmetric and $c > 0$, then the graph of

$$x^t A x = c$$

is an ellipsoid.

Our theorem now follows.

**Theorem 10.6** *Let $A$ be an $n \times n$ nonsingular matrix. Then $L(x) = Ax$ maps ellipsoids to ellipsoids.*

**Proof.** We argue this theorem in two parts.

Part a. We show that the image of an ellipsoid is on an ellipsoid. To do this, let $E$ be an ellipsoid. Then $E$ is the graph of $x^t B x = c$ where $B$ is a positive definite symmetric matrix and $c$ a positive scalar. Since $B$ is positive definite, $B = R^t R$ for some nonsingular matrix $R$.

Let $x \in E$. Then

$$x^t R^t R x = c. \tag{10.8}$$

Now let $L(x) = y$. Then $A^{-1}y = x$. Substitution into (10.8) leads to

$$y^t \left(A^{-1}\right)^t R^t R A^{-1} y = c. \tag{10.9}$$

Since $\left(A^{-1}\right)^t R^t R A^{-1} = \left(RA^{-1}\right)^t \left(RA^{-1}\right)$, this matrix is positive definite symmetric. Thus, $y$ is on the ellipsoid defined by (10.9).

Part b. We show the image of the ellipsoid $E$ is all of the ellipsoid defined by (10.9). For this let $y$ be on the ellipsoid defined by (10.9). We need to show there is an $x \in E$ such that $L(x) = y$. This part will be left as an exercise. ∎

An example showing a consequence of this theorem follows.

**Example 10.8** *Let $L(x) = Ax$ where $A = \begin{bmatrix} 2 & 1 \\ 1 & 2 \end{bmatrix}$. Then $L$ maps the unit circle into an ellipse. By Rayleigh's Principle,*

$$\max_{\|x\|_2 = 1} \|L(x)\|_2 = \max_{\|x\|_2 = 1} \left(x^t A^t A x\right)^{\frac{1}{2}}$$

$$= \left(\max_{\|x\|_2 = 1} x^t A^2 x\right)^{\frac{1}{2}}$$

$$= \lambda_1, \text{ the largest eigenvalue of } A.$$

And the value is achieved at $u_1$, a corresponding eigenvector of unit length. Thus, $L(u_1)$ is the major axis of the image ellipse.

Similarly, the length of the minor axis is $\lambda_2$, the smallest eigenvalue of $A$ and is achieved at $L(u_2)$, $u_2$ a corresponding eigenvector of unit length. Thus, $L(u_2)$ is the minor axis of the ellipse.

Since $\lambda_1 = 3$, $u_1 = \begin{bmatrix} \frac{1}{\sqrt{2}} \\ \frac{1}{\sqrt{2}} \end{bmatrix}$ and $L(u_1) = 3\begin{bmatrix} \frac{1}{\sqrt{2}} \\ \frac{1}{\sqrt{2}} \end{bmatrix}$ and since $\lambda_2 = 1$,

$u_2 = \begin{bmatrix} -\frac{1}{\sqrt{2}} \\ \frac{1}{\sqrt{2}} \end{bmatrix}$ and $L(u_2) = \begin{bmatrix} -\frac{1}{\sqrt{2}} \\ \frac{1}{\sqrt{2}} \end{bmatrix}$, we see the image ellipse in Figure 10.8.

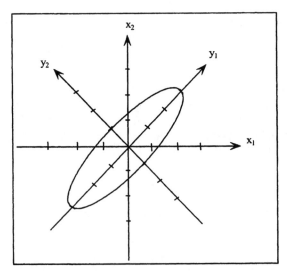

FIGURE 10.8.

## 10.2.1  Optional (Optimization)

In elementary calculus, we saw that if $f(x)$ was a function of one variable, and $x_0$ a critical point, then $f''(x_0) > 0$ implied the critical point was at a local minimum while $f''(x_0) < 0$ assured a local maximum. We outline a corresponding such test for a function $f(x, y)$ of two variables. (It can be extended to more variables.)

If $(x_0, y_0)$ is a critical point of $f(x, y)$, then

$$\frac{\partial}{\partial x} f(x_0, y_0) = 0$$
$$\frac{\partial}{\partial y} f(x_0, y_0) = 0.$$

Thus writing $f(x,y)$ in a series about $(x_0, y_0)$, we have $f(x,y) = f(x_0, y_0) +$
$\frac{1}{2}(x - x_0, y - y_0) H \begin{bmatrix} x - x_0 \\ y - y_0 \end{bmatrix} + R(x,y)$ where

$$H = \begin{bmatrix} \frac{\partial^2 f}{\partial x^2}(x_0, y_0) & \frac{\partial^2 f}{\partial y \partial x}(x_0, y_0) \\ \frac{\partial^2 f}{\partial x \partial y}(x_0, y_0) & \frac{\partial^2 f}{\partial y^2}(x_0, y_0) \end{bmatrix}$$

is called the *Hessian* of $f(x,y)$ at $(x_0, y_0)$. Under rather mild conditions on $f$, $H$ is symmetric.

If $H$ is positive definite, which we can easily check by Theorem 10.5, then

$$q(x,y) = (x - x_0, y - y_0) H \begin{bmatrix} x - x_0, \\ y - y_0 \end{bmatrix} > 0$$

except when $x = x_0$ and $y = y_0$. And it can be shown that $q(x,y) > R(x,y)$ for $(x,y)$ close to (not equal) $(x_0, y_0)$. So $f(x,y) - f(x_0, y_0) = q(x,y) + R(x,y) > 0$ for all such $(x,y)$, and thus $f(x_0, y_0)$ is a local minimum of $f$. (See Figure 10.9.)

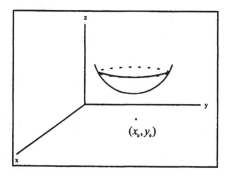

FIGURE 10.9.

If $H$ is negative definite, then $f(x_0, y_0)$ is a local maximum. We give an example.

**Example 10.9** *Let* $f(x,y) = 2x^2 + xy + 3y^2 - 6x - 13y + 6$.

    i. *We find the critical points of* $f$.

        *Setting the partial derivatives of* $f$ *equal to 0, we have*

$$\begin{array}{rrrl} 4x & +y & -6 & = 0 \\ x & +6y & -13 & = 0. \end{array}$$

*or*

$$4x + y = 6$$
$$x + 6y = 13.$$

*The solution to these equations is* $(1, 2)^t$.

ii. *We decide if f has a local maximum or local minimum at* $(1,2)^t$.

*To do this, we calculate the Hessian of f at* $(1,2)^t$. *We get*

$$H = \begin{bmatrix} \delta \frac{\delta^2 f}{\delta x^2}(1,2) & \delta \frac{\delta^2 f}{\delta y \delta x^2}(1,2) \\ \frac{\delta^2 f}{\delta x \delta y}(1,2) & \frac{\delta^2 f}{\delta y^2}(1,2) \end{bmatrix}$$

$$= \begin{bmatrix} 4 & 1 \\ 1 & 6 \end{bmatrix},$$

*which is positive definite. Thus, there is a local minimum at* $(1,2)^t$.

# Exercises

1. Graph by changing coordinates using the basis $Y$ that provides the eigenvalue description.

   (a) $5x_1^2 + 2x_1x_2 + 5x_2^2 = 5$    (b) $x_1^2 + 2x_1x_2 + x_2^2 = 1$

2. Graph by using the basis $Y$ that provides the eigenvalue description. Describe each shape.

   (a) $q(x_1, x_2) = 4x_1^2 + 2x_1x_2 + 4x_2^2$
   (b) $q(x_1, x_2) = x_1^2 + 4x_1x_2 + x_2^2$

3. Using the hypothesis of Rayleigh's Principle, prove that $\min_{\|x\|_2=1} h(x) = \lambda_n$.

4. Decide which matrices are positive definite.

   (a) $\begin{bmatrix} -1 & 2 \\ 2 & 1 \end{bmatrix}$    (b) $\begin{bmatrix} 1 & 2 & 1 \\ 2 & 1 & 1 \\ 1 & 1 & 3 \end{bmatrix}$    (c) $\begin{bmatrix} 3 & 1 & 2 \\ 1 & 3 & 2 \\ 2 & 2 & 2 \end{bmatrix}$

5. Demonstrate the Inclusion Principle for $\begin{bmatrix} 3 & 1 & 2 \\ 1 & 3 & 2 \\ 2 & 2 & 2 \end{bmatrix}$.

6. Let $L(x) = \begin{bmatrix} 3 & 2 \\ 2 & 3 \end{bmatrix} x$. Then $L$ maps the unit circle in $R^2$ into an ellipse in $R^2$. Find the ellipse as in Example 10.8.

7. Prove that the sum of two $n \times n$ positive definite matrices is positive definite.

8. Give the details for the proof of Theorem 10.2.

9. Let $A$ be a $3 \times 3$ Hermitian matrix where $a_{11} = 0$. Prove that $A$ has a nonnegative and a nonpositive eigenvalue. (Hint: Apply the Inclusion Principle to $A_1$, $A_2$, and $A$ where $A_k$ is the submatrix of $A$ in the first $k$ rows and $k$ columns of $A$.)

10. Let $A$ be an $n \times n$ positive definite Hermitian matrix. Prove that the submatrix in rows $2, \ldots, n-1$ and columns $2, \ldots, n-1$ is positive definite. (Actually, any submatrix sharing the same rows and columns of $A$ is positive definite.)

11. Let $A$ be an $n \times n$ positive definite Hermitian matrix. Suppose we can obtain an echelon form $E$ by only applying the add operation $\alpha R_i + R_j$ where $i < j$. Prove that $A$ is positive definite if and only if the entries on the main diagonal of $E$ are positive.

12. (Optional) Let

$$f(x,y) = 4x^2 + 2x + 4y^2 + 4y + 2.$$

(a) Find the critical points of $f$.

(b) Analyze the critical points to see if they yield local maximum or local minimum values of $f$.

13. (Optional) Repeat Exercise 12(b) for

$$f(x,y) = \sin x + \cos y + xy$$

for the critical point $(0,0)$.

14. (MATLAB) Graph and describe each shape.

(a) $q(x) = x^t A x$ where $A = \begin{bmatrix} 2 & 1 \\ 1 & 2 \end{bmatrix}$

(b) $q(x) = x^t A x$ where $A = \begin{bmatrix} 1 & 2 \\ 2 & 1 \end{bmatrix}$

(c) $q(x) = x^t A x$ where $A = \begin{bmatrix} -3 & 1 \\ 1 & -3 \end{bmatrix}$

# 11
# Graphics and Topology

In this chapter we will show how matrices can be used in computer graphics and, to some extent, how special pieces (nonsingular matrices, diagonalizable matrices, etc.) of matrix space can be viewed. So in some sense, both topics deal with pictures.

## 11.1  Two Projection Matrices

In this section we study two special maps: the projection map and the perspective projection map, which are maps from $R^n$ into $R^n$. We first develop the projection map.

Recall that we have defined and used the orthogonal projection matrix. This matrix projected Euclidean n-space orthogonally onto a subspace of itself. We now extend this notion to allow projections at various angles.

**Definition 11.1** *Let $P$ be an $n \times n$ matrix. If $P$ is similar to a diagonal matrix $D$ (thus, $P = RDR^{-1}$ for some $n \times n$ matrix $R$) whose main diagonal consists of 0's and 1's, then $P$ is called a <u>projection</u> (or idempotent) <u>matrix</u>.*

To give a grid view of $L(x) = Px$, we start by letting $Y = \{r_1, \ldots, r_n\}$, where $r_1, \ldots, r_n$ are the columns of $R$, a basis for Euclidean n-space. Recall that $Y$ determines axes for the Y-coordinate system and that $R = [r_1 \ldots r_n]$ converts coordinates,

$$Ry = x,$$

where $y$ is the coordinate vector of $x$ in the Y-coordinate system.

Now, in the Y-coordinate system $L_Y(y) = Dy$ is easily seen as a projection of the space. For example, $L_Y(y) = \begin{bmatrix} 1 & 0 \\ 0 & 0 \end{bmatrix} y$ projects $R^2$ onto the $y_1$-axis, parallel to the $y_2$-axis; while $L_Y(y) = \begin{bmatrix} 1 & 0 & 0 \\ 0 & 1 & 0 \\ 0 & 0 & 0 \end{bmatrix}$ projects $R^3$ into the $y_1 y_2$-plane parallel to $y_3$. (See Figure 11.1.)

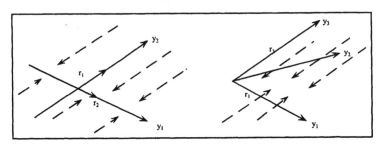

FIGURE 11.1.

To link $L(x) = Px$ and $L_Y(y) = Dy$, we need a theorem whose proof is described in Chapter 3, Section 3.

**Theorem 11.1** *Let $P$ be a projection matrix with $P = RDR^{-1}$ and $R = [r_1 \ldots r_n]$. If $Y = \{r_1, \ldots, r_n\}$, then $L_Y(y) = Dy$ in the Y-coordinate system gives the same map as $L(x) = Px$.*

The theorem makes clear that a projection matrix $P$ behaves as does $D$, using axes determined from $Y$. To help clarify the theorem, we give an example.

**Example 11.1** *Let $P = \begin{bmatrix} \frac{4}{3} & -\frac{2}{3} \\ \frac{2}{3} & -\frac{1}{3} \end{bmatrix}$. Then we can factor*

$$P = \begin{bmatrix} 2 & 1 \\ 1 & 2 \end{bmatrix} \begin{bmatrix} 1 & 0 \\ 0 & 0 \end{bmatrix} \begin{bmatrix} \frac{2}{3} & -\frac{1}{3} \\ -\frac{1}{3} & \frac{2}{3} \end{bmatrix}.$$

*So $D = \begin{bmatrix} 1 & 0 \\ 0 & 0 \end{bmatrix}$. Now $L(x) = Px$, described naturally, is the same as $L_y(y) = Dy$ in the Y-coordinate system, where $Y = \left\{ \begin{bmatrix} 2 \\ 1 \end{bmatrix}, \begin{bmatrix} 1 \\ 2 \end{bmatrix} \right\}$.*

*Note $L_Y$ projects onto the $y_1$-axis (determined from $\begin{bmatrix} 2 \\ 1 \end{bmatrix}$) parallel to the $y_2$-axis (determined from $\begin{bmatrix} 1 \\ 2 \end{bmatrix}$). See Figure 11.2.*

Another example may help.

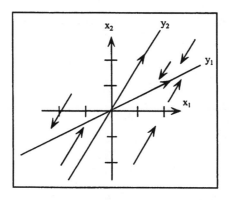

FIGURE 11.2.

**Example 11.2** *Suppose we want to construct a projection that collapses $R^2$ into the line $y = x$ parallel to $\begin{bmatrix} 0 \\ 1 \end{bmatrix}$. We take a vector on the line, say, $\begin{bmatrix} 1 \\ 1 \end{bmatrix}$. Then set*

$$Y = \left\{ \begin{bmatrix} 0 \\ 1 \end{bmatrix}, \begin{bmatrix} 1 \\ 1 \end{bmatrix} \right\} \quad and$$
$$D = \begin{bmatrix} 0 & 0 \\ 0 & 1 \end{bmatrix}.$$

*Now $R = \begin{bmatrix} 0 & 1 \\ 1 & 1 \end{bmatrix}$ and $P = RDR^{-1} = \begin{bmatrix} 1 & 0 \\ 1 & 0 \end{bmatrix}$. And, $L(x) = Px$ is the desired projection. See Figure 11.3.*

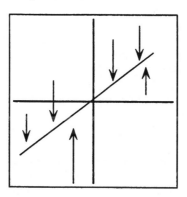

FIGURE 11.3.

It is possible to tell if $P$ is a projection matrix without factoring.

**Theorem 11.2** *P is a projection matrix if and only if $P^2 = P$. (This says $L(L(x)) = P(Px) = Px = L(x)$.)*

**Proof.** The direct implication is proved by setting $P = RDR^{-1}$ and showing $P^2 = P$. Thus we need only argue the converse.

Suppose

$$P^2 = P. \tag{11.1}$$

Factor $P$, by Jordan's theorem, so

$$P = RJR^{-1}$$

where $J$ is a Jordan form for $P$. Substituting into (11.1), and simplifying yields

$$J^2 = J.$$

Since $J$ is block diagonal,

$$J_k^2 = J_k \tag{11.2}$$

for each Jordan block $J_k$ of $J$. Viewing the main diagonals of these blocks, we see from (11.2) that any eigenvalue $\lambda$ must satisfy

$$\lambda^2 = \lambda$$

so $\lambda = 1$ or $0$. And, all Jordan blocks must be $1 \times 1$. (If not, view the $1, 2$-entry for a contradiction) Thus $P$ is similar to a diagonal matrix with main diagonal entries 0's and 1's. ∎

Notice in the examples that the projections there are "slanted." As you might expect, "orthogonal" projections require the basis $Y$, and thus the matrix $R$, to be orthogonal.

**Definition 11.2** *A projection matrix $P$ is an orthogonal projection if $P = QDQ^t$ for some orthogonal matrix $Q$ and $D$ a diagonal matrix whose main diagonal consists of 0's and 1's.*

We can also tell, without factoring, if $P$ is an orthogonal projection.

**Theorem 11.3** *Let $P$ be a projection matrix. Then $P$ is an orthogonal projection if and only if $P^t = P$.*

**Proof.** The proof is left as an exercise with the following hint: If $P^t = P$, $P$ is normal. ∎

The second kind of projection map we consider is the perspective projection map. In this kind of map, Euclidean n-space is projected toward a 'point at infinity.' (Think of looking at railroad tracks to see how lines are intended to go.)

**Definition 11.3** *An* $(n+1) \times (n+1)$ *matrix* $A$ *is called a* perspective projection matrix *if it can be partitioned*

$$A = \begin{bmatrix} B & 0 \\ b & 1 \end{bmatrix} \ or \ A = \begin{bmatrix} B & b \\ 0 & 1 \end{bmatrix}$$

*where* $B$ *is an* $n \times n$ *matrix.*

As a use for this matrix, note that the difference equation

$$x_{k+1} = Bx_k + b \tag{11.3}$$

can be written

$$\begin{bmatrix} x_{k+1} \\ 1 \end{bmatrix} = \begin{bmatrix} B & b \\ 0 & 1 \end{bmatrix} \begin{bmatrix} x_k \\ 1 \end{bmatrix}. \tag{11.4}$$

The convergence of (11.4) depends on the eigenvalues and Jordan blocks of $A = \begin{bmatrix} B & b \\ 0 & 1 \end{bmatrix}$. Concerning the eigenvalues of $A$, we use the notion of the *spectrum of a matrix* $C$, namely,

$$\sigma(C) = \{\lambda : \lambda \text{ is an eigenvalue of C}\}.$$

The following lemma is easily proved.

**Lemma 11.1** *If* $A$ *is a perspective projection matrix then*

$$\sigma(A) = \sigma(B) \cup \{1\}.$$

Perspective projection matrices also arise in graphics. In art, a painter might hold up his thumb to help envision a vanishing point behind the canvas. The drawing diminishes the back (top, bottom, and sides) to provide perspective. A draftsman will do this by perhaps initially establishing a vanishing point for a drawing, perhaps putting a point in the upper right corner of the drafting paper. (See Figure 11.4.)

The same results can be achieved in computer graphics using mathematics. We project to the xy-plane (our canvas). However, since we are projecting, rather than drawing, the point we need is in front of the object. (So we get the eye view.) We label the axes as in Figure 11.5.

We choose a vanishing point $e$, $e = (0, 0, d)^t$ on the $z$-axis and use it to obtain perspective. To calculate where $e$ places the point $p$, $p = \begin{bmatrix} x \\ y \\ z \end{bmatrix}$, in the $xy$-plane, say, at $\begin{bmatrix} x^* \\ y^* \end{bmatrix}$, we use the line determined by $e$ and $p$, namely

$$\alpha p + (1 - \alpha) e \tag{11.5}$$

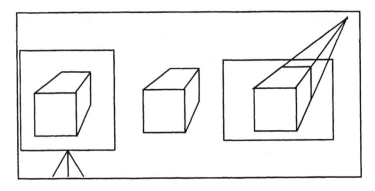

FIGURE 11.4.

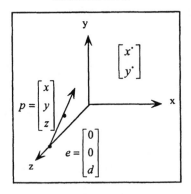

FIGURE 11.5.

where $-\infty < \alpha < \infty$. We choose $\alpha$ so that $z = 0$, i.e.

$$\alpha z + (1 - \alpha)\, d = 0.$$

Solving yields

$$\alpha = \frac{d}{d - z}\ .$$

Now, using this $\alpha$ in (11.5), we have that

$$x^* = \frac{d}{d - z}x$$

$$y^* = \frac{d}{d - z}y.$$

Note that

$$\begin{bmatrix} 1 & 0 & 0 & 0 \\ 0 & 1 & 0 & 0 \\ 0 & 0 & 0 & 0 \\ 0 & 0 & -\frac{1}{d} & 1 \end{bmatrix} \begin{bmatrix} x \\ y \\ z \\ 1 \end{bmatrix} = \begin{bmatrix} x \\ y \\ 0 \\ \frac{d-z}{d} \end{bmatrix}. \qquad (11.6)$$

Thus, if we define $\pi$ on those vectors in $R^4$ with nonzero last entry, such that $\pi$ normalizes the vector to have last entry 1, then we have

$$\pi \begin{bmatrix} x \\ y \\ 0 \\ \frac{d-z}{d} \end{bmatrix} = \begin{bmatrix} \frac{dx}{d-z} \\ \frac{dy}{d-z} \\ 0 \\ 1 \end{bmatrix} = \begin{bmatrix} x^* \\ y^* \\ 0 \\ 1 \end{bmatrix}.$$

(Vectors that are scalar multiples of each other are said to have homogeneous coordinates. Hence, $\pi$ maps vectors into vectors with homogeneous coordinates and having last entry 1.) Putting together, if

$$A = \begin{bmatrix} 1 & 0 & 0 & 0 \\ 0 & 1 & 0 & 0 \\ 0 & 0 & 0 & 0 \\ 0 & 0 & -\frac{1}{d} & 1 \end{bmatrix},$$

then the perspective projection map $\pi \circ A$ maps vectors into the $xy$-plane with the perspective of a vanishing point at $e$, a distance $d$ from the origin.

## 11.1.1  Optional (Drawing Pictures Using Projection Maps)

In this optional we will draw a box to be viewed on a computer screen. We do this with a perspective projection map and a projection map.

Part a. Perspective projection.

The eye view we take of the box is from a vertex, say, $f$, to the farthest vertex, say, $h$, from $f$. Thus we place the line through $h$ and $f$ on the $z$-axis so that $f$ is at $(0,0,3)^t$ and $h$ at $(0,0,1)^t$. So the vertices of the box that we view are

$$a = (-2,0,3)^t$$
$$b = (-2,-2,3)^t$$
$$c = (0,-2,3)^t$$
$$d = (2,0,1)^t$$
$$e = (2,2,1)^t$$
$$f = (0,0,3)^t$$
$$g = (0,2,1)^t.$$

(This box is not a cube.) To outline the box, we intend to draw the edges in the following sequence:

$$a - b - c - d - e - f - c - f - a - g - e.$$

Listing the x, y, and z coordinates of this sequence, we have the following.
$x = [-2, -2, 0, 2, 2, 0, 0, 0, -2, 0, 2]$;
$y = [0, -2, -2, 0, 2, 0, -2, 0, 0, 2, 2]$;
$z = [3, 3, 3, 1, 1, 3, 3, 3, 3, 1, 1]$;
Now we position our eye at $e = (0, 0, 10)$, so $d = 10$.
$d = 10$;
Computing our perspective projection on the $xy$-plane, we have
$s = d* \text{ ones}(1, 11)$;
$x1 = d * x./ (s - z)$;
$y1 = d * y./ (s - z)$;
Now, plotting in the $xy$-plane
$w = \text{zeros } (1, 11)$;
$\text{plot3 } (x1, y1, w)$
To view the picture from the $z$-axis, we use the following.
view $(0, 90)$     % Tilts axes so that the $z$-axis points
                     toward us.
axis equal     % Puts tick marks so they are equal.
axis off     % Removes appearence of axes.
The picture is below in Figure 11.6.

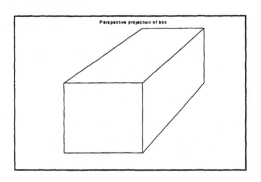

FIGURE 11.6.

Of course, if we increase $d$, the back square will appear to increase in size, so this can be adjusted to suit the viewer.

Part b. Projection.

To contrast, suppose we simply project our box on the $xy$-plane. We use $x$, $y$, and $w$ from the previous program. And, we add
$\text{plot3}(x, y, w)$
view $(0, 90)$
axis equal
axis off
The picture is in Figure 11.7.

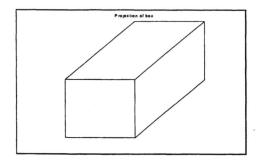

FIGURE 11.7.

Notice that in this picture, the back square appears larger than the front square. However, measurement shows they are the same.

Perspective is important in drawing. Our eyes expect it. And when it is missing, we see (perceive) a distortion.

## Exercises

1. Find $P$ that projects $R^2$ onto the line $y = 2x$ parallel to $(1,1)^t$.

2. Find $P$ that projects $R^3$ onto the plane $x + y + z = 0$ parallel to $(1,1,0)^t$.

3. Find the orthogonal projection of $R^3$ onto plane $x + y + z = 0$.

4. Is $P = \begin{bmatrix} 1 & 1 \\ 1 & 1 \end{bmatrix}$ a projection matrix?

5. Find $P$ that projects $R^3$ onto the line $x = t \begin{bmatrix} 1 \\ 1 \\ 1 \end{bmatrix}$ (parametrically described), and parallel to the $xy$-plane. (All points project parallel to the $xy$-plane.)

6. Find the projection of $R^4$ onto the $x_1 x_2$-plane parallel to the plane $x_1 + x_2 + x_3 + x_4 = 0$.

7. Prove that if $Q = [q_1 \ldots q_r]$ is an $n \times r$ matrix whose columns form an orthonormal set, then $QQ^t$ is an orthogonal projection of $R^n$ onto span $\{q_1, \ldots, q_r\}$. (Write in the form $RDR^{-1}$.)

8. Prove Theorem 11.3.

9. Prove Lemma 11.1.

10. Find the perspective projection map into $R^2$ using the vanishing point $(0,0,6)^t$.

11. Let $d$ be the vanishing point of a projective projection map into $R^2$. If $d$ increases, what would happen to the picture of a house?

12. (MATLAB)

    (a) Find the perspective projection matrix $P$ that projects $R^3$ into $R^2$ with vanishing point $e = (0, 0, 20)^t$.

    (b) Change the box in Optional into a cube. Use $P$ to project the cube into $R^2$.

    (c) Does the picture "look right"? If not, how should the vanishing point be changed? (Art, even on computers requires experience and sense.)

## 11.2   Manifolds and Topological Sets

In working with matrices, it is very helpful to have some sense or feel for the special sets of matrices: the nonsingular matrices as well as the matrices that have distinct eigenvalues. As shown in Chapter 5, Section 2, a matrix relatively close to singular matrices has a large condition number, $c(A)$. And, as given in Chapter 9, Section 2, if a matrix is close to matrices with multiple eigenvalues, that matrix might have large eigenvalue and eigenvector condition numbers. In this section, we provide results intended to give that sense. Most of the work is based on the following definition.

**Definition 11.4** *The concepts described below concern the set of $m \times n$ matrices with matrix norm $\|\cdot\|$. Let $\epsilon > 0$ and $A$ an $m \times n$ matrix. A ball $\mathbf{B}$ about $A$ of radius $\epsilon$ is defined as*

$$\mathbf{B} = \{B : B \text{ is an } m \times n \text{ matrix and } \|B - A\| < \epsilon\}.$$

Let $K$ be a set of $m \times n$ matrices.

   i. $K$ is *open* if for each $A \in K$ there is some ball $\mathbf{B}$ about $A$ such that $\mathbf{B} \subseteq K$.

   ii. $K$ is *closed* if whenever $A_1, A_2...$ are in $K$ and the sequence converges to some $A$, called a *limit point* of $K$, then $A \in K$.

It is left an exercise to show that the compliment of an open set is closed and vice versa.

Open sets $K$ are important since sufficiently small errors made in estimating or calculating an $A \in K$ results in a matrix in $K$. Not always so in closed sets. However, in closed sets $K$, limits of convergent sequences in $K$ must be in $K$.

Two special closed and open sets of matrices follow.

**Theorem 11.4** *In the space of $n \times n$ matrices,*

  *(a) The set of singular matrices is closed.*

  *(b) The set of nonsingular matrices is open.*

**Proof.** There are two parts.

Part a. The determinant is a continuous function. Hence if $A_1, A_2, \ldots$ is a sequence of singular matrices which converge to $A$, then $\det A = \lim_{k \to \infty} \det A_k = \lim_{k \to \infty} 0 = 0$. Thus $A$ is singular. Hence, the set of singular matrices is closed.

Part b. Since the set of nonsingular matrices is the compliment of the set of singular matrices, the result follows. ∎

Results concerning diagonalizable matrices follow.

**Theorem 11.5** *In the space of $n \times n$ matrices,*

  *(a) The set of matrices that have multiple eigenvalues (at least one eigenvalue of multiplicity 2 or more) is closed.*

  *(b) The set of matrices that have distinct eigenvalues is open.*

**Proof.** There two parts.

Part a. Let $A_1, A_2, \ldots$ be a sequence of matrices which have multiple eigenvalues. Suppose the sequence converges to $A$. By the continuous dependence of eigenvalues, $A$ cannot have distinct eigenvalues since if $A$ had distinct eigenvalues, we could find small nonintersecting disks, say, of radius $\epsilon$, about them. But then, for some $\delta$, if $\|A_k - A\|_\infty < \delta$, the eigenvalues of $A_k$ would have to be within $\epsilon$ of those of $A$. Thus, the set of matrices that have multiple eigenvalues is closed.

Part b. Left as exercise. ∎

Continuing with the definition,

  iii. $K$ is *dense* if for each matrix $A$ in the space, and each scalar $\epsilon > 0$, the ball about $A$ of radius $\epsilon$ contains a matrix from $K$.

Thus, any matrix can be approximated arbitrarily close using matrices in a dense set. Two such sets of matrices follow.

**Theorem 11.6** *In the space of $n \times n$ matrices,*

  *(a) The set of matrices with distinct eigenvalues is dense.*

  *(b) The set of nonsingular matrices is dense.*

**Proof.** There are two parts.

Part a. Let $A$ be an $n \times n$ matrix. Write

$$A = PJF^{-1}$$

where $J$ is the Jordan form of $A$. Let

$$\delta = \min |\lambda_i - \lambda_j|$$

where the minimum is over all distinct eigenvalues $\lambda_i, \lambda_j$ of $A$. Let $\epsilon$ be a variable, $0 < \epsilon < \frac{\delta}{n}$. Define

$$D = diag\,(\epsilon, 2\epsilon, \ldots, n\epsilon)\,.$$

And set

$$B = P\,(J + D)\,P^{-1}.$$

Note that $B$ has distinct eigenvalues for all $\epsilon$ and that

$$\lim_{\epsilon \to 0} B = A.$$

Thus, there are matrices with distinct eigenvalues arbitrarily close to $A$.

Part b. This part is similar to Part a. (Note, nonsingular is equivalent to nonzero eigenvalues.) ∎

We might mention that neither the nonsingular matrices nor the matrices with distinct eigenvalues are convex sets, so they don't have a convex set dimension. Intuitively, however, it is nice to have some notion of dimension of the sets we studied. Thus, we will need an extended definition for dimension. And, we will do this only for the real numbers.

For some intuition on this, let $X$ be a nonempty subset of $m \times n$ matrices. We will say the $X$ has dimension $k$ if at each $A \in X$, there is an open set containing $A$ which looks something like an open set in $R^k$. (Say we can lay an open set in $R^k$ one-to-one, on the open set containing $A$.) So around any point in $X$, it looks like $R^k$. (See Figure 11.8.)

For a mathematical description, let $X$ be a nonempty subset of $m \times n$ matrices. Using the matrix norm $\|\cdot\|_F$, we can define ball, open and closed sets in $X$ (rather than in the whole set of $m \times n$ matrices) as we did in the space of $m \times n$ matrices. And using the vector norm $\|\cdot\|_2$, we can define those same notions in $R^k$.

Now suppose that at each $A \in X$ there is an open set $W$ (open in $X$) containing $A$, an open set $V$ in $R^k$, and a function $f$.

$$f : V \to W$$

which is

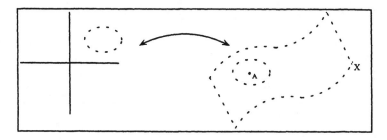

FIGURE 11.8.

1. One-to-one and onto, and such that both

2. $f$ and $f^{-1}$ are continuous.

Then we say that $X$ is a $k$-manifold. And, we add, all $k$-manifolds are assigned the dimension $k$.

Giving some dimension to nonsingular matrices, we have the following.

**Theorem 11.7** *Let $U$ be the set of $n \times n$ nonsingular matrices. Then $U$ is an $n^2$-manifold.*

**Proof.** Define $f : R^{n^2} \to R^{n \times n}$ by

$$f(a_{11}, \ldots, a_{1n}, a_{21}, \ldots, a_{2n}, \ldots, a_{n1}, \ldots, a_{nn}) = \begin{bmatrix} a_{11} \ldots a_{1n} \\ \cdots \\ a_{n1} \ldots a_{nn} \end{bmatrix}.$$

Then $f$ is one-to-one and onto.

Let $V = f^{-1}(U)$. (That is, $V = \{x : f(x) \in U\}$.) To show that V is open, let $x \in V$. Then $f(x) \in U$. Since $U$ is open, there is a ball $\mathbf{B}(f(x), r)$ about $f(x)$ of radius $r$ such that

$$\mathbf{B}(f(x), r) \subseteq U.$$

If $\mathbf{B}(x, r)$ is the ball about $x$ of radius $r$ in $R^{n^2}$, then $f : \mathbf{B}(x, r) \to \mathbf{B}(f(x), r)$. Thus, it follows that $\mathbf{B}(x, r) \subseteq V$. As $x$ was arbitrarily chosen, $V$ is open.

Finally, it is clear (checking to see if $x_k \to x$, then $f(x_k) \to f(x)$ and if $B_k \to B$, then $f^{-1}(B_k) \to f^{-1}(B)$) that $f$ and $f^{-1}$ are continuous. Thus, $U$ is an $n^2$-manifold. ∎

As an additional result we have the following.

**Theorem 11.8** *The set of $n \times n$ matrices, having distinct eigenvalues, is an $n^2$-manifold.*

**Proof.** Exercise. ∎

## 11.2.1    Optional (Rank k Matrices)

In this optional we look at special subsets of $m \times n$ matrices having various rank conditions. The first of these is Rank $\geq k$ defined by

$$\text{Rank} \geq k = \{A : \text{rank}\, A \geq k\}.$$

We show two properties about the set.

**Theorem 11.9** *In the space of $m \times n$ matrices,*

(a) *Rank $\geq k$ is an open set, and*

(b) *Rank $\geq k$ is an $mn$-manifold.*

**Proof.** There are two parts.

Part a. Let $A \in \text{Rank} \geq k$. Then $A$ has a $k \times k$ submatrix $C$ such that $C$ is nonsingular. To simplify the argument, we will suppose that $C$ is in the upper left corner of $A$.

Let $U$ denote the set of the $k \times k$ nonsingular matrices. Since $U$ is open, there is a ball about $C$ of radius $r$ such that

$$\mathbf{B}(C, r) \subseteq U.$$

Now note that if $R$ is an $n \times n$ matrix and $S$ the submatrix in the $k \times k$ upper left corner of $R$, then

$$\|A - R\|_F < r$$

implies that

$$\|C - S\|_F < r.$$

Thus, if

$$R \in \mathbf{B}(A, r)$$

then $S \in \mathbf{B}(C, r)$, and so $S$ is nonsingular. From this it follows that all matrices in $\mathbf{B}(A, r)$ have rank at least $k$. Thus

$$\mathbf{B}(A, r) \subseteq \text{Rank} \geq k$$

and so, Rank $\geq k$ is open.

Part b. Mimicking the proof of Theorem 12.7 yields this result. ∎

From this theorem it follows that, using that the compliment of an open set is closed,

$$\text{Rank} \leq k = \{A : \text{rank}\, A \leq k\}$$

is a closed set.

An additional result concerns

$$\text{Rank } k = \{A : \text{rank } A = k\}.$$

This set is neither open nor closed in $R^{m \times n}$. However, Rank $k$ is a manifold.

**Theorem 11.10** Rank $k$ *is a* $k^2 + k(n-k) + k(m-k)$ *manifold.*

**Proof.** We will prove this theorem for $k = 2$ and $3 \times 3$ matrices. The extension to the general case will be clear.

Let $A \in \text{Rank } 2$ and suppose that the $2 \times 2$ nonsingular submatrix $C$ in $A$ is in the upper left corner. Then, partitioning $A$, we have

$$A = \left[\begin{array}{c|c} C & y \\ \hline x & \alpha \end{array}\right]$$

where $x$ is a $1 \times 2$ vector, $y$ a $2 \times 1$ vector, and $\alpha$ a scalar. If $C = [c_1 c_2]$, $c_1$ and $c_2$ the columns of $C$, then

$$y = \alpha_1 c_1 + \alpha_2 c_2$$

for some scalars $\alpha_1$ and $\alpha_2$. And, if $A = \begin{bmatrix} a_1 \\ a_2 \\ a_3 \end{bmatrix}$, where $a_1$, $a_2$, and $a_3$ are

the rows of $A$, then $a_3 = \gamma_1 a_1 + \gamma_2 a_2$ for some scalars $\gamma_1$ and $\gamma_2$.

Now, set $f : R^8 \to R^{3 \times 3}$ by

$$f(c_{11}, c_{12}, c_{21}, c_{22}, \alpha_1, \alpha_2, \gamma_1, \gamma_2) =$$
$$\begin{bmatrix} c_{11} & c_{12} & \alpha_1 c_{11} + \alpha_2 c_{12} \\ c_{21} & c_{22} & \alpha_1 c_{21} + \alpha_2 c_{22} \\ \begin{pmatrix} \gamma_1 c_{11} + \\ \gamma_2 c_{21} \end{pmatrix} & \begin{pmatrix} \gamma_1 c_{12} + \\ \gamma_2 c_{22} \end{pmatrix} & \begin{pmatrix} \gamma_1(\alpha_1 c_{11} + \alpha_2 c_{12}) + \\ \gamma_2(\alpha_1 c_{21} + \alpha_2 c_{22}) \end{pmatrix} \end{bmatrix}$$

Let $V = f^{-1}(\text{Rank } 2)$. Then

$$f : V \to \text{Rank } 2$$

is one-to-one and continuous. Also, $f^{-1}$ is continuous. Thus, Rank $2$ is a $4 + 2(3-2) + 2(3-2)$-manifold. ∎

From this theorem, we get a view of the $n \times n$ singular matrices. This set can be seen as the union of Rank $k$ sets for $k < n$. And, since Rank $k$ is a $k^2 + 2k(n-k)$-manifold, we see the singular matrices as a union of manifolds, the largest dimension of which is $n^2 - 1$ obtained by the set Rank $(n-1)$.

# Exercises

1. Write out the definitions for ball, open, close, and dense sets in

   (a) $X$, a subset of $m \times n$ matrices.
   (b) $R^k$.

2. Show that the set of $2 \times 2$ matrices is a 4-manifold.

3. Show that the set of $2 \times 2$ symmetric matrices is a 3-manifold.

4. Show that the orthogonal matrices in the $2 \times 2$ matrices form a manifold of dim 1.

5. Prove that the compliment of an open set in $R^{m \times n}$ is closed and vice versa.

6. Explain why Theorem 11.5, part (b) is true.

7. Show that $k^2 + 2k(n-k)$, where $1 \le k \le n-1$ is largest when $k = n-1$. And, at $k = n-1$, this value is $n^2 - 1$. (Hint: Use $f(x) = x^2 - 2x(n-x)$ and apply calculus techniques.)

8. Prove that the set of $n \times n$ orthogonal matrices is closed.

9. Prove Theorem 11.6, part b.

10. Prove Theorem 11.8.

11. (MATLAB) Let $S = \left\{ A : A = \begin{bmatrix} a & b \\ c & \frac{bc}{a} \end{bmatrix} \text{ where } a, b, c \in R \right\}$. Graph all rank 1 matrices in $S$ such that $a > 0$ and $b = a + c$ over $1 \le a \le 10$, $1 \le c \le 10$.

# Appendix A: MATLAB

In this appendix, we go over some of the basics of the MATLAB software package. More appears, as it is needed, in the text.

**Numbers:** The arithmetic operations for numbers, as with calculators, are $+$, $-$, $*$, and $/$.

**Matrices:** To enter a matrix, say $A = \begin{bmatrix} 1 & 2 & 3 \\ 4 & 5 & 6 \end{bmatrix}$, type in

$A = \begin{bmatrix} 1 & 2 & 3; & 4 & 5 & 6 \end{bmatrix}$

The semicolon indicates the beginning of a new row. If we don't want the matrix to appear on the screen, we can use a semicolon at the end of the command, as in

$A = \begin{bmatrix} 1 & 2 & 3; & 4 & 5 & 6 \end{bmatrix}$;

**Arithmetic:** If $A$ and $B$ have been entered, we can do *arithmetic* with them by using the commands

$A + B, A - B$

| | |
|---|---|
| $A * B$ | for the matrix product |
| $\alpha * A$ | for the scalar product |
| $\text{inv}(A)$ | for $A^{-1}$ |
| $A \backslash b$ | for the solution to $Ax = b$ |
| $A \wedge 2$ | for $A^2$ |
| $A'$ | for $A$ transpose |

Sometimes we need an *element-wise operation*. Placing a period in front of the operation provides that result.

$A.\backslash B$    gives $\left[\frac{a_{ij}}{b_{ij}}\right]$

$A.\wedge 2$    gives $\left[a_{ij}^2\right]$

**Functions:** MATLAB provides a large list of functions of matrices. These functions provide us with numerical calculations that would require a great deal of time if done by hand. For example, if $A$ has been entered, we can get

rref($A$)                    for the reduced row echelon form

det $(A)$, rank $(A)$, …

Probably, some good advice is, if we want something, say rank $A$, type in what seems natural. Usually, this is correct.

**Graphics:** We break this up into two parts.

**Part 1. 2-D Graphics.** To plot a function, say $f(t)$, it is required that we decide at what points we want to see the graph. We enter these points by indicating where the interval is to start, where it is to end, and the number of points desired in the interval. For example, since the variable is $t$,

$t$ =linspace$(0, 1, 5)$    gives $t = \left[0, \frac{1}{4}, \frac{2}{4}, \frac{3}{4}, 1\right]$

Now to plot $f(t)$, we use

plot$(t, f(t))$.

MATLAB will connect the points

$(0, f(0))$    $\left(\frac{1}{4}, f\left(\frac{1}{4}\right)\right)$    $\left(\frac{1}{2}, f\left(\frac{1}{2}\right)\right)$    $\left(\frac{3}{4}, f\left(\frac{3}{4}\right)\right)$    $(1, f(1))$

with segments. (Remember, $t$ is a matrix so if $f(t) = t^2$, we would write plot$(t, t. \wedge 2)$.)

Curves described parametrically can be graphed in the same way. For example

plot$(t. \wedge 2, t. \wedge 5)$

graphs $(t^2, t^5)$ over our interval.

**Part 2. 3-D Graphics.** To plot a function, say $f(s, t)$, we need two intervals

$s$ =linspace$(-1, 3, 100)$ ;

$t$ =linspace$(2, 6, 100)$ ;

Now to get the grid on which we will plot $f(s, t)$, we use

$[s, t]$ =meshgrid$(s, t)$ ;

Thus, $[s, t]$ provides the matrix of points in $s \times t$, all ordered pairs $(s_i, t_j)$ where $s_i$ is in $s$ and $t_j$ is in $t$.

To graph $f(s, t)$, we can use

mesh$(s, t, f(s, t))$

which plots $(s, t, f(s, t))$ and connects with rectangular-like sheets.

Curves can be graphed using plot3 as in

plot3$(t, t. \wedge 2, t. \wedge 5)$

Sometimes we need to put more than one graph on the screen. To superimpose graphs, use the command *hold* after the first graphing command has been entered.

**Programming:** Calculations which are iterative (a sequence of calculations) can done using *for loops, while loops* and perhaps incorporating *if* statements. We briefly go over each of these.

**Part 1.** For loop. If a calculation needs to be done for say $n = 1, 2, \ldots, r$, we can do them by using a program such as

```
for n = 1 : r
    calculation
end
```

As an example, to add the first ten natural numbers, we would use

```
S = 0;
for n = 1 : 10
    S = S + n
end
```

(If you do not want $S$ printed on each pass through the loop, end it with a semicolon. Then add $S$ after *end* as in *end, S.*

**Part 2.** While loop. The *while loop* works with a relation, such as $>$ or $\geq$, which is a bit different from the *for loop*. In general, we use

```
while (relation)
    calculation
end
```

For example, to add the first ten natural numbers, we might have

```
S = 0;
c = 1;
while c < 11
    S = S + c;
    c = c + 1;
end, S
```

**Part 3.** If and else. Sometimes, a decision needs to be made which determines our next calculation. And, often this occurs within a loop. For example, if we want an upper triangular matrix of 1's we can use

```
A = zeros(3, 3);
for i = 1 : 3
    for j = 1 : 3
        if i < j
            A(i, j) = 1;
        end
    end
end
```

A bit more complicated example use *else* as well.

```
A = [ 1  -2  0;  3  0  -4;  0  -1  2 ];
for i = 1 : 3
   for j = 1 : 3
      if A (i, j) > 0
            A (i, j) = 1;
      elseif A (i, j) < 0
            A (i, j) = -1;
      else
            A (i, j) > 0;
      end
   end
end
```

**Help:** If assistance is needed with a command, type in *help* and the name of the command as in *help det*.

# Answers to Selected Exercises

## Chapter 1

2. (a) $\overline{z+w} = \overline{(a+c)+(b+d)\,i} = (a+c)-(b+d)\,i = a-bi+c-di = \bar{z}+\bar{w}$

4. (b) $[-a_2 \ \ a_1+a_2]$

5. $\left[\begin{array}{ccc} \lambda_1 p_1 & \lambda_2 p_2 & \lambda_3 p_3 \end{array}\right]$

6. (a) Using the first column of $T$, and backward multiplication, show that the first column of $X$ is $\left(t_{11}^{-1},0,0\right)^t$. Continue to the second column.

7. (a) In arithmetic, if $ab = ac$, then $a\,(b-c) = 0$. So if $a \neq 0$, $b-c = 0$ or $b = c$. The missing arithmetic property is: nonzero constants have inverses.

10. Note that $\det A \det B = 1$, (Show this.) so $\det A \neq 0$. Thus, $A^{-1}$ exists. Solve for $B$.

11. (a) Show $\left(A^{-1}\right)^{-1}$ satisfies the inverse equation for $A^{-1}$ so it's the inverse of $A^{-1}$.

    (b) Use induction on $m$.

21. If $B = \operatorname{adj} A$, then $b_{ij} = c_{ji} = (-1)^{i+j} \det A_{ji}$. Now argue that $\det A_{ji}$ is rational.

22. $x = \frac{1}{\det A} \begin{bmatrix} c_{11} & c_{21} & c_{31} \\ c_{12} & c_{22} & c_{32} \\ c_{13} & c_{23} & c_{33} \end{bmatrix} \begin{bmatrix} b_1 \\ b_2 \\ b_3 \end{bmatrix}$ so

$$
\begin{aligned}
x_1 &= \frac{1}{\det A}(b_1 c_{11} + b_2 c_{21} + b_3 c_{31}) \\
&= \frac{1}{\det A} \det \begin{bmatrix} b_1 & a_{12} & a_{13} \\ b_2 & a_{22} & a_{23} \\ b_3 & a_{32} & a_{33} \end{bmatrix}.
\end{aligned}
$$

24. $\det A = (-1)^t \alpha_1 \ldots \alpha_r \det E$ where $t$ =number of interchanges used and where $\alpha_i R_i$ was used $r$ times. Explain why the last row of $E$ is 0. (Note: The determinant section follows the section on systems. However, the determinant results used here could have preceded that section. And thus our use of determinant results here to prove a systems result is legitimate.)

# Chapter 2, Section 1

1. (c)  Use the identity $(-1+1) = 0$ and so $(-1+1)x = 0x$. Now simplify.

3. (a) (iii) $0t^2 + 0t + 0$ which can also be written as 0.

(iv) $z$ where $z(t) = 0$ for all $t$.

8. Let $x \in W$. Since $W$ is closed under scalar multiplication, $0x \in W$. Since $0x = 0$, $0 \in W$.

11. Let $S$ be the subspace. If $S \neq \{0\}$, let $x \in S$, $x \neq 0$. Then $\alpha x \in S$ for all scalars $\alpha$. So $S$ contains a line through the origin. If $S$ contains nothing else, $S$ is that line. Continue.

13. (a)  Choose an arbitrary vector, say $(a, b, c, )^t$ in $R^3$. Show that $\alpha_1 (1, 1, 0)^t + \alpha_2 (1, -2, -1)^t + \alpha_3 (-1, 2, 2)^t = (a, b, c)^t$ has a solution

18. (a) Let $x, y \in U \cap W$. Then $x \in U$ and $y \in U$. Since $U$ is a subspace, $x + y \in U$. Similarly, $x + y \in W$. Thus, $x + y \in U \cap W$.

# Chapter 2, Section 2

3. Rearrange the pendent equation $\alpha_1 u + \alpha_2 (u + v) + \alpha_3 (u + v + w) = 0$ and set the coefficients of $u$, $v$, and $w$ to 0. (Explain why this can be done.) Now solve that system of equations for $\alpha_1, \alpha_2, \alpha_3$. Give the conclusion.

5. Let $S = \{x_1, \ldots, x_n\}$ be linearly independent. By reindexing, if necessary, let $\hat{S} = \{x_1, \ldots, x_r\}$ be the chosen subset of $S$. Now suppose $\hat{S}$ is linearly dependent and $(\beta_1, \ldots, \beta_r)$ a nontrivial solution to its pendent equation. Extend to a nontrivial solution to the pendent equation for $S$.

6. Consider $\alpha f(t) + \beta g(t) = 0$. Differentiation yields $\alpha f'(t) + \beta g'(t) = 0$. Thus,

$$\begin{bmatrix} f(t) & g(t) \\ f'(t) & g'(t) \end{bmatrix} \begin{bmatrix} \alpha \\ \beta \end{bmatrix} = \begin{bmatrix} 0 \\ 0 \end{bmatrix}.$$

So if there is a single $t$ such that $W(f(t), g(t)) \neq 0$, for that $t$, $\begin{bmatrix} f(t) & g(t) \\ f'(t) & g'(t) \end{bmatrix}$ is nonsingular. That is enough to show $\alpha = \beta = 0$.

8. If $x, y \in N(A)$, then $Ax = 0$ and $Ay = 0$. Thus, $A(x+y) = 0$ and so $x + y \in N(A)$.

13. Let $\{x_1, \ldots, x_n\}$ be a basis for $V$. Consider $\alpha_1 y_1 + \cdots + \alpha_m y_m = 0$. Write each $y_i$ as a linear combination of $x_1, \ldots, x_n$ and substitute these into the equation. Rearrange, set coefficients of the $x_i$'s to 0. Note the number of solutions here.

14. (a) $\dim W = 3$

16. Try $y = e^{rt}$ and determine $r$ so $y$ works.

18. Remove vectors, one at a time, from the set until no dependent vectors are left. Explain why this set is a basis.

19. (a) Let $(\beta_1, \beta_2)$ be a solution to the pendent equation. So $\beta_1 x_1 + \beta_2 x_2 = 0$. If $\beta_2 \neq 0$, we can solve for $x_2$ showing $x_2 \in \text{span}\{x_1\}$. Thus $\beta_2 = 0$. Now we have $\beta_1 x_1 = 0$. Since $x_1$ is linearly independent $\beta_1 = 0$.

20. Suppose $S = \{x_1, \ldots, x_n\}$ is linearly dependent. Then some vector, say $x_n$, in $S$ is dependent, so $\text{span}\, S \backslash \{x_n\} = \text{span}\, S$. Continue to a contradiction to $\dim V = n$.

21. (c) Let $x, y$ be in the parallelepiped. Then $x = \alpha_1 x_1 + \cdots + \alpha_n x_n$ and $y = \beta_1 y_1 + \cdots + \beta_n y_n$ where $0 \leq \alpha_k, \beta_k \leq 1$. Thus, $\alpha x + (1 - \alpha) y = (\alpha \alpha_1 + (1 - \alpha) \beta_1) x_1 + \cdots + (\alpha \alpha_n + (1 - \alpha) \beta_n) x_n$. Note that $0 \leq \alpha \alpha_k + (1 + \alpha) \beta_k \leq \alpha + (1 - \alpha) = 1$. So the parallelepiped is convex.

## Chapter 2, Section 3

1. (a) $L(x+y) = L\left(\begin{bmatrix} x_1 + y_1 \\ x_2 + y_2 \end{bmatrix}\right) = \begin{bmatrix} (x_1 + y_1) + 2(x_2 + y_2) \\ 2(x_1 + y_1) - (x_2 + y_2) \end{bmatrix}$

$= \begin{bmatrix} (x_1 + 2x_2) + (y_1 + 2y_2) \\ (2x_1 - x_2) + (2y_1 - y_2) \end{bmatrix} = \begin{bmatrix} x_1 + 2x_2 \\ 2x_1 - x_2 \end{bmatrix} + \begin{bmatrix} y_1 + 2y_2 \\ 2y_1 - y_2 \end{bmatrix}$

$= L(x) + L(y)$.

(c) $L(f(t) + g(t)) = (f(t) + g(t))' + (f(t) + g(t))$

$= (f'(t) + f(t)) + (g'(t) + g(t)) = L(f(t)) + L(g(t))$.

3. $L$ rotates the $x_1$-axis counter clockwise a bit and the $x_2$-axis clockwise a little, stretching both.

7. $L_1 \circ L_2 (x_1, x_2) = L_1 \left( \begin{bmatrix} x_1 - x_2 \\ x_2 + 1 \end{bmatrix} \right) = \begin{bmatrix} (x_1 - x_2) + (x_2 + 1) \\ (x_1 - x_2) - (x_2 + 1) \end{bmatrix}$

$= \begin{bmatrix} x_1 + 1 \\ x_1 - 2x_2 - 1 \end{bmatrix}$.

9. (a) $A = \begin{bmatrix} 2 & 1 \\ 1 & 2 \end{bmatrix}$.

13. Try $y = at + b$. Plug in and determine $a$ and $b$.

15. Suppose $(\beta_1, \ldots, \beta_n)$ is a solution to the pendent equation for

$$L(x_1), \ldots, L(x_n).$$

Then $\beta_1 L(x_1) + \cdots + \beta_n L(x_n) = 0$ or $L(\beta_1 x_1 + \cdots + \beta_n x_n) = 0$. Since $L$ is one-to-one, $\beta_1 x_1 + \cdots + \beta_n x_n = 0$. Thus $\beta_1 = \cdots = \beta_n = 0$, and so $L(x_1), \ldots, L(x_n)$ are linearly independent.

16. $L_1 \circ L_2 (x + y) = L_1 (L_2(x) + L_2(y)) = L_1 (L_2(x)) + L_1 (L_2(y)) = L_1 \circ L_2(x) + L_1 \circ L_2(y)$.

# Chapter 3, Section 1

2. Let $E$ be an echelon form of $A$ obtained by interchange and add operations. Then $\det A = (-1)^t \det E$ and since $\det A = 0$, $\det E = 0$. Thus, $E$ has a row of 0's and hence there is a free variable in the solution to $Ax = 0$. (There are other proofs as well.)

4. Arguing by contradiction, suppose $x_1, \ldots, x_r, u_i$ is linearly dependent for each $i$. Show that this means $u_i \in \text{span}\{x_1, \ldots, x_r\}$ for each $i$. If $x \in V$, $x = \alpha_1 u_1 + \cdots + \alpha_n u_n$ for some scalars $\alpha_1, \ldots, \alpha_n$. Show $x \in \text{span}\{x_1, \ldots, x_r\}$, and explain why $V \subseteq \text{span}\{x_1, \ldots, x_r\}$. But this means $\dim V = r$, $r < n$, a contradiction.

9. No. Find an $A$ and $B$ such that $\text{rank}(A + B) \neq \text{rank} A + \text{rank} B$.

10. Look at $3 \times 3$ matrices with lots of 0's.

# Chapter 3, Section 2

3. Define $A = PDF^{-1}$.

4. Find different sets of eigenvectors for $P$.

7. Use that if $A = PDF^{-1}$, $A - \lambda I = P(D - \lambda I) P^{-1}$ and that

$$\text{rank}(A - \lambda I) = \text{rank}(D - \lambda I).$$

8. Suppose $A$ has a real eigenvalue $\lambda$. Then $Ax = \lambda x$. But, $A$ rotates while $\lambda$ stretched, etc.

11. A diagonal matrix.

12. (a)   Let $\varphi(\lambda) = \det(A - \lambda I)$. If $\varphi(\lambda_1) = 0$ and $\lambda_1$ is a complex number, then $0 = \varphi(\lambda_1) = \det(\bar{A} - \bar{\lambda}_1 I) = \det(A - \bar{\lambda}_1 I) = \varphi(\bar{\lambda}_1)$. Now note that $(\lambda_1 - \lambda)(\bar{\lambda}_1 - \lambda)$ is a real polynomial. So $\varphi(\lambda)/(\lambda_1 - \lambda)(\bar{\lambda}_1 - \lambda)$ is a real polynomial, namely

$$\varphi_1(\lambda) = (\lambda_3 - \lambda) \cdots (\lambda_n - \lambda).$$

Continue, by working with $\varphi_1(\lambda)$, to see that complex conjugate eigenvalues pair up. And, put everything together.

# Chapter 3, Section 3

1. (a)  Can't diagonalize. The eigenvalues are $\lambda_1 = 2$, $\lambda_2 = 2$, but the corresponding eigenspace has dimension 1. So, we can't find linearly independent eigenvectors $p_1, p_2$ to form a nonsingular $P$.

4. Note that $P^{-1}A = DP^{-1}$. Then $\hat{p}_i A = \lambda_1 \hat{p}_i$ where $\hat{p}_i$ is the $i$-th row of $P^{-1}$. So, the rows of $P^{-1}$ will give left eigenvectors.

5. No.  Find a counter example.

7. Show $\det(A - \lambda I) = \det(A^t - \lambda I)$ by using $\det B = \det B^t$.

9. If $A = PBP^{-1}$ and $Ax = \lambda x$, $PBP^{-1}x = \lambda x$. Rearrange this to $B(P^{-1}x) = \lambda(P^{-1}x)$. Conclude.

10. If $A$ is singular $Ax = 0$ has a nontrivial solution, say $y$. Then $Ay = 0y$ so 0 is an eigenvalue of $A$. The converse still needs to be argued.

11. Note that $\begin{bmatrix} I & A \\ 0 & I \end{bmatrix}$ has an inverse.

# Chapter 3, Section 4

1. (a) $\begin{bmatrix} 2 & 1 & 0 \\ 0 & 2 & 0 \\ 0 & 0 & 3 \end{bmatrix}, \begin{bmatrix} 2 & 0 & 0 \\ 0 & 2 & 0 \\ 0 & 0 & 3 \end{bmatrix}.$

2. (a) $P = \begin{bmatrix} 0 & \frac{1}{2} \\ 1 & 0 \end{bmatrix}$   $J = \begin{bmatrix} 3 & 1 \\ 0 & 3 \end{bmatrix}$

3. There are 2 linearly independent eigenvectors for $\lambda$ (from the 1,1 and 3,3 entries of $J$). So dim (eigenspace for $\lambda$) = 2.

4. Solve $AP = PB$ for $P$.

8. Use $(A - \lambda I)p_1 = 0$ and $(A - \lambda I)p_2 = p_1$.

9. Find $R$ such that $A = R \begin{bmatrix} 2 & 1 \\ 0 & 2 \end{bmatrix} R^{-1}$. Then

$$A = R \begin{bmatrix} \epsilon^{-1} & 0 \\ 0 & 1 \end{bmatrix} \begin{bmatrix} 2 & \epsilon \\ 0 & 2 \end{bmatrix} \begin{bmatrix} \epsilon & 0 \\ 0 & 1 \end{bmatrix} R^{-1}.$$

Thus, $P = R \begin{bmatrix} \epsilon^{-1} & 0 \\ 0 & 1 \end{bmatrix}$.

13. If $A = PJF^{-1}$, $A^{-1} = PJ^{-1}P^{-1}$. So $A^{-1}$ is similar to $J^{-1}$. Conclude.

# Chapter 4, Section 1

3. $\lim_{k \to \infty} (A_k + B_k) = \lim_{k \to \infty} \left[ a_{ij}^{(k)} + b_{ij}^{(k)} \right] = \left[ \lim_{k \to \infty} \left( a_{ij}^{(k)} + b_{ij}^{(k)} \right) \right]$

$= \left[ \lim_{k \to \infty} a_{ij}^{(k)} + \lim_{k \to \infty} b_{ij}^{(k)} \right] = [a_{ij} + b_{ij}] = A + B.$

8. Suppose $f$ is continuous. Let $A \in R^2$ and $A_1, A_2, \ldots$ be a sequence that converges to $A$. Then $\lim_{k \to \infty} f(A_k) = f(A) = \begin{bmatrix} f_1(A) \\ f_2(A) \end{bmatrix}$. Also,

$\lim_{k \to \infty} f(A_k) = \begin{bmatrix} \lim_{k \to \infty} f_1(A_k) \\ \lim_{k \to \infty} f_2(A_k) \end{bmatrix}$, so both $f_1$ and $f_2$ are continuous at

$A$. Since $A$ was chosen arbitrarily, $f_1$ and $f_2$ are continuous in $R^2$. (The converse still needs to be proved.)

9. (a) $\lim_{t \to 0} A(t) = \begin{bmatrix} \lim_{t \to 0} (2t - 1) & \lim_{k \to 0} e^t \\ \lim_{t \to 0} \frac{t}{t-1} & \lim_{k \to 0} 0 \end{bmatrix} = \begin{bmatrix} -1 & 1 \\ 0 & 0 \end{bmatrix} = A(0).$

Thus $A(t)$ is continuous at $t = 0$.

14. $m_1 x_1'' = -k_1 x_1 + k_2 (x_2 - x_1)$

$m_2 x_2'' = -k_2 (x_2 - x_1) - k_3 x_2$

15. Same as in Optional.

# Chapter 4, Section 2

1. (a) $\lim_{k \to 0} A^k = \lim_{k \to 0} A \left( \begin{bmatrix} 1 & -4 \\ 1 & 3 \end{bmatrix} \begin{bmatrix} .6 & 0 \\ 0 & -.1 \end{bmatrix} \begin{bmatrix} \frac{3}{7} & \frac{4}{7} \\ -\frac{1}{7} & \frac{1}{7} \end{bmatrix} \right)^k$

$= \begin{bmatrix} 1 & -4 \\ 1 & 3 \end{bmatrix} \begin{bmatrix} \lim_{k \to 0} (.6)^k & 0 \\ 0 & \lim_{k \to 0} (-.1)^k \end{bmatrix} \begin{bmatrix} \frac{3}{7} & \frac{4}{7} \\ -\frac{1}{7} & \frac{1}{7} \end{bmatrix}$

$= \begin{bmatrix} 1 & -4 \\ 1 & 3 \end{bmatrix} \begin{bmatrix} 0 & 0 \\ 0 & 0 \end{bmatrix} \begin{bmatrix} \frac{3}{7} & \frac{4}{7} \\ -\frac{1}{7} & \frac{1}{7} \end{bmatrix} = \begin{bmatrix} 0 & 0 \\ 0 & 0 \end{bmatrix}.$

2. (a) $x = \alpha_1 2^k \begin{bmatrix} 1 \\ 1 \end{bmatrix} + \alpha_2 4^k \begin{bmatrix} 1 \\ -1 \end{bmatrix}$.

4. (a) $\alpha_2 4^k \begin{bmatrix} 1 \\ -1 \end{bmatrix}$ (The vector could be any eigenvector for $\lambda = 4$.)

11. Substitute $PDF^{-1}$ for $A$ and proceed as in the introduction to this section.

# Chapter 4, Section 3

1. (a) $x = \alpha_1 e^{-t} \begin{bmatrix} 1 \\ 1 \end{bmatrix} + \alpha_2 e^{-3t} \begin{bmatrix} 1 \\ -1 \end{bmatrix}$. (The choice of eigenvectors can be different.)

8. $y(t) = e^{-t} \begin{bmatrix} 1 \\ 1 \end{bmatrix}$.

9. Find the Taylor series for $\sin \tau$. Replace $\tau$ by $At$. Then differentiate termwise.

12. $y(t) = \sin\left(\sqrt{A}t\right) \begin{bmatrix} \alpha_1 \\ \alpha_2 \end{bmatrix} + \cos\left(\sqrt{A}t\right) \begin{bmatrix} \beta_1 \\ \beta_2 \end{bmatrix}$. Define $\sqrt{A}$, etc.

13. If $\left|a_{ij}^{(k)}\right| \leq (nm)^k$ then $A^{k+1} = AA^k$ so $a_{ij}^{(k+1)} = \sum_{r=1}^{n} a_{ir} a_{rg}^{(k)} \leq \sum_{r=1}^{n} m(nm)^k$. (The proof by induction would still need a formal write-up.)

# Chapter 5, Section 1

5. (a) Yes (b) no. Find $x$'s that support these answers.

10. (a) $\|x + ty\|^2 = (x + ty, x + ty) = (x, x) + 2t(x, y) + t^2(y, y)$. Substitute $\|x\|^2 = (x, x)$, $(x, y) = \sum_{r=1}^{n} x_k y_k$.

11. (a) $\|x + y\|_2^2 = (x + y, x + y) = \|x\|_2^2 + (x, y) + (y, x)$

$+ \|y\|_2^2 \leq \|x\|_2^2 + 2\|x\|_2 \|y\|_2 + \|y\|_2^2 = \left(\|x\|_2^2 + \|y\|_2^2\right)$ Now take the square root of both sides for the triangle inequality.

12. (a) $\|x\|_2 \leq \|x\|_1$ so $\|x\|_2 \leq .001$.

# Chapter 5, Section 2

1. (a) Let $m_s = \max\{m_1, \ldots, m_r\}$, then $m_k \le m_s$ for all $k$ so $cm_k \le cm_s$. Thus,

$$\max\{cm_1, \ldots, cm_r\} = cm_s.$$

   (b) Let $n_s = \max\{n_1, \ldots, n_r\}$ and $m_t = \max\{m_1, \ldots, m_r\}$. Continue.

2. (b) $\|I\| = \max_{\|x\|=1} \|Ix\| = \max_{\|x\|=1} \|x\| = 1.$

7. (a) Plot $Ax$ for $x = e_1, -e_1, e_2, -e_2$. (The vertices of $C_0$.) and connect with edges.

   (b) Graph $\pm Ae_1, \pm Ae_2$, and connect with segments.

9. (b) $\left\| L\left(\begin{bmatrix} 1 \\ 1 \end{bmatrix}\right) - L\left(\begin{bmatrix} 2 \\ 2 \end{bmatrix}\right) \right\|_1 = \left\| \begin{bmatrix} -1 \\ -3 \end{bmatrix} \right\|_1 = 4,$

   $\left\| \begin{bmatrix} 1 \\ 1 \end{bmatrix} - \begin{bmatrix} 2 \\ 2 \end{bmatrix} \right\|_1 = 2.$

11. Use that $\sum_{k=1}^{n} |a_{kj}| = \sum_{k=1}^{n} 1 \, |a_{kj}| \le \left(\sum_{k=1}^{n} 1^2\right)^{\frac{1}{2}} \left(\sum_{k=1}^{n} |a_{kj}|^2\right)^{\frac{1}{2}}$ by the Cauchy-Schwarz inequality. Now, if $\|A\|_1 = \sum_{k=1}^{n} |a_{kr}|$,

$$\|A\|_1 \le \sqrt{n} \left(\sum_{k=1}^{n} |a_{kj}|^2\right)^{\frac{1}{2}} \le \sqrt{n} \, \|A\|_F.$$

12. (d) $c(AB) = \|AB\| \left\|(AB)^{-1}\right\| = \|AB\| \left\|B^{-1}A^{-1}\right\|$

   $\le \|A\| \|B\| \|B^{-1}\| \|A^{-1}\| = \|A\| \|A^{-1}\| \|B\| \|B^{-1}\| = c(A) \, c(B).$

13. (a) Note that $x = A^{-1}b$ and $y = A^{-1}c$. So $\|x - y\| \le \|A^{-1}\| \|b - c\|$. Also $\|A\| \|x\| \ge \|b\|$. So $\|x\| \ge \frac{\|b\|}{\|A\|}$. Thus $\frac{\|x-y\|}{\|x\|} \le \frac{\|A^{-1}\| \|b-c\|}{\|b\|/\|A\|}$, etc.

16. (a) $|123.4 - x| < 10^{-3}(123.4) = .1234$. So $x = 123.4\pm$ a number less than $.1234$. So $x$ differs from $123.4$ by a number starting in the fourth digit of $123.4$. (Multiplying by $10^{-3}$ shifts the decimal 3 places to the left, causing a number which starts in the 4th digit of $123.4$.)

# Chapter 5, Section 3

2. (b) Write out the expressions for $\|AB\|_F$ and $\|A\|_F \|B\|_F$ and compare.

3. For the triangle inequality, $\|x + y\|_R = \|R(x + y)\| = \|Rx + Ry\| \leq \|Rx\| + \|Ry\|$ (since $\|\cdot\|$ is a norm) $= \|x\|_R + \|y\|_R$.

6. (a) The eigenvalues are 4, −1.

8. (b) $\begin{bmatrix} 1 & 1 \\ 0 & 1 \end{bmatrix}$

10. (a) $x = (I - A)^{-1} b = \frac{1}{.77} \begin{bmatrix} .9 & .2 \\ .2 & .9 \end{bmatrix} \begin{bmatrix} 1 \\ 1 \end{bmatrix} = \begin{bmatrix} 1.429 \\ 1.429 \end{bmatrix}$ using Neumann's formula. Rate, using the 1-norm, is $\|A\|_1^k = .3^k$.

# Chapter 5, Section 4

4. For Example 5.10:

(1) If $x \neq 0$, then $(x, x) = x_1 \bar{x}_1 + \cdots + x_n \bar{x}_n = |x_1|^2 + \cdots + |x_n|^2 > 0$ since some $x_i \neq 0$.

(2) $(x, y) = x_1 \bar{y}_1 + \cdots + x_n \bar{y}_n = \overline{y_1 \bar{x}_1} + \cdots + \overline{y_n \bar{x}_n}$
$= \overline{y_1 \bar{x}_1 + \cdots + y_n \bar{x}_n} = \overline{(y, x)}$.

(3) $(\alpha x, y) = (\alpha x_1) \bar{y}_1 + \cdots + (\alpha x_n) \bar{y}_n = \alpha (x_1 \bar{y}_1 + \cdots + x_n \bar{y}_n) = \alpha (x, y)$.

5. For (iii): $(x, \alpha y) = \overline{(\alpha y, x)}$ by (2), $= \overline{\alpha (y, x)}$ by (3), $= \bar{\alpha} \overline{(y, x)} = \bar{\alpha} (x, y)$ by (2).

6. $(0, x) = (0x, x) = 0(x, x) = 0$. Give reasons.

9. $\|\alpha_1 u_1 + \cdots + \alpha_m u_m\|_2^2 = (\alpha_1 u_1 + \cdots + \alpha_m u_m, \alpha_1 u_1 + \cdots + \alpha_m u_m)$
$= \alpha_1 \bar{\alpha}_1 (u_1, u_1) + \cdots + \alpha_m \bar{\alpha}_m (u_m, u_m) = |\alpha_1|^2 + \cdots + |\alpha_m|^2$. Give reasons.

14. First apply Gram-Schmidt to $\begin{bmatrix} 1 \\ 1 \\ 0 \end{bmatrix}, \begin{bmatrix} 1 \\ 1 \\ 1 \end{bmatrix}$ to get $u_1, u_2$. Then use the corresponding Fourier sum.

16. (a) The line is span $\left\{ \begin{bmatrix} 1 \\ 2 \end{bmatrix} \right\}$. So $u_1 = \begin{bmatrix} \frac{1}{\sqrt{5}} \\ \frac{2}{\sqrt{5}} \end{bmatrix}$ and $P = u_1 u_1^t = \begin{bmatrix} \frac{1}{5} & \frac{2}{5} \\ \frac{2}{5} & \frac{4}{5} \end{bmatrix}$.

(b) $P \begin{bmatrix} 1 \\ 1 \end{bmatrix} = \begin{bmatrix} \frac{3}{5} \\ \frac{6}{5} \end{bmatrix}$.

17. $P^2 = (UU^t)(UU^t) = UIU^t = UU^t$. Give reasons.

19. No. Show.

# Chapter 6, Section 1

6. (a) The columns of $U$ form a linearly independent set so $U$ is non-singular. (There are other ways of showing this.)

   (b) Since $U$ is nonsingular and $U^H U = I$, we can multiply this equation, on the right, by $U^{-1}$ to get $U^H = U^{-1}$.

10. Apply Theorem 6.5.

11. Cos of the angle between $x, y$ is $\frac{(x,y)}{\|x\|_2 \|y\|_2}$. Write out the equivalent expression for $Qx, Qy$. Then manipulate one of these to the other.

13. Let $a = \frac{1}{2} \left( \begin{bmatrix} 3 \\ 4 \end{bmatrix} + \begin{bmatrix} 5 \\ 0 \end{bmatrix} \right)$, the average of the vectors. Since we want to reflect about span $\{a\}$, take $u$ to be orthogonal to $a$.

16. (b) Under rotations or reflections the flag points away from the origin.

17. Let $\hat{S} = \{L(x) : x \in S\}$. Prove $\|L(x) - L(z)\| = r$, for all $L(x) \in \hat{S}$. (This proves $L : S \rightarrow \hat{S}$.) And prove that if $\|w - L(z)\| = r$, then $w = L(x)$ for some $x$ where $\|x - z\| = r$. (This proves onto.)

# Chapter 6, Section 2

1. (a) $Q = \begin{bmatrix} \frac{1}{\sqrt{2}} & -\frac{1}{\sqrt{2}} \\ \frac{1}{\sqrt{2}} & \frac{1}{\sqrt{2}} \end{bmatrix}$. (There could be others.)   $T = \begin{bmatrix} 3 & 1 \\ 0 & 0 \end{bmatrix}$.

3. Let $A(x + y) = (\alpha + \beta i)(x + iy)$. Then $Ax = \alpha x - \beta y$, $Ay = \beta x + \alpha y$. Thus $L(x) = Ax$ maps span $\{x, y\} \rightarrow$ span $\{x, y\}$. Apply Gram-Schmidt to $x, y$ to get $u_1, u_2$. Then $Au_1 = r_1 u_1 + r_2 u_2$, $Au_2 = s_1 u_1 + s_2 u_2$. So $A[u_1 u_2] = [u_1 u_2] \begin{bmatrix} r_1 & s_1 \\ r_2 & s_2 \end{bmatrix}$. Extend $u_1 u_2$ to an orthonormal basis for $R^4$ say $u_1, u_2, u_3, u_4$. Set $Q = [u_1 u_2 u_3 u_4]$ and show $AQ = QT$

8. Let $x, y$ be linearly independent vectors which are not orthogonal. Let $P = [xy], D = \begin{bmatrix} \alpha & 0 \\ 0 & \beta \end{bmatrix}$ where $\alpha \neq \beta$. Form $A = PDP^{-1}$. (A particular such example will be fine.)

10. No. An example still needs to be given.

11. Argue by contradiction. Use that $TT^H = T^H T$.

# Chapter 7, Section 1

1. There are infinitely many solutions.

2. If the equation of the line is $mx + b = y$, solve

$$
\begin{aligned}
m & +b = 1 \\
2m & +b = 1 \\
2m & +b = 2
\end{aligned}
$$

4. (b)  The line is span $\left\{ \begin{bmatrix} 1 \\ 1 \end{bmatrix} \right\}$. So, $P = \begin{bmatrix} \frac{1}{\sqrt{2}} \\ \frac{1}{\sqrt{2}} \end{bmatrix} \begin{bmatrix} \frac{1}{\sqrt{2}} & \frac{1}{\sqrt{2}} \end{bmatrix} = \begin{bmatrix} \frac{1}{\sqrt{2}} & \frac{1}{\sqrt{2}} \\ \frac{1}{\sqrt{2}} & \frac{1}{\sqrt{2}} \end{bmatrix}$. And the closest point is $P \begin{bmatrix} 2 \\ 1 \end{bmatrix} = \begin{bmatrix} \frac{3}{2} \\ \frac{3}{2} \end{bmatrix}$.

8. No.

9. Since $A^t A = A^2$ and the eigenvalues of $A^2$ are $\lambda_1^2, \ldots, \lambda_n^2$, it follows that $\sigma_1 = \sqrt{\lambda_1^2} = |\lambda_1|, \ldots, \sigma_n = \sqrt{\lambda_n^2} = |\lambda_n|$.

16. (a)  $A^H AV = V\Sigma^H U^H U\Sigma V^H V = V\Sigma^H \Sigma$, so $A^H Av_i = \sigma_i^2 v_i$ for all $i$.

# Chapter 7, Section 2

1. $0^t$ since it satisfies the equation (i)-(iv) of a pseudo-inverse.

3. (a)  $A^+ = \begin{bmatrix} \frac{1}{\sqrt{2}} & -\frac{1}{\sqrt{2}} \\ \frac{1}{\sqrt{2}} & \frac{1}{\sqrt{2}} \end{bmatrix} \begin{bmatrix} \frac{1}{3} & 0 \\ 0 & 1 \end{bmatrix} \begin{bmatrix} \frac{1}{\sqrt{2}} & \frac{1}{\sqrt{2}} \\ -\frac{1}{\sqrt{2}} & \frac{1}{\sqrt{2}} \end{bmatrix} = \begin{bmatrix} \frac{2}{3} & -\frac{1}{3} \\ -\frac{1}{3} & \frac{2}{3} \end{bmatrix}$.

4. Show the matrix satisfies (i)-(iv) of a pseudo-inverse.

5. (a)  1.

7. 1.  The supporting work must still be given.

9. (a)  $L(y) = \begin{bmatrix} 2y_1 \\ 0 \end{bmatrix}$ where $Z = Y = \left\{ \begin{bmatrix} \frac{1}{\sqrt{2}} \\ \frac{1}{\sqrt{2}} \end{bmatrix} \begin{bmatrix} -\frac{1}{\sqrt{2}} \\ \frac{1}{\sqrt{2}} \end{bmatrix} \right\}$. So $L$ collapses (projects) $R^2$ into the $y_1$-axis and then doubles this axis and sends to the $z_1$-axis.

10. Change the problem to finding the closest orthogonal matrix $Q$ to $\Sigma$. Then show, by looking at the terms in $\|Q - \Sigma\|_F$, that such a matrix is $I$. (This is the idea. Organization is still required.)

11. It says that the condition number is squared, perhaps doubling the number of additional digits in error when we solve the normal equations $A^t Ax = A^t b$.

# Chapter 8, Section 1

1. (c) $P = \begin{bmatrix} 0 & 1 & 0 \\ 0 & 0 & 1 \\ 1 & 0 & 0 \end{bmatrix}$

3. (a)

$$\begin{bmatrix} 1 & 0 & 0 \\ 1 & 1 & 0 \\ 1 & 2 & 1 \end{bmatrix} \begin{bmatrix} 1 & 1 & 1 \\ 0 & 1 & 3 \\ 0 & 0 & 2 \end{bmatrix}$$

5. Use $3R_2 + R_1$ so $E = \begin{bmatrix} 1 & 3 \\ 0 & 1 \end{bmatrix}$.

6. (a)

$$\begin{bmatrix} 1 & 0 \\ * & * \end{bmatrix} \begin{bmatrix} 2 & 4 \\ \times & * \end{bmatrix}, \quad \begin{bmatrix} 1 & 0 \\ \ell_{21} & 1 \end{bmatrix} \begin{bmatrix} 2 & 4 \\ 0 & \times \end{bmatrix} = \begin{bmatrix} 2 & 4 \\ 4 & 11 \end{bmatrix}.$$

9. (a) Just multiply $\begin{bmatrix} 1 & 0 & 0 \\ \ell_{21} & 1 & 0 \\ \ell_{31} & \ell_{32} & 1 \end{bmatrix} \begin{bmatrix} 1 & 0 & 0 \\ \hat{\ell}_{21} & 1 & 0 \\ \hat{\ell}_{31} & \hat{\ell}_{32} & 1 \end{bmatrix}$. (See why it

works.)

(b) From Chapter 1, we know that the product of two lower triangular matrices is lower triangular. Now, if $\ell = [\ell_{k1} \dots \ell_{kk-1}\, 1\, 0 \dots 0]$ and $u = [0 \dots 0\, 1\, \ell_{j+1,j} \dots \ell_{nj}]^t \dots$

11. Use the first row of $L$ to find the first row of $X$. Continue.

12. Note that $A = \begin{bmatrix} e_1 A \\ \cdots \\ e_n A \end{bmatrix}$. Now apply an elementary operation and

see what happens with the $e_i A$ rows.

# Chapter 8, Section 2

6. $(w, u) = 0$ is $u_1 w_1 + \cdots + u_n w_n = 0$ or in augmented form $[u_1 \dots u_n | 0]$. So, there are $n - 1$ free variables and so $\dim W = n - 1$.

8. (c) If $u$ has length 1,

$$\begin{aligned}
HH^t &= (I - 2u\,u)^t\,(I - 2u\,u^t) \\
&= I - 4u\,u^t + 4u\,u^t u\,u^t \\
&= I - 4u\,u^t + 4u\,u^t \quad \text{(Note } u\,^t u = 1.\text{)} \\
&= I.
\end{aligned}$$

11. The $QR$ decomposition, with partial pivoting, produces an $R$ of the form $F = \begin{bmatrix} R_{11} & R_{12} \\ 0 & 0 \end{bmatrix}$ where $R_{11}$ is upper triangular and nonsingular. Since $\operatorname{rank} A = \operatorname{rank}(QR) = \operatorname{rank} R$, $R_{11}$ must be $r \times r$.

12. Note that $Hu = -u$ so $-1$ is an eigenvalue of $H$. And, since $Aw = w$ for all $w \in W$, and $\dim W = n-1$, there are $n-1$ linearly independent eigenvectors, say $w_1, \ldots, w_{n-1}$ for the eigenvalue 1.

14. Use $|\det A| = |\det(QR)| = |\det Q \det R| = |\det R| = |r_{11} \cdots r_{nn}|$. Since $a_i = Qr_i$ where $a_i$ and $r_i$ are the $i$-th columns of $A$ and $R$, respectively, $\|a_i\|_2 = \|Qr_i\|_2 = \|r_i\| \geq |r_{ii}|$. Now, put together.

15. Note that $a_i = r_{1i}q_1 + \cdots + r_{ii}q_i$, so $a_i - (r_{1i}q_1 + \cdots + r_{i,i-1}q_{i-1}) = r_{ii}q_i$. Now, using $Q^t A = R$ (so $q_k^t a_i = r_{ki}$), show that $r_{i1}q_1 + \cdots + r_{i,i-1}q_{i-1}$ is the Fourier sum for $a_i$ and thus the closest vector in span $\{a_1, \ldots, a_{i-1}\}$ to $a_i$. Finish.

17. Apply Givens matrices to $Q$ to obtain an upper triangular matrix $R$ where $r_{11} > 0, \ldots, r_{n-1,n-1} > 0$. Now $R$ must be orthogonal so $R$ is diagonal with $r_{11} = 1, \ldots, r_{n-1,n-1} = 1$ and $r_{nn} = \pm 1$. Since the determinant of a Givens matrix is 1, $\det Q = \det R = r_{11} \cdots r_{nn}$. Finish.

# Chapter 9, Section 1

3. Check the trace.

5. Yes

6. (b) $|\lambda - a_{ii}| \leq \sum_{\substack{k=1 \\ k \neq i}}^{n} |a_{ki}|$, etc.

# Chapter 9, Section 2

3. (a) $\lambda_1'(0) = .5$.

5. (a) $\frac{1}{s_1} = 1$. So $|\lambda_1'(0)| \leq 1$.

# Chapter 10, Section 1

3. $\begin{bmatrix} 1 & 1 \\ 1 & 1 \end{bmatrix} = \begin{bmatrix} \frac{1}{\sqrt{2}} & -\frac{1}{\sqrt{2}} \\ \frac{1}{\sqrt{2}} & \frac{1}{\sqrt{2}} \end{bmatrix} \begin{bmatrix} 2 & 0 \\ 0 & 0 \end{bmatrix} \begin{bmatrix} \frac{1}{\sqrt{2}} & \frac{1}{\sqrt{2}} \\ -\frac{1}{\sqrt{2}} & \frac{1}{\sqrt{2}} \end{bmatrix}$

$= \left( \begin{bmatrix} \frac{1}{\sqrt{2}} & -\frac{1}{\sqrt{2}} \\ \frac{1}{\sqrt{2}} & \frac{1}{\sqrt{2}} \end{bmatrix} \begin{bmatrix} \sqrt{2} & 0 \\ 0 & 0 \end{bmatrix} \right) \left( \begin{bmatrix} \sqrt{2} & 0 \\ 0 & 0 \end{bmatrix} \begin{bmatrix} \frac{1}{\sqrt{2}} & \frac{1}{\sqrt{2}} \\ -\frac{1}{\sqrt{2}} & \frac{1}{\sqrt{2}} \end{bmatrix} \right)$

$= RR^t$ where $R = \begin{bmatrix} 1 & 0 \\ 1 & 0 \end{bmatrix}$.

8. The model is $\begin{bmatrix} 8 & 0 \\ 0 & 4 \end{bmatrix} x'' + \begin{bmatrix} \frac{1}{2} & -\frac{1}{4} \\ -\frac{1}{4} & \frac{1}{4} \end{bmatrix} x = 0.$

9. (a) Using Theorem 10.1, the following equations are equivalent.

$$\det(\lambda B - A) = 0$$
$$\det\left(\lambda\left(P^H\right)^{-1}P^{-1} - \left(P^H\right)^{-1}DP^{-1}\right) = 0$$
$$\det\left(\left(P^H\right)^{-1}(\lambda I - D)P^{-1}\right) = 0$$
$$\det(\lambda I - D) = 0.$$

The last equation has solutions $\lambda_1, \ldots, \lambda_n$.

## Chapter 10, Section 2

4. Use Theorem 10.5.

6. The eigenvalues are 5 and 1 with corresponding orthonormal eigen-
vectors $\begin{bmatrix} \frac{1}{\sqrt{2}} \\ \frac{1}{\sqrt{2}} \end{bmatrix}$ and $\begin{bmatrix} \frac{1}{\sqrt{2}} \\ -\frac{1}{\sqrt{2}} \end{bmatrix}$, respectively. So, for the image ellipse,
the major axis is 5 $\begin{bmatrix} \frac{1}{\sqrt{2}} \\ \frac{1}{\sqrt{2}} \end{bmatrix}$, and the minor axis is 1 $\begin{bmatrix} \frac{1}{\sqrt{2}} \\ -\frac{1}{\sqrt{2}} \end{bmatrix}$. Sketch
from this.

7. Let $A$ and $B$ be the two positive definite matrices. Then $h(x) = x^H(A+B)x = x^H Ax + x^H Bx \geq 0$ with equality only when $x = 0$. Thus, $A + B$ is positive definite. (Now show $A + B$ is Hermitian.)

11. Observe that by applying these elementary operations, $\det A_k = \det E_k$ where $E_k$ is the $k \times k$ submatrix in the upper left corner of $E$. Tell why.

## Chapter 11, Section 1

2. A basis for the plane is $\begin{bmatrix} 1 \\ -1 \\ 0 \end{bmatrix}, \begin{bmatrix} 1 \\ 0 \\ -1 \end{bmatrix}$. Add $\begin{bmatrix} 1 \\ 1 \\ 0 \end{bmatrix}$ and form
$R = \begin{bmatrix} 1 & 1 & 1 \\ -1 & 0 & 1 \\ 0 & 1 & 0 \end{bmatrix}$. Set $P = R\begin{bmatrix} 1 & 0 & 0 \\ 0 & 1 & 0 \\ 0 & 0 & 0 \end{bmatrix}R^{-1}$.

4. Test by $P^2 = P$.

5. A basis for the line is $\begin{bmatrix} 1 \\ 1 \\ 1 \end{bmatrix}$, for the plane is $\begin{bmatrix} 1 \\ 0 \\ 0 \end{bmatrix}\begin{bmatrix} 0 \\ 1 \\ 0 \end{bmatrix}$. Set
$R = \begin{bmatrix} 1 & 1 & 0 \\ 1 & 0 & 1 \\ 1 & 0 & 0 \end{bmatrix}$ and $D = \begin{bmatrix} 1 & 0 & 0 \\ 0 & 0 & 0 \\ 0 & 0 & 0 \end{bmatrix}$ and $P = RDR^{-1}$.

7. Try extending $q_1, \ldots, q_r$ to $q_1, \ldots, q_n$ an orthonormal basis. Set $\hat{Q} = [q_1, \ldots, q_n]$. Then,

$$QQ^t = \hat{Q} D \hat{Q}^t$$

where $D = \text{diag}\, (1, \ldots, 1, 0, \ldots, 0)$ where there are $r$ 1's in $D$.

# Chapter 11, Section 2

1. (a) Let $A \in X$. A ball $\mathbf{B}$ of radius $\epsilon$ about $A$ is

$$\mathbf{B} = \{B : B \in X \text{ and } \|B - A\| < \epsilon\}.$$

3. Use $f(a, b, c) = \begin{bmatrix} a & b \\ b & c \end{bmatrix}$. To show $f$ is continuous (it should be clear), let $(a_1, b_1, c_1)(a_2, b_2, c_2) \cdots \to (a, b, c)$. Then the image sequence is $\begin{bmatrix} a_1 & b_1 \\ b_1 & c_1 \end{bmatrix}, \begin{bmatrix} a_2 & b_2 \\ b_2 & c_2 \end{bmatrix}, \cdots$ which converges to $\begin{bmatrix} a & b \\ b & c \end{bmatrix}$. Since $f(a, b, c) = \begin{bmatrix} a & b \\ b & c \end{bmatrix}$, $f$ is continuous, etc.

4. Use $f(\theta) = \begin{bmatrix} \cos\theta & -\sin\theta \\ \sin\theta & \cos\theta \end{bmatrix}$ for $0 < \theta < 2\pi$ and $\pi < \theta < 3\pi$. This covers the Givens matrices. Now do the Householder matrices.

8. Let $Q_1, Q_2, \ldots$ be a sequence of orthogonal matrices that converge to $A$. Then

$$Q_k^t Q_k = I.$$

Taking the limit gives

$$A^t A = I.$$

So, $A$ is orthogonal. Hence, the set of orthogonal matrices is closed.

# Bibliography

**A. Matrix Theory**

1. Horn, R., Johnson, C.R. (1985): Matrix anaysis. Cambridge University Press, NY

2. Franklin, J. (1968): Matrix theory. Prentice-Hall, Inc., Englewood Cliff, NJ

**B. Numerical Linear Algebra**

1. Datta, B.N. (1995): Numerical linear algebra and applications. Brooks/Cole Publishing Co., Pacific Grove, CA

2. Golub, G.H, Van Loan, C.F. (1996): Matrix computations, 3rd ed. John Hopkins University Press, Baltimore

**C. MATLAB**

1. Coleman, T.F., Van Loan, C. (1988): Handbook for matrix computations. Siam, Philadelphia

2. Pratap, R. (1996): Getting started with MATLAB. Harcourt Brace College Publishers, Orlando

3. Dongarra, J., Bunch, J.R., Moler, C.B., Steward, G.W. (1978): LINPACK User's Guide. Siam, Philadelphia

4. Hanselman, D., Littlefield, B. (1997): The Student Edition of MATLAB, version 5. Prentice Hall, Upper Saddle River, NJ

# Index